PROBLEMS IN MANAGEMENT OF LOCALLY ABUNDANT WILD MAMMALS

Academic Press Rapid Manuscript Reproduction

Proceedings of the Management of Locally Abundant Wild Mammals: A Workshop to Examine the Need for and Alternatives to the Culling of Wild Animals, Sponsored by the International Fund for Animal Welfare in Yarmouth Port, Massachusetts, September 29 - October 3, 1980

PROBLEMS IN MANAGEMENT OF LOCALLY ABUNDANT WILD MAMMALS

Edited by

PETER A. JEWELL
SIDNEY HOLT

St. Johns' College
Cambridge, United Kingdom

Associate Editor

DONNA HART

International Fund for Animal Welfare
Yarmouth Port, Massachusetts

1981

ACADEMIC PRESS
A Subsidiary of Harcourt Brace Jovanovich, Publishers
New York London
Paris San Diego San Francisco São Paulo Sydney Tokyo Toronto

ACADEMIC PRESS, INC.
111 Fifth Avenue, New York, New York 10003

United Kingdom Edition published by
ACADEMIC PRESS, INC. (LONDON) LTD.
24/28 Oval Road, London NW1 7DX

Library of Congress Cataloging in Publication Data

Management of Locally Abundant Wild Mammals: a Workshop to Examine the Need for and Alternatives to the Culling of Wild Animals (1980 : Yarmouth Port, Mass.)
Problems in management of locally abundant wild mammals.

"Proceedings of the Management of Locally Abundant Wild Mammals: a Workshop to Examine the Need for and Alternatives to the Culling of Wild Animals, sponsored by the International Fund for Animal Welfare in Yarmouth Port, Massachusetts, September 29-October 3, 1980"--Verso t.p.
Includes index.
1. Wildlife management--Congresses. 2. Mammal populations --Congresses. I. Jewell, Peter Arundel. II. Holt, Sidney Joseph, Date. III. Hart, Donna. IV. International Fund for Animal Welfare. V. Title. VI. Title: Culling of Wild Animals.
SK352.M36 1980 639.9'79 81-20529
ISBN 0-12-385280-3 AACR2

PRINTED IN THE UNITED STATES OF AMERICA

81 82 83 84 9 8 7 6 5 4 3 2 1

CONTENTS

Contributors ix
Foreword xi
Preface xiii

I. INTRODUCTION 1

II. WHAT IS OVERABUNDANCE?

Overpopulation 7
Graeme Caughley

III. CASE STUDIES

Management of Locally Abundant Mammals—The South African Experience 21
John Hanks, W. D. Densham, G. L. Smuts, J. F. Jooste, S. C. J. Joubert, P. le Roux, and P. leS. Milstein

Stability Properties of Semiarid Savannas in Southern African Game Reserves 57
Brian H. Walker

An Outline of a Management Plan for Kasungu National Park, Malawi 69
R. H. V. Bell

The Management of Elephant and Other Large Mammals in Zimbabwe 91
David H. M. Cumming

The Problems of Overpopulation of Deer in North America 119
David R. Klein

The White Rhino Overpopulation Problem and a Proposed Solution 129
Norman Owen-Smith

Locally Abundant Marine Mammals—Problems and Attempted Solutions 151
John Harwood and David M. Lavigne

The Case of the Sea Otter 167
James A. Estes

The Leopard—Problems of an Overabundant, Threatened, Terrestrial Carnivore 181
Stephen Cobb

The Northern Fur Seal—An Example of Complexity 193
Douglas G. Chapman

IV. THEORY AND MANAGEMENT

Notes on Some Topics in Theoretical Ecology, in Relation to the Management of Locally Abundant Populations of Mammals 205
Robert M. May and John R. Beddington

Large Mammal Feeding Strategies and Related Overabundance Problems 217
Richard M. Laws

Responses of Vegetation to the Abundance of Mammalian Herbivores 233
Imanuel Noy-Meir

Environmental Carrying Capacity and the Evidence for Overabundance 247
A. R. E. Sinclair

V. DIAGNOSIS

The Debate about the Vicuna Population in Pampa Galeras Reserve 259
K. C. Otte and R. K. Hofmann

The Vicuna of the Pampa Galeras National Reserve—The Conservation Issue 277
S. K. Eltringham and W. J. Jordan

The Elephant Seal 291
Burney J. Le Boeuf

The Pacific Walrus 303
G. Carleton Ray

VI. THE WORKSHOP REPORT 317

Index 355

CONTRIBUTORS

Numbers in parentheses indicate the pages on which the authors' contributions begin.

John R. Beddington (205), *International Institute for Environment and Development, London W1, England*

R. H. V. Bell (69), *Department of National Parks and Wildlife, Kasungu National Park, Kasungu, Malawi*

Graeme Caughley (7), *Division of Wildlife Research, CSIRO, Lyneham, Canberra, A. C. T. 2602, Australia*

Douglas G. Chapman (193), *Center for Quantitative Science, University of Washington, Seattle, Washington 98195*

Stephen Cobb (181), *Department of Zoology, University of Nairobi, Nairobi, Kenya*

David H. M. Cumming (91), *Department of National Parks and Wildlife Management, Causeway, Salisbury, Zimbabwe*

W. D. Densham (21), *Natal Parks Board, Pietermaritzburg, South Africa*

S. K. Eltringham (277), *Department of Applied Biology, University of Cambridge, CB2 3DX, England*

James A. Estes (167), *National Fish and Wildlife Laboratory, Center for Coastal Marine Studies, University of California, Santa Cruz, California 95064*

John Hanks (21), *Institute of Natural Resources, Pietermaritzburg, South Africa*

John Harwood (151), *Natural Environment Research Council, Sea Mammal Research Unit, Cambridge CB3 0ET, England*

R. K. Hofmann (259), *Proyecto Vicuña, Nazca, Peru*

J. F. Jooste (21), *Cape Department of Nature and Environmental Conservation, Stellenbosch, South Africa*

W. J. Jordan (277), *People's Trust for Endangered Species, Guildford, Surrey GU1 3EH, England*

S. C. J. Joubert (21), *National Parks Board of Trustees, Skukuza, Kruger National Park, South Africa*

David R. Klein (119), *Alaska Cooperative Wildlife Research Unit, University of Alaska, Fairbanks, Alaska 99701*

David M. Lavigne (151), *Department of Zoology, University of Guelph, Guelph, Ontario, N1G 2W1, Canada*

Richard M. Laws (217), *British Antarctic Survey, Natural Environment Research Council, Cambridge CB3 7PX, England*

Burney J. Le Boeuf (291), *Crown College, University of California, Santa Cruz, California 95064*

P. le Roux (21), *Orange Free State Division of Nature Conservation, Bloemfontein, South Africa*

Robert M. May (205), *Department of Biology, Princeton University, Princeton, New Jersey 08544*

P. leS. Milstein (21), *Transvaal Nature Conservation Division, Pretoria, South Africa*

Imanuel Noy-Meir (233), *Botany Department, The Hebrew University, Jerusalem, Israel*

K. C. Otte (259), *Proyecto Vicuña, Nazca, Peru*

Norman Owen-Smith (129), *Center for Resource Ecology, Department of Botany, University of the Witwatersrand, Johannesburg 2001, South Africa*

G. Carleton Ray (303), *Department of Environmental Sciences, University of Virginia, Charlottesville, Virginia 22903*

A. R. E. Sinclair (247), *Institute of Animal Resource Ecology, University of British Columbia, Vancouver, British Columbia V6R 1W5, Canada*

G. L. Smuts (21), *Natal Parks Board, Pietermaritzburg, South Africa*

Brian H. Walker (57), *Center for Resource Ecology, Department of Botany, University of the Witwatersrand, Johannesburg 2001, South Africa*

FOREWORD

The suggestion for the international workshop of which this book is the outcome was made by Sidney Holt at a meeting of the Species Survival Commission of the International Union for the Conservation of Nature and Natural Resources (IUCN) in Cambridge, England, in mid-1979. One of us (PS) having drafted a position paper for IUCN on the implications of habitat destruction by elephants in certain East African national parks, had been requested to broaden the statement to cover other situations, and specifically that of the vicuna in the Pampa Galeras Reserve in the Andes of Peru. Discussions of this by the Species Survival Commission revealed sharply how far we were from clear definition of such problems, and still further from their solution. Meanwhile, the International Fund for Animal Welfare was beginning to question the basis for the widespread belief that intervention of one kind or the other was, in fact, necessary.

The workshop that emerged from subsequent organization and consultation was an important first step on the path toward reasonable and humane treatment of certain animals and ecosystems that are under human care.

We feel the organizations with which we are associated have been well served by those scientists who agreed to contribute their time and energy to this project. They and others will inevitably continue the field studies and refine the scientific concepts. It is now up to those of us who are engaged in public interest organizations and in management of protected areas to clarify our minds on policy issues, to ensure that scientific advice is made available, and to monitor the effects of human interventions of all kinds in living systems, as well as of natural events on those systems.

Brian D. Davies
Executive Director, International Fund for Animal Welfare

Peter Scott
Chairman, World Wildlife Fund

PREFACE

The question of the degree to which protected animals may be allowed to increase in number, with possible detriment to human interests and to their own habitat, is one which is of increasing concern to many people. Farmers, pastoralists, and fishermen may be worried about competition from such animal populations for space and fodder; managers of reserves, parks, and industries based on exploitation of wild living resources may wish to see any "surplus" put to economic use; persons and groups concerned with conservation or animal welfare may have strong views about so-called culling or other programs for environmental management; and local and national authorities may have to make decisions aiming to strike a balance between diverse interests. All look, to some extent, to scientists to help them with their problems. But it has become evident that such help is limited by both the bounds of present biological knowledge and the fact that such knowledge is but one of many factors that determine management policies.

The workshop, the results of which are given in this book, was sponsored by the International Fund for Animal Welfare with a view to reexamining the scientific basis for possible management aimed at restraining local increase in numbers of protected species. This limited objective was set with the expectation that subsequent meetings, with somewhat different participation, would look at practical aspects of reducing numbers or restraining growth where, for whatever reason, it had to be concluded that such action was desirable. It was expected also that further meetings would take up the related but nevertheless distinct problem of exploitation of animal populations which interact directly with other resources, such as seals or dolphins and commercially useful fish.

This book contains all the material, including selected case studies, prepared as background for the workshop and revised by the authors afterwards in the light of the discussions there. It contains also a report of the workshop, prepared by a steering committee, based on drafts written during the course of the meeting by various groups of participants, all of whom were consulted subsequently. This report thus represents a general consensus of the participants, although many of them would perhaps have reservations about details and the form of expressing certain generalizations. As far as we know, this subject has not previously been addressed in a comprehensive way, so the book will, we hope, be of wide interest to other scientists, to managers, to members of various public interest groups, and to users of natural living resources. It is admittedly highly technical, but an attempt has been made to produce a text which is comprehensible to those who are not specialists. We believe that although many scientific questions remain unanswered, and the workshop drew attention to areas of considerable uncertainty, a degree of sifting and synthesis of ideas and findings has been possible, which leads to practical implications for environmental management. Not least of the results of this project is a better identification of the types of information needed for the definition of both management policy and its implementation.

Apart from acknowledging the farsightedness of the sponsoring organization in recognizing the need for a scientific review of the problem of "local overabundance" and their timely action to facilitate it, thanks are due to all those scientists who contributed to the workshop, and hence to this publication, by writing working documents or making case studies, by participating in the workshop, and by assisting in various ways to convert its results into this book. Included in this acknowledgment and thanks is recognition of the input of Dan Botkin, who chaired the workshop session on diagnostic characteristics.

Our special thanks go to Dan Morast of the International Fund for Animal Welfare for coordinating the workshop, to Ruth Walicki, also of IFAW, for acting as workshop secretary, and to Mary Ann Pitchford and Donna Perdun for typing the manuscript.

Sidney Holt
On behalf of the editors and members of the Steering Committee

INTRODUCTION

Is wildlife ever too abundant? A great many species of animals, once widespread and numerous, have been so reduced by man's activities in changing the face of the earth that they now survive in only a few small populations. Among these animals are many of the world's larger and most attractive mammals. All too many of these species are threatened with extinction.

While this is so, it is an irony that these small populations may become, apparently, locally overabundant and create intense problems. The "elephant problem" is one of the best known examples. Conservationists are faced with the dilemma of a species in urgent need of protection over most of its range, which occurs in such abnormally large numbers in certain limited areas that the need to cull, perhaps by killing, must be considered.

Is there any way out of this dilemma? What advice can scientists who work in this field, particularly ecologists, offer?

I. BACKGROUND

The International Fund for Animal Welfare, which has engaged in several operations to translocate wild animals as an alternative to killing them, organized and funded a workshop to pool the appropriate experiences of ecologists. (The suggestion for this workshop was made at a meeting of the Species Survival Commission of the International Union for the Conservation of Nature and Natural Resources held in Cambridge, England in mid-1979.)

As first viewed, the central task of the workshop was to consider the dilemma of local *overabundance* of *threatened* mammals, but it was recognized that this concept could lead to too narrow an approach. For one thing, to speak of "over" abundance prejudges the issue. Furthermore, while threatened species present the real worries, many species that are not endangered do become excessively abundant locally and do provide

ISBN 0-12-385280-3

valuable case histories. Hence the broader title of the workshop -- "The Management of Locally Abundant Wild Mammals: A workshop on the need for and alternatives to the culling of wild animals." The purpose of the workshop was defined as a forum for the discussion of the problems which can arise when man-made changes permit a species to become so numerous locally it seems to have an adverse effect on its food supply, its competitors, or even the species composition of the ecosystem. This objective statement hides a host of problems that are of pressing urgency for those concerned with the conservation of wild animals and their welfare. In particular the question of culling -- in whatever form it might take -- was the central issue. As used in this volume, culling is defined as the selective removal of animals from a population. Removal may be by killing or by live capture and subsequent transportation of animals elsewhere (either translocation to other areas of natural habitat or for confinement in zoos and parks).

There are now many examples of national parks or conservation areas where the culling of wild mammals has been carried out as part of a management policy. Large numbers of certain species have been killed to reduce their population density and their impact on some other facet of the environment. Such action is an extremely onerous and unpalatable task for conservationists. In some places even relatively rare or threatened mammals have been culled by killing in the very reserves originally set aside for their protection. Examples are hippopotamus in Uganda, nyala in South Africa, elephant in many African countries, vicuna in Peru, and grey seals in Britain. Even if they were not classified as rare, many other spectacular mammals have been involved. This drastic intervention, in attempts to manage wildlife areas, presents those in charge with an agonizing dilemma.

II. ABUNDANCE AND OVERABUNDANCE

Why should such a contrary situation arise? As a result of conservation efforts in recent decades many species and populations of animals, the survival of which had been threatened by human activities, have increased in numbers. The activities that originally created the threats to survival were high levels of hunting and changing land use leading to reduction in the area or quality of the animals' habitats. Either or both of these factors were present in any particular case and to save the animal populations from extinction, reduction or cessation of hunting was imposed and protected areas were created.

Many ecosystems, as a result of human activities, have been perturbed from an equilibrium. Adjustment to new equilibria, or even the process of returning to the old equilibria, has thrown up its own imbalances that may at times have seemed to imperil the ecosystem itself. Many early endeavors in conservation failed to set the protection of the whole ecosystem as their objective. Sometimes the preservation of a particular population of a single species, often one of the more spectacular ones, provided the false start.

Ironically, many of the problems that have arisen occur where action directed towards conservation was successful. Moreover, these emergent problems of management imply that successful control of illegal hunting has been achieved, and possibly such successful protection of the habitat that, at least temporarily, it can sustain rising numbers of animals.

The human interests affected are of several different kinds. They range from damage to fishing gear by protected whales, through plunder of livestock by carnivores, to the destruction of vegetation and soil cover by large numbers of herbivores. In cases like the last-mentioned, the human interest involved will often be the wish to ensure the survival of the habitat or ecosystem, so the problem may be perceived also as damage to the animals' own "interests." Quite commonly, it is difficult to know whether the apparent local abundance is a reflection of overall increase in numbers or densities of animals, is only a local increase, or reflects a change in the behavior of the animals such that their presence (or the effects of their presence) is more visible. Whatever the characteristics of overabundance, there is usually a human response calling for "remedies" which more often than not include culling.

III. THE NECESSITY FOR ACTION

It may be quite uncertain whether culling, i.e., selective removal, will in fact significantly lower local abundance and effectively ameliorate the conflict of known interests, including undesired changes in the environment. A proposal to cull by killing may also be seen simply as an easy (and only short-term) "solution" to a social or ecological problem. Even if culling could provide a solution, it may be that other management actions could have an equivalent effect; such actions would include manipulation of the habitat in some way, or "scaring" or "herding" which would cause the animals to disperse or to remove themselves, perhaps selectively, from the localities of high density. Physically transporting animals to another locality within a protected area may also be an option when there are either geographical or behavioral impediments to natural dispersal around the area of high density.

Perhaps combinations of such actions would be more effective than any single one of the above. If culling turns out to be necessary, then the removal by means other than killing will be a matter of intense interest to conservation groups. Whether culling is done just once or over a limited number of years (while relatively stable "recovered" population and habitat are attained), or whether it is to be carried out as a continuous management practice, may also be a matter of concern, as well as having different ecological consequences. Finally, it seems generally agreed that the specific interests of humans living near or in the localities of overabundance must receive special consideration in management programs.

The regular, annual killing of animals as part of a management policy can give rise to special dangers because this culling may come to include an economic use of the culled animals -- dead or alive -- and the true purpose of the cull may become obscured.

One further facet of the problem requires mention. Conservationists have often devoted attention to animal populations and ignored plant communities. *Plants are wildlife.* A shattered baobab tree is a sorry and arresting sight, and a devastated grove of mature acacias is a real loss. Many pastures are highly tolerant of grazing, but not able to sustain the loss of soil and sheet erosion. These wild flora assets cannot be replaced indefinitely and their protection must be given an appropriate value in the equation.

IV. SCOPE OF THE WORKSHOP

The central tasks of the workshop were four in number:

(1) To assess the extent to which past predictions have been fulfilled in practice, and whether understanding of the dynamics of living systems is now adequate for useful prediction.

(2) Identification of the circumstances which allow a species to become so abundant the adverse effects outlined can occur.

(3) Identification of characteristics which are diagnostic of the early stages in the development of these effects.

(4) Consideration of what options for control are available when these effects occur or are feared, and how these options can be assessed.

Supplementary questions concern whether experiences in a limited number of places with a few species can be generalized to provide useful guidelines for management, and whether present techniques of census and assessment are adequate for maintaining the managed systems and hence for testing the predictions. These questions are particularly important where management (including culling) proposals are made with the intention of avoiding damage from high local abundance rather than correcting an already undesirable situation. However, even where conflict is perceived, it is usually not certain at the outset into which category the problem falls. Determination of the true nature of the problem -- from the array of opinions that will invariably be offered by various persons involved -- is itself a major scientific task. Even determination of the real rather than the apparent degree of impact of high local abundance on human values and on the environment may not be a trivial task, since it can involve not only extension of the geographical range of survey and of the range of organisms monitored, but also study of short-term trends with a view to longer-term predictions. Guidelines are needed by managers and organizations regarding the research necessary to make these determinations.

It is evident the scientific questions involved in both correctly identifying problems and making valid predictions of solutions to the more important of these, can be extremely difficult ones. The phenomena of species or population extinction, of catastrophic processes, and multi-species dynamics are still poorly understood. While experimental approaches to management are invaluable, the time scales involved may be so long that a very substantial input from theory is necessary if sense is to be made of the experimental results. The time factor also gives great importance to comparative studies -- of species, of systems, and of management regimes. The chapters in this volume are designed to facilitate such comparisons.

OVERPOPULATION

Graeme Caughley

Division of Wildlife Research
CSIRO
Lyneham, A.C.T.
Australia

"Overpopulation" may be defined rigorously as too many animals, but the rigor ends there. It is not a single neat phenomenon but a set of them. Before management activities can be planned to cope appropriately with a case of overpopulation, we must know not so much why the area is overpopulated but rather in what sense it is overpopulated. I will outline four senses in which animals may be too numerous and will suggest that these are often confused, one with another, leading as often as not to inappropriate management.

I. CLASSES OF "OVERPOPULATION"

The classification to follow is not exhaustive but it provides a simple framework embracing most cases of overpopulation.

A. *Class 1: The Animals Threaten Human Life or Livelihood*

Into this class go all cases where the animals are too numerous only because their presence is inconvenient. Three examples will suffice: World Health Organization (WHO) is convinced that the tropics are overpopulated by the *Anopheles* mosquito; most sheepmen in Australia consider that the inland plains are overpopulated by kangaroos; and most sheepmen in the Midwest of the United States perceive an overabundance of coyotes. Obvious in each example is the reason the animals are seen as too numerous, and equally obviously that insight may

ISBN 0-12-385280-3

be denied to those who do not have to contend with children dying of malaria, who have never had to scratch a living from a drought-prone grazing property, or who have not found their prize ram deprived of important appendages.

But no delicate ecological balance is at risk with Class 1 problems. No deeply felt conservation ethic is threatened.

B. *Class 2: The Animals Depress the Densities of Favored Species*

Few people enjoy all species equally. Some organisms are favored and others treated with indifference. When it is apparent that reducing the density of a dreary species will enhance the density of an interesting species, many will favor doing just that. An example is the continuing wrangle over elk and aspen in Yellowstone National Park. The aspen is a beautiful tree, and it is held at low density by the browsing of elk. People fond of these trees favor thinning out the elk; people fond of elk oppose that action. It is more complicated than that because fire also affects aspen regeneration (Gruell and Loope, 1974; Bartos and Mueggler, 1979), and artificial feeding of elk in winter perturbs that population's dynamics (Beetle, 1979), but the above simplistic summary captures the essence of that dispute.

A second example of Class 2 overpopulation concerns a practice common in many African national parks thirty years ago -- the shooting of African hunting dogs. It was argued, probably correctly, that a reduction in the density of *Lycaon* would increase the density of herbivores, and so *Lycaon* was shot on sight. More recently, lions and hyenas have been culled in Kruger National Park to increase the density of wildebeest and zebra (Smuts, 1975).

C. *Class 3: The Animals Are Too Numerous for Their Own Good*

Animals held at low density by hunting tend to be larger, fatter, healthier and more fecund than those at the unharvested equilibrium density. Some people approve strongly of this and argue that, for their own good, animals should be hunted each year to hold density down. That argument is used as an ecological justification of sports hunting (see Dasmann, 1971).

In its commonest form, however, the argument goes that the object of the hunting is to hold the population "within carrying capacity" (see Lang and Wood, 1976). Implicit or tacit is the assumption that the system comprising the population and its resources is incapable of holding an equilibrium unaided and so must be managed constantly. That philosophy underlies most of the culling of herbivores in national parks.

D. *Class 4: The System of Plants and Animals Is Off Its Equilibrium*

This class of overpopulation differs from the last where the equilibrium density itself is seen as "overpopulation." Here the population really is above its equilibrium density. The population may be erupting into new habitat, it may be reacting to a surge of plant growth occasioned by a run of good seasons, it may be responding to a reduction of hunting, or -- a most uncommon occurrence -- it may be locked into a stable limit cycle with its food supply.

II. MISCLASSIFICATION

Class 4 problems are the only ones that need detain us. Class 1 contains only conflicts with human interests, real enough but not of overwhelming ecological interest. Class 2 problems are purely technical, calling for a manipulation of the system to provide an outcome (higher diversity of animal species, more tree cover, less tree cover, and so on) more acceptable than the equilibrium to which the system would otherwise converge. Class 3 "overpopulation" is a rejection of the ecological that-which-is in favor of an opinion as to what is best for the population. Ecological considerations are irrelevant because the opinion is a judgment of value. The Class 4 problem of a displaced equilibrium is the only one to which ecology is central and the only one about which a relatively objective and scholarly discussion is possible. That is not to say the problems of Class 1, 2, and 3 are illusionary but that no large ecological principle is involved. Disputes about cases tend to reflect the personal background and system of values of the protagonists.

The importance of Class 1, 2, and 3 problems to us is that they are usually presented erroneously as examples of Class 4, and management actions are then represented as ways of achieving a goal dictated, not by personal preference, but by ecological necessity. The elk/aspen debate is a good example. It is carried out on an elevated plane, personal likes and dislikes not being mentioned. They are replaced by "zootic climax," "balance of nature," "disclimax," "dynamics of the ecosystem," and other polysyllabic swear-words.

Most cases of "overpopulation" I have looked at personally fall easily into Class 1, 2, or 3, but the majority were viewed officially as the displaced equilibrium of Class 4. The commonest cause of that confusion is the application of range management notions of carrying capacity (the standing crop of animals at equilibrium with a maximum sustained harvest) to an unharvested population of herbivores at equilibrium. Hence the composition of the vegetation does not conform to the criteria of "good range conditions," the "proper use of factor" is exceeded, and therefore the population is above carrying capacity. And so it is, but only according to the special criteria by which a farmer judges the density of stock that will maximize his net revenue. These have no relevance to the equilibrium that ecologists call "carrying capacity."

III. MANAGEMENT OF DISPLACED EQUILIBRIA

The commonest form of Class 4 overpopulation that exercises conservationists is an eruption of terrestrial herbivores past the level to which they will eventually stabilize. Usually the eruption reflects the response of the population to abundant food, as when a new population is started by translocation of a nucleus to an area presently lacking that species. Sometimes it reflects the relaxation of hunting pressure, as when an area is dedicated as a national park. Over the next twenty years or so the population climbs to its equilibrium density, overshoots it, crashes, and then converges to relative stability. The density of edible plants follows a reciprocal trajectory. That is close to general for temperate zones. The water-driven systems of arid zones are less predictable.

An eruption is the pattern of growth typical of most herbivore populations. It reflects not a breakdown of regulatory mechanisms but rather the predictable demographic response of a population presented with more food than needed to ensure an animal's replacement in the next generation (Caughley, 1976a). The overshoot results from an animal's supply of food being reduced by the grazing and browsing of the previous generation (Caughley, 1979), an effect represented by a lag term in one-species mathematical models of population growth. While the population is rising, falling, and stabilizing, plant composition and density changes progressively and often dramatically until an equilibrium is achieved after about 35 to 40 years.

A. *Can an Eruption be Contained?*

Caughley (1976a) modeled a herbivore population and its food as if it constituted a prey/predator system, the plants being the prey and the herbivores the predators. Crête *et al.* (1980) tested that model against the dynamics of moose in Quebec, concluding that despite its simplicity it predicted quantitatively the effects of their experimental manipulations. Consequently this model may serve as a useful device to explore the effect on eruption of various forms of management.

Computer simulation indicates that the overshoot of animals, and the reciprocal undershoot of plants, can be eliminated by a regime of harvesting initiated when the animal population is about halfway along its upswing (Caughley, 1976a). The rate of harvesting is reduced progressively until the density of edible plants settles gently onto its equilibrium. Eruptive oscillations can be dampened in this way but the final outcome, in terms of the equilibrium densities of plants and animals, is unaffected. Note that the treatment must begin before "overpopulation" has occurred. An attack on a population at peak density will not avert an undershooting by the plants. If the aim of management is to guide the system to equilibrium without the trauma of oscillation, the method of cropping is irrelevant to that result. The animals may be harvested by shooting, poisoning, translocation, or by any other method that appeals.

B. Is Containment Desirable?

It is one thing to suggest that an eruption can be contained; quite another to advocate it. I have heard the following arguments in favor of flattening an eruption:

(1) Since the animal population crashes because many of its members starve to death, it is more humane to shoot them first.
(2) The crash wastes protein that could be harvested without interfering with the ecological outcome.
(3) Having dead bodies lying around a national park is disturbing to the public.
(4) "I am a wildlife manager. I am here to manage."

Excepting perhaps the last, these arguments are sensible and logically valid. But whether we support them or not depends on the shape of our souls, not upon the nature of nature.
To balance up, here are arguments I have heard against managing eruption:

(1) Man should intervene as little as possible in natural systems, particularly in national parks whose *raison d'etre* is the preservation of relatively undisturbed ecological processes.
(2) The expense is damnable.
(3) Given the current state of the art in wildlife management, the inevitable ham-fistedness of an attempt at fine-tuning will cause more problems than it solves.

C. Is Containment Necessary?

The previous subsection presented arguments for and against action that would change an eruption's path but not its outcome -- an equilibrium between plants and animals. This subsection addresses a more important question: whether dreadful things would result if we did nothing. Ecosystem theory provides a rogues' gallery of apocalypses, bottlenecks, and black holes into which a theoretical system might be sucked should its parameters get out of kilter. In particular it might be conceived that an eruption could affect the long-term equilibrium by changing vegetative composition and soil quality. If such dark outcomes were likely we would act swiftly to intervene.

The commonest dreads are that species of plants will be lost during an eruption, that erosion will be severe and irreversible, or that the system will not converge upon an equilibrium but instead will gyrate out of control or shake itself apart.

These predictions of potential disaster have been tested several hundred times by the liberation of various ungulates in New Zealand. It is an extreme test because that vegetation had not been browsed or grazed since the extinction of moas in the sixteenth century, and it had never previously been used by mammals. If violent eruptions lead occasionally to ecological malfunction it is here that we should see the incipient or manifest breakdown of an unfenced system. Some of these eruptions have been studied in greater or lesser detail (Riney *et al.*, 1959; Riney, 1964; Howard, 1965; Caughley, 1970), and the outcome at least is known for most of the others. In all cases the composition of the vegetation that came to equilibrium with the alien grazing and browsing differed radically from its previous state. But no system fell apart, no plant species was lost although many were driven to low density, and only one liberation (moose in Fiordland) entered what appeared to be a limit cycle to extinction.

These animals have long been blamed for causing erosional aggradation of river beds and consequent flooding. But recent research (Griffiths, 1979 and personal communication; Adams, 1980) reveals that the high rates of erosion characteristic of the New Zealand mountains (up to 13,000 tons/km^2/yr in large catchments and up to 50,000 tons in some tributaries) may be entirely natural. The rates can be predicted closely from rainfall alone (up to 10 m/yr). Rock type and plant cover account for no significant portion of the residual variance. The immense shingle screes providing "visible proof" of accelerated erosion have now been identified as relatively ancient geomorphic features whose surfaces are to be dated in centuries rather than decades.

To summarize, the deer and other introduced herbivores changed the composition of the New Zealand vegetation, but not irreversibly. Those animals may well consititute an overpopulation in the sense of Class 2 but they do not now constitute a threat to themselves or their environment in the sense of Class 4.

That conclusion probably holds in a general way for Class 4 overpopulation of most species in most countries. I do not know of any system dislocated permanently by a bout of overpopulation. The phenomenon is temporary and its remission spontaneous. Most treatments of overpopulation are justified by a dire prediction of what might have happened had the treatment been withheld. A more convincing case would be made by demonstrating that the effects of untreated overabundance are irreversible. The snowshoe hare, which overpopulates every ten years and which modifies the vegetation severely at each peak, is an obvious candidate for such a study.

Two exceptions to my optimistic prognosis are worth noting: domestic stock and African elephants. The dynamics of domestic stock are imposed upon them by man, and man sometimes holds stock at densities higher than would be possible if the stock arranged their own bionomics. Severe overgrazing may result and its effects may sometimes be irreversible.

The African elephant is susceptible to Class 4 overpopulation, but the causes postulated (compression according to Laws *et al.*, 1975; limit cyclicity according to Caughley, 1976b) are clearly not those that trigger eruptions of other herbivores. Whatever the cause of the elephant problem it is essentially a special case and we would be unwise to extrapolate from it to the more common and mundane examples of overabundance.

Is containment of an eruption necessary? That is a scientific question and I interpret the evidence available as implying that it is seldom or never necessary. Is containment of an eruption desirable? That is not a scientific question. I can boast no qualifications that would make my opinion any more valuable than those of my two immediate neighbors, a garage mechanic on one side and an Air Vice-marshall on the other.

IV. POSTSCRIPT

The above comments, save for minor editorial tidying, were written a couple of months before the workshop was held and before I had seen the other submissions. The course of the meeting swifly demonstrated that I had been remiss, or at least premature, in my cavalier dismissal of overpopulation Classes 2 and 3, particularly as they relate to management of national parks. They would have been given greater stress had I written the draft during the 1960's because then they occasioned much debate. Subsequently those issues have faded considerably. The main reason for that was a convergence of opinion on what national parks were for. The prevalent attitude is that national parks should be manipulated as little as possible. They form one half of an experiment testing the effect of people intervening in ecological processes. The parks are the control, the rest of the world the treatment. The control is never perfect because each park carries a memory of the state it was in before it became a park, and because what happens outside parks affects conditions within parks. But the intent remains valid for all that. Not that this aim is necessarily spelled out in the definition of the function of this or that park. Such definitions are usually fuzzy (as when they evoke the preservation of "pristine" or "natural" states) and often silly (as when they demand simultaneous maximization of several linked characteristics). Yet the notion toward which these unfocused definitions grope is clear enough: the maintenance of ecological processes protected as far as possible from fiddling by people.

The trend over the last decade, in most areas, has been a reduction of human intervention in national parks, a largely unstated policy reinforced by observing that a phasing out of manipulative management has not caused ecological whiplash injuries. I do not know of any present culling of indigenous mammals in any national park in Australia, southern Asia, China, South America, or in Africa north of about the Zambesi. In some of these regions there is considerable de facto culling by illegal hunting, in others little or none. In contrast to even ten years ago, culling in now extremly uncommon in the national parks of North America. I have no recent information from the U.S.S.R. or Europe, and almost certainly I have missed isolated instances of culling in the other regions; but that does not weaken the force of my general conclusion: culling of indigenous mammals is now a rare practice in national parks and becoming rarer.

That was the context shaping my original draft. Unfortunately I did not realize, or more likely had forgotten, that national parks in southern Africa are managed very closely. Culling and translocation are integral to that management. Although that region occupies only 2.3% of the earth's land area it occupied a somewhat higher proportion of the workshop's discussion time. Consequently I take this opportunity to redress what was, in terms of that discussion, a serious imbalance in my original draft.

Management of national parks in southern Africa contrasts markedly with that in the rest of the world. There are two reasons. One reflects the southern African view of what national parks are for. The other is more complicated: a conceptual model, carried in the minds of the managers, of the ecological consequences of not doing what they are doing.

Bell (this volume) states clearly the defined function of national parks in Malawi: "... to preserve in perpetuity representative examples of plant and animal communities." That definition tacitly underlies management of national parks in much of southern Africa. In contrast to the imprecise definitions alluded to previously, this one is internally consistent and completely unambiguous. It begs no question. But note that the definition itself creates overpopulation in the sense of Class 2, because many communities do not exist at unmanaged levels of browsing and grazing. Bell understands that very well and makes it explicit in his draft plan of management for Kasungu National Park. It calls for culling in some areas but not in others to preserve a diversity of plant associations. Bell stresses that the decision as to which communities will be favored is aesthetic, not scientific. There can be no logical argument against Bell's view of a national park's function. One may agree with it, or disagree with it, but that decision is not reached by a dispassionate sifting of facts. Whether one decides to create or retain a desired state by manipulative management, or whether one prefers to let the system choose its own course, the decision is a judgment of value. No telling points are scored by detractors referring to the first option as "landscape gardening" and to the second as the "let-her-rip system."

These divergent aims supply one reason for the disparity in management practices between southern Africa and elsewhere. The second reason is an untested conviction, I think unique to southern Africa, that without continuous culling and translocation there would likely be a breakdown of the system to a lower equilibrium, a loss of genetic variability, and devastating soil erosion (Hanks *et al.*, this volume). These are seen as consequences of constraining animals within a boundary fence, of parks being too small and of animals being too large. The "too small" category apparently includes parks as big as 20,000 km^2 (Kruger) and the "too large" category includes animals weighing as little as 70 kg (impala and reed buck). Table I is a list of national parks that have been established for more than ten years in New South Wales, Australia, together with their areas. Only some are fenced, the fences grading in quality from stock-proof to kangaroo-proof. In addition to the national parks, New South Wales contains 123 nature reserves averaging 3,540 ha, extremes being 1 ha (Moon Island) and 65,000 ha (Pilliga). My Table I is for comparison with Hanks' *et al.* (this volume) Table I which lists wildlife areas in South Africa. Indigenous mammals are culled in all the parks and reserves listed by Hanks, in none of the parks I listed or the nature reserves I mentioned. In spite of that these systems persist.

TABLE I. National Parks of New South Wales That Have Been Established for Ten Years or More

Park	*Area(ha)*	*Largest indigenous mammal*
Barrington Tops	*16,325*	*Wallaroo*
Blue Mountains	*198,881*	*Wallaroo*
Bouddi	*1,111*	*Swamp wallaby*
Brisbane Water	*7,903*	*Swamp wallaby*
Cocoparra	*8,356*	*Western grey kangaroo*
Dharug	*14,186*	*Swamp wallaby*
Dorrigo	*3,909*	*Red-neck wallaby*
Gibraltar Range	*17,273*	*Wallaroo*
Heathcote	*2,251*	*Red-neck wallaby*
Kanangra Boyd	*67,881*	*Wallaroo*
Kinchega	*44,005*	*Red kangaroo*
Kosciusko	*627,218*	*Eastern grey kangaroo*
Ku-ring-gai Chase	*14,709*	*Swamp wallaby*
MacQuarie Pass	*1,064*	*Swamp wallaby*
Morton	*109,968*	*Wallaroo*
Mount Kaputar	*31,700*	*Wallaroo*
Mount Warning	*2,138*	*Wallaroo*
New England	*23,351*	*Eastern grey kangaroo*
Royal	*14,944*	*Red-neck wallaby*
Warrumbungle	*18,175*	*Wallaroo*

I conclude with two points, one trivial, the other not. I agree that a cessation of culling in southern African parks might bring on all the disruption and degradation so confidently predicted. Southern Africans have deemed it is too dangerous to find out. That is a cautious and responsible sovereign decision, entailing a preemptive strike against a perceived surfeit of population dynamics. Sovereignty would be transcended only if they advanced their management system as a general panacea, but they do not. My colleagues in southern Africa are not proselytizers. They see their chosen option as local hazard management, their own choice for their own area.

My second point is not trivial. Southern Africa is irrelevant to the hypothesis that culling is necessary in national parks. The wildlife managers there are locked into a system of continuous culling from which they cannot break free -- culling is institutionalized in those parks. There is no chance, at present, of testing that hypothesis in that region. Such a test would first require a complete mental reorientation, the crossing of a mountain range of the mind. I do not lightly dismiss that possibility. Did they not themselves traverse the Drakensberg?

ACKNOWLEDGMENTS

I am grateful to John Beddington, Jeff Short, and Alan Newsome for criticizing this draft, to the International Fund for Animal Welfare for funding my attendance at the workshop, and to Peter Jewell who urbanely chaired its progress through difficult bouts of informality.

REFERENCES

Adams, J. (1980). High sediment yields from major rivers of the western Southern Alps, New Zealand. *Nature (London) 287*, 88-89.

Bartos, D. L. and Mueggler, W. F. (1979). Influence of fire on vegetation production in the aspen ecosystem in western Wyoming. *In* "North American Elk: Ecology, Behavior and Management" (M.S. Boyce and L. D. Hayden-Wing, eds.), pp. 75-88. Univ. Wyoming Press, Laramie.

Beetle, A. A. (1979). Jackson Hole elk herd: a summary of 25 years of study. *In* "North American Elk: Ecology, Behavior and Management" (M. S. Boyce and L. D. Hayden-Wing, eds.), pp. 259-262. Univ. Wyoming Press, Laramie.

Caughley, G. (1970). Eruption of ungulate populations with emphasis on Himalayan tahr in New Zealand. *Ecology 51,* 53-72.
Caughley, G. (1976a). Wildlife management and the dynamics of ungulate populations. *In* "Applied Biology" Vol. I. (T. H. Coaker, ed.), pp. 183-246. Academic Press, New York.
Caughley, G. (1976b). The elephant problem -- an alternative hypothesis. *East Afr. Wildl. J. 14,* 265-283.
Caughley, G. (1979). What is this thing called carrying capacity? *In* "North American Elk: Ecology, Behavior and Management" (M. S. Boyce and L. D. Hayden-Wing, eds.), pp. 2-8. Univ. Wyoming Press, Laramie.
Crête, M., Taylor, R. J., and Jordon, P. A. (1980). Optimization of moose harvest in southwestern Quebec. *J. Wildl. Manage.* (In press).
Dasmann, W. (1971). "If Deer Are to Survive." Stackpole Books, Harrisburg.
Griffiths, G. A. (1979). High sediment yields from major rivers in the western Southern Alps, New Zealand. *Nature (London) 282,* 61-63.
Gruell, G. E. and Loope, L. L. (1974). Relationships among aspen, fire, and ungulate browsing in Jackson Hole, Wyoming. *USDA Forest Service Inter. Region Publ.* 33 pp.
Howard, W. E. (1965). Control of introduced mammals in New Zealand. *New Z. Dep. Sci. Ind. Res. Inf. Ser. 45,* 1-95.
Lang, L. M. and Wood, G. W. (1976). Manipulation of the Pennsylvania deer herd. *Wildl. Soc. Bull. 4,* 159-165.
Laws, R. M., Parker, I. S. C., and Johnstone, R. C. B. (1975). "Elephants and Their Habitats." Oxford Unv. Press (Clarendon), London and New York.
Riney, T. (1964). The impact of introduced large herbivores on the tropical environment. *Int. Union Conserv. Nat. Nat. Resour. Pubs. New Ser. 4,* 261-273.
Riney, T., Watson, J. S., Bassett, C., Turbott, E. G., and Howard, W. E. (1959). The Lake Monk expedition. *New Z. Dep. Sci. Ind. Res. Bull. 135,* 1-75.
Smuts, G. L. (1975). Why we killed those lions. *Afr. Wildl. 29(3),* 30-33.

MANAGEMENT OF LOCALLY ABUNDANT MAMMALS -- THE SOUTH AFRICAN EXPERIENCE

John Hanks

Institute of Natural Resources
Pietermaritzburg
South Africa

W. D. Densham
G. L. Smuts

Natal Parks Board
Pietermaritzburg
South Africa

J. F. Jooste

Cape Department of Nature and Environmental Conservation
Stellenbosch
South Africa

S. C. J. Joubert

National Parks Board of Trustees
Skukuza
Kruger National Park
South Africa

P. le Roux

Orange Free State Division of Nature Conservation
Bloemfontein
South Africa

P. leS. Milstein

Transvaal Nature Conservation Division
Pretoria
South Africa

ISBN 0-12-385280-3

I. INTRODUCTION

In the past ten years South Africa's human population has increased by 5,000,000 to 23,771,970, a 27% increase in the ten years since the last population census was completed in 1970 (Reilly, 1980). This total does not include the independent homelands of Transkei, Bophuthatswana, and Venda. As more and more land is being called into use for human settlement and for food production to accommodate legitimate human demands associated with this high rate of population growth, it is inevitable that the national parks and game reserves of South Africa (the "designated areas") will become increasingly isolated from one another as "islands" in a "sea" of expanding agriculture and human settlements.

With increasing fragmentation and isolation of wildlife populations and their habitats, the theory of island biogeography becomes very relevant in considerations of wildlife management. The theory predicts that island biotas tend to approach dynamic equilibrium in the number of species present. The forces that bring about this dynamic equilibrium are immigration and extinction. For a particular island, the precise number of species at equilibrium depends mostly on island size (the area effect) and the distance from the source of immigrants (the distance effect). Within South Africa, existing designated areas are already experiencing a process of attrition, and this is being coupled with an increase in the distances between viable populations as the process of fragmentation continues. An obvious deduction from the species/area/distance relationship of island biogeography is that any further attrition or isolation will lead to a loss of species, and the eventual attainment of a new, lower equilibrium (Diamond and May, 1976; Miller, 1978, 1979; Miller and Harris, 1977; Polunin and Eidsvik, 1979; Soulé and Frankel, 1980; Terborgh, 1974; Vuilleumier, 1970). The question of concern for this volume is the speed at which biotas collapse to lower equilibria once the area is reduced. If the collapse rate is geological in scale, that is, occurring over millions of years, then we need not be too concerned, because evolution might produce new species as fast as old ones are being lost. If the rate is historical, however, in decades or a few centuries, our descendants will live in a noticeably less diverse world (Soulé and Frankel, 1980). Such possibilities for collapse emphasize the need for management strategies and objectives in wildlife sanctuaries to be related not only to the immediate environmental changes brought about by locally overabundant mammals and/or fire (e.g., Dodds and Patton, 1968; Laws, 1969, 1970; Laws *et al.*, 1975; Naylor *et al.*, 1973; Thomson, 1975), but more to the realities of long-term genetic conservation in a situation of spatial isolation.

The designated wildlife areas of South Africa discussed in this chapter in terms of their locally abundant mammals are all relatively small, with the exception of Kruger National Park (see Table I), and they are isolated from one another. Locally abundant mammals have presented problems of habitat degradation in nearly every area and are usually a consequence of both a lack of predation and the prevention of dispersion caused by fences around the reserves, or by people and associated agriculture on or close to the reserve boundaries discouraging emigration. Animal removal programs, both by live capture and culling (dead removal), have been introduced to prevent or reduce the habitat degradation that results from persistent overabundance. In the Kruger National Park, however, elephant and buffalo removal is related more to reducing competition for water with other animals than to habitat degradation. In most cases the population reduction programs are part of a management plan related to the specific management objectives of the area concerned, but with few exceptions these management objectives have not given adequate consideration to long-term genetic conservation. The proposed goal for nature reserves suggested by Soulé and Frankel (1980) comes very close to meeting these long-term requirements: "The purpose of a nature reserve is to maintain, hopefully for perpetuity, a highly complex set of ecological, genetic, behavioural, evolutionary and physical processes and the coevolved, compatible populations which participate in these processes." The significance of this goal in relation to the objectives of this workshop will be referred to later.

Holt *et al.* (1979) have defined three central tasks prior to the beginning of this workshop. First, to identify the types of situation which produce a necessity for population reduction or control. Secondly, to produce appropriate diagnostic characteristics so that such problems are identified at an early stage. Thirdly, to deal both in general and practical terms with the actual details of the control action required. With these central tasks in mind, the conservation authorities in South Africa listed in Table I were asked to provide the following information for the purposes of this review:

(1) Number of each species culled (dead removal) in each designated area in South Africa during the last five years.

(2) Number of each species live captured in each designated area in South Africa during the last five years.

(3) Criteria used to reach decisions to remove animals (e.g., habitat degradation, poor physiological condition of animals). Special consideration should be given to the identification of diagnostic characteristics that can be used to help management staff recognize the need for population reduction in the future.

TABLE I. *The Conservation Authorities in South Africa and, Where Relevant, the Designated Wildlife Areas under Their Control Referred to in the Text*[a]

	Natal Parks, Game and Fish Preservation Board	Area (ha)
1	Chelmsford Public Resort Nature Reserve	6014
2	Coleford Nature Reserve	1272
3	Enseleni Nature Reserve	293
4	False Bay Park	2247
5	Giant's Castle Game Reserve	34284
6	Himeville Nature Reserve	105
7	Hluhluwe Game Reserve	23067
8	Itala Nature Reserve	26328
9	Kenneth Stainbank Nature Reserve	214
10	Kosi Bay Nature Reserve	20
11	Krantzkloof Nature Reserve	532
12	Loteni Nature Reserve	3984
13	Midmar Public Resort Nature Reserve	2831
14	Mkuzi Game Reserve	25091
15	Moor Park Nature Reserve	264
16	Ndumu Game Reserve	10117
17	Nxwala State Land	5503
18	Queen Elizabeth Park Nature Reserve	93
19	Royal Natal National Park	8094
20	St. Lucia Game Reserve	36826
21	St. Lucia Park	12545
22	Spioenkop Public Resort Nature Reserve	4562
23	Umfolozi Game Reserve	47753
	National Parks Board	
	Kruger National Park	1,950,000

	Nature Conservation Division of the Transvaal Provincial Administration	
1	*Doorndraai Dam Nature Reserve*	7392
2	*Hans Merensky Nature Reserve*	4153
3	*Langian Nature Reserve*	4766
4	*Loskop Dam Nature Reserve*	14800
5	*Nylsvley Nature Reserve*	3120
6	*Ohrigstad Dam Nature Reserve*	2564
7	*Percy Fyfe Nature Reserve*	2939
8	*Rustenburg Nature Reserve*	2898
9	*Roodeplaat Dam Nature Reserve*	1667
10	*S. A. Lombard Nature Reserve*	3663
11	*Suikerbosrand Nature Reserve*	13337
	Cape Department of Nature and Environmental Conservation	
1	*Andries Vosloo Kudu Reserve*	6493
2	*De Hoop Nature Reserve*	18763
3	*Hester Malan Nature Reserve*	4676
4	*Oviston Nature Reserve*	13000
5	*Rolfontein Nature Reserve*	4749
6	*Thomas Baines Nature Reserve*	975
7	*Vrolijkheid Nature Conservation Station*	1827
	Orange Free State Division of Nature Conservation	
1	*Erfenis Dam Nature Reserve*	343
2	*Gamefarm Tussen-die-Riviere*	22000
3	*Verwoerd Dam Nature Reserve*	23525
4	*Willem Pretorius Game Reserve*	8278

[a] *In each case the numbers refer to the location of the areas on the maps (Figs. 1-5).*

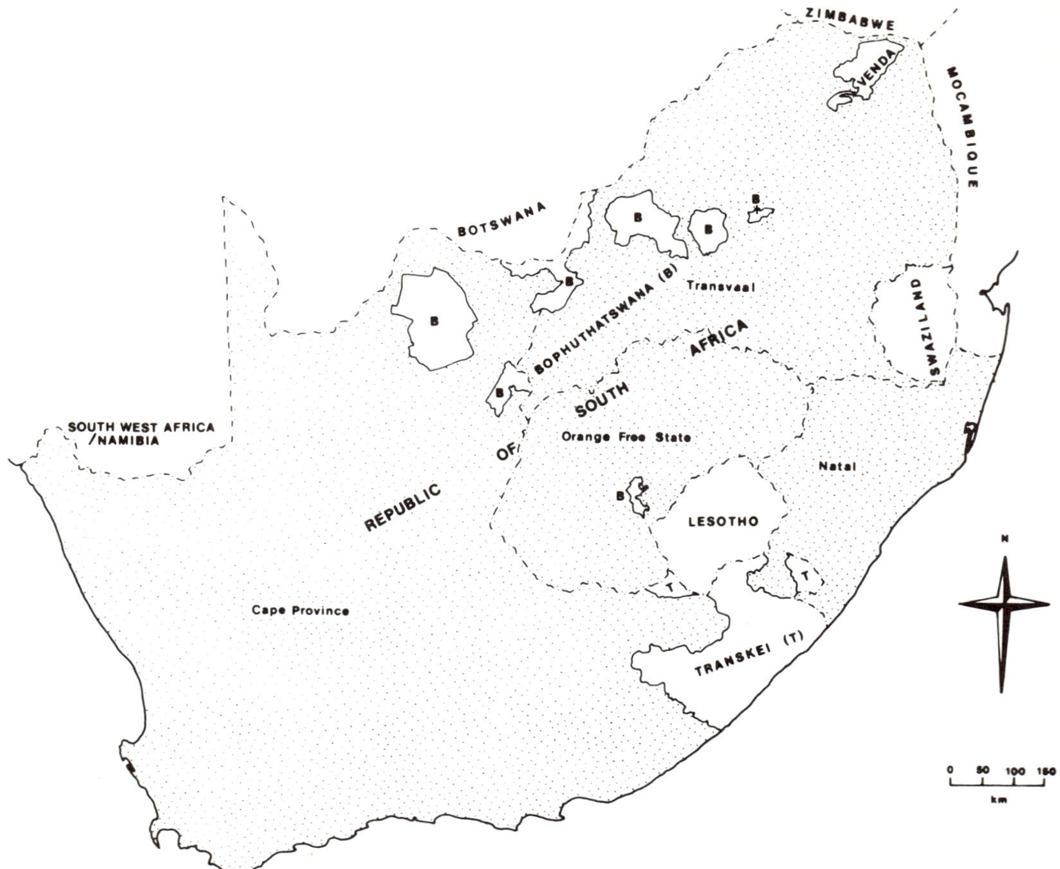
ZIMBABWE
MOCAMBIQUE
VENDA
BOTSWANA
B
Transvaal
SWAZILAND
BOPHUTHATSWANA (B)
REPUBLIC OF SOUTH AFRICA
Orange Free State
SOUTH WEST AFRICA
/NAMIBIA
Natal
LESOTHO
T
Cape Province
TRANSKEI (T)
N
0 50 100 150
km

Figure 1.

(4) Subsequent monitoring programs that were introduced in each case to assess the effectiveness of animal removals in relation to specified management goals.

(5) Methods used in animal removal programs.

(6) Destination and/or use of removed animals.

(7) Revenue generated from animal removal programs (if any) and use to which this revenue was put.

(8) Suggestions (if any) on alternatives to animal removal as a method of management.

(9) Comprehensive list of publications related to any aspect of animal removal programs.

The information received forms the basis of this review of management of locally abundant mammals in South Africa. The mammals referred to in the text are baboon *(Papio ursinus)*, blesbok *(Damaliscus dorcas phillipsi)*, bontebok *(Damaliscus dorcas dorcas)*, buffalo *(Syncerus caffer)*, bushbuck *(Tragelaphus scriptus)*, grey duiker *(Sylvicapra grimmia)*, red duiker *(Cephalophus natalensis)*, eland *(Taurotragus oryx)*, elephant *(Loxodonta africana)*, gemsbok *(Oryx gazella)*, giraffe *(Giraffa camelopardalis)*, hare *(Lepus* sp.), red hartebeest *(Alcelaphus buselaphus)*, hippopotamus *(Hippopotamus amphibius)*, spotted hyena *(Crocuta crocuta)*, impala *(Aepyceros melampus)*, black-backed jackal *(Canis mesomelas)*, kudu *(Tragelaphus strepsiceros)*, lion *(Panthera leo)*, vervet monkey *(Cercopithecus pygerythrus)*, nyala *(Tragelaphus angasi)*, oribi *(Ourebia ourebi)*, roan *(Hippotragus equinus)*, mountain reedbuck *(Redunca fulvorufula)*, southern reedbuck *(Redunca arundinum)*, black rhino *(Diceros bicornis)*, white rhino *(Ceratotherium simum)*, sable *(Hippotragus niger)*, springbok *(Antidorcas marsupialis)*, steenbok *(Raphicerus campestris)*, tsessebe *(Damaliscus lunatus)*, warthog *(Phacochoerus aethiopicus)*, waterbuck *(Kobus ellipsiprymnus ellipsiprymnus)*, black wildebeest *(Connochaetus gnou)*, blue wildebeest *(Connochaetus taurinus)*, Burchell's zebra *(Equus burchelli)* and mountain zebra *(Equus zebra)*. Nomenclature follows Meester and Setzer (1971).

II. NATURE CONSERVATION AUTHORITIES IN SOUTH AFRICA

Fig. 1 is a map of four provincial boundaries in the Republic of South Africa and the location of the three independent homelands of Transkei, Bophuthatswana, and Venda. The National

Fig. 1. The four provincial boundaries in the Republic of South Africa and the location of the three independent homelands of Transkei, Bophuthatswana, and Venda.

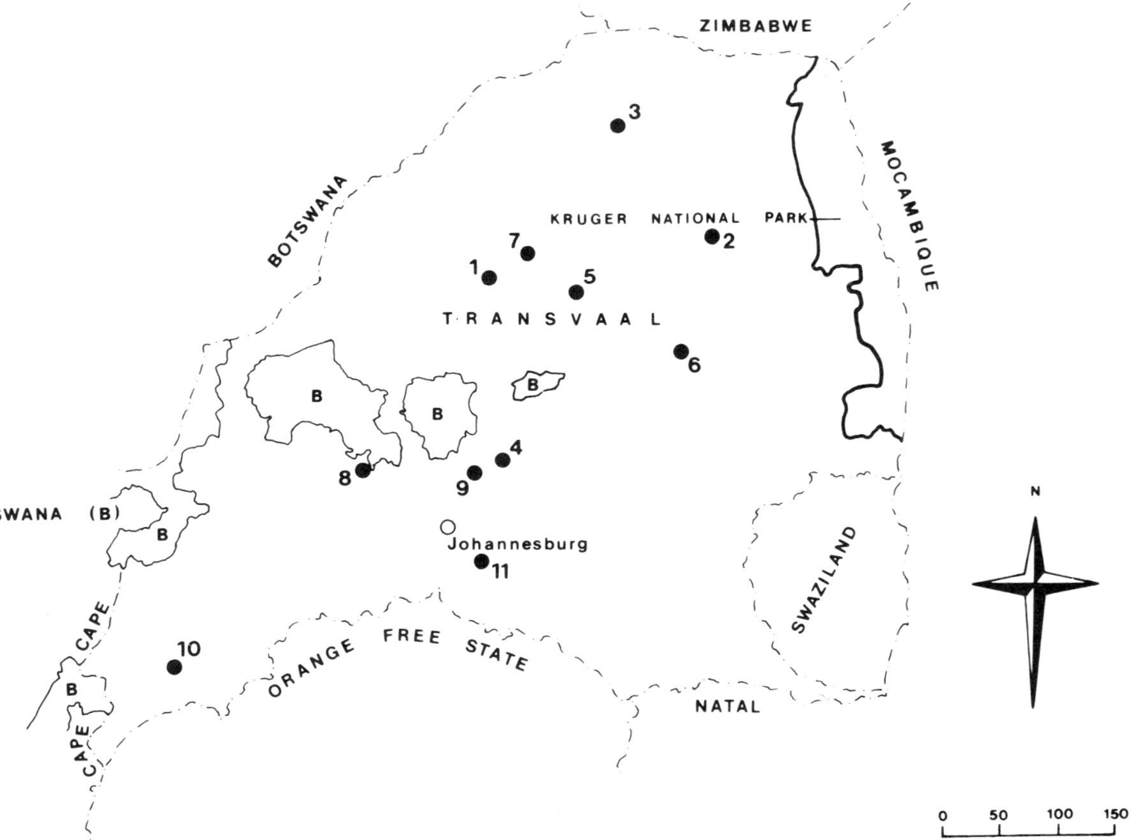
ZIMBABWE
MOCAMBIQUE
BOTSWANA
KRUGER NATIONAL PARK
TRANSVAAL
BOPHUTHATSWANA (B)
B
Johannesburg
SWAZILAND
CAPE
ORANGE FREE STATE
NATAL
1
2
3
4
5
6
7
8
9
10
11
N
0
50
100
150
km

Figure 2.

Parks Board is the only organization concerned with nature conservation on a Republic-wide basis. Nine national parks have been established by the National Parks Board, but in this chapter information is restricted to the Kruger National Park in Transvaal (Fig. 2 and Table I). Each province has in addition its own regional authority in charge of nature conservation. In Transvaal the Nature Conservation Division of the Transvaal Provincial Administration was established in 1949, and at present the Division controls 39 nature reserves throughout the province. In addition, the province has over 500 private nature reserves, with a total area of 1,292,392 ha. The designated areas under the direct control of the provincial administration in which animal removal programs took place during the period of this review are listed in Table I (Fig. 2). In the Cape, the provincial administration established the Department of Nature and Environmental Conservation in 1952. The Department is responsible for 26 provincial nature reserves and 59 subsidized nature reserves, which together with 36 private reserves, cover a total area of 146,343 ha. The seven provincial nature reserves in which management of locally abundant mammals occurred are listed in Table I (Fig. 3). The Orange Free State has a Division of Nature Conservation which was established in 1965. Animal removal programs described in this chapter took place in 4 of the 13 designated areas in the province (Fig. 4 and Table I). In Natal, nature conservation falls under the control of a statutory body known at the Natal Parks, Game and Fish Preservation Board. From 1975 to 1979, locally abundant mammals were removed from 23 of the 52 designated areas under the Board's control (Fig. 5 and Table I). The minor conservation authorities in South Africa described by Hey (1977) are not covered by this chapter as their activities in the field of management of mammal populations are, on the whole, relatively insignificant.

III. THE MANAGEMENT OF LOCALLY ABUNDANT MAMMALS IN SOUTH AFRICA

The data presented from each of the five main nature conservation authorities are not always strictly comparable. Time precluded a more comprehensive analysis for each area, and in addition the detailed information required was not always

Fig. 2. The location of the Kruger National Park in Transvaal, and the 11 provincial nature reserves in which management of locally abundant mammals occurred. For identification of numbers, see Table I.

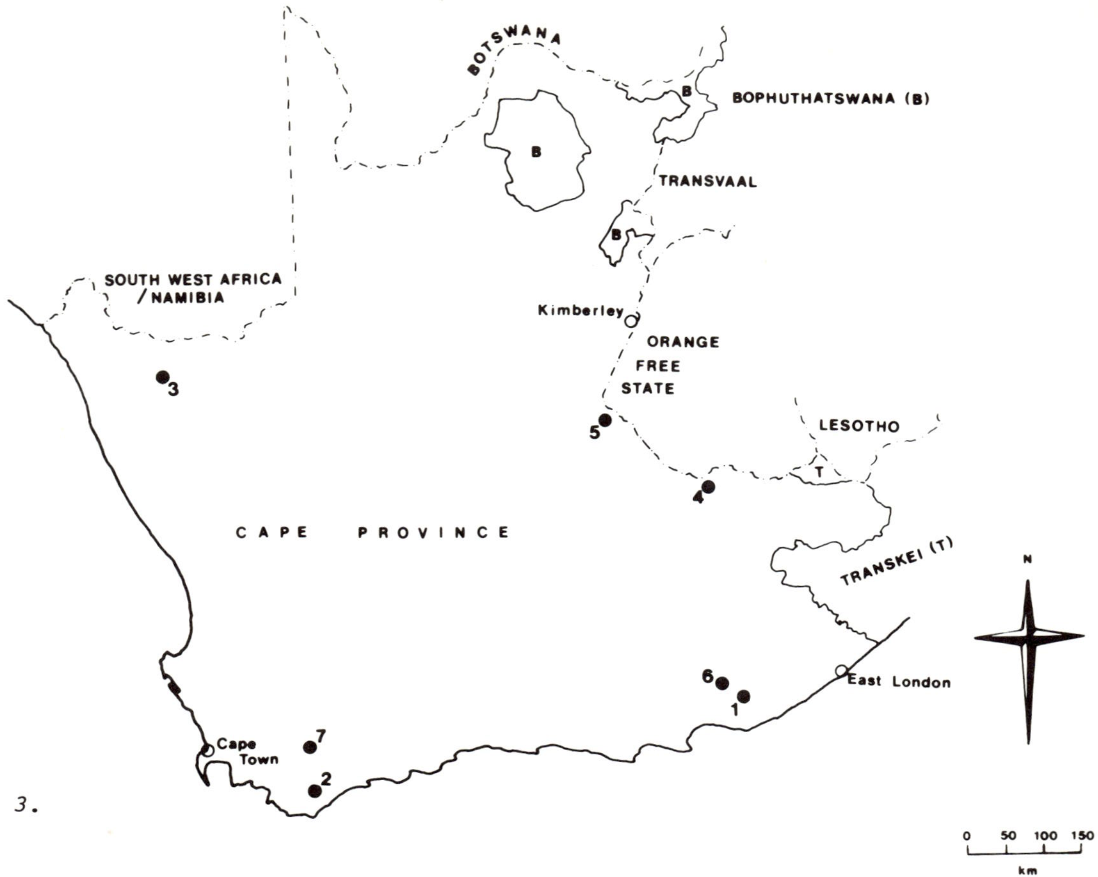
BOTSWANA
BOPHUTHATSWANA (B)
B
B
B
TRANSVAAL
SOUTH WEST AFRICA
/NAMIBIA
Kimberley
ORANGE
FREE
STATE
LESOTHO
T
CAPE PROVINCE
TRANSKEI (T)
N
East London
Cape
Town
1
2
3
4
5
6
7
0 50 100 150
km

Figure 3.

available. Furthermore, each authority has its own nature conservation ordinance, resulting in significant differences in the nomenclature used to describe designated areas and in their systems of management.

(1) Number of each species culled in designated areas.

(a) Natal Parks Board: It is the policy of the Natal Parks Board (NPB) to remove as many animals alive as possible. Animals are culled when difficult terrain, disturbance of other species, and high capture costs and/or time of capture per individual animal make capture operations impracticable. Veterinary restrictions on the translocation of certain species is a further factor inhibiting live capture programs. From 1975 to 1979 inclusive, NPB culled 23,183 mammals from 23 designated areas (Table II). Seventeen species were involved.

(b) Kruger National Park: In the five-year period under review in the Kruger National Park, culling operations were restricted to buffalo, elephant, and hippo; a total of 11,306 were killed (Table III). Culling of impala, blue wildebeest, and Burchell's zebra was terminated five years ago when the populations started declining and/or habitat conditions improved. Culling of lion and spotted hyena populations has recently been initiated in an attempt to reduce the decline in the wildebeest and zebra populations.

(c) Transvaal Nature Conservation Division: The culling data are available in several categories, but with the emphasis the province has placed on data from private nature reserves, the information differs from other provincial authorities. Nevertheless, these data provide a valuable insight into the role of private nature reserves in South Africa. Table IV shows the number of each species for which hunting permits were issued to private nature reserves in Transvaal in 1977, and these totals are more or less equivalent to the mean of the past five years. Permits were issued each year for a total of nearly 8,000 animals from 15 species (Table IV), but this would be an overestimate of the numbers killed, since many owners of private nature reserves do not shoot the maximum numbers allocated to them. Table V shows the number of each species for which cropping permits were issued in Transvaal from 1976-1979, a total of 26,186 individual animals, from

Fig. 3. The location of the seven provincial nature reserves in the Cape Province in which management of locally abundant mammals occurred. For identification of numbers, see Table I.

TABLE II. Number of Each Species Culled from the Natal Parks System from 1975 to 1979

Species	1975	1976	1977	1978	1979	Total
Blesbok	8	32	54	24	32	150
Buffalo	88	46	5	26	107	272
Eland	2	1	0	2	2	7
Giraffe	0	0	0	2	0	2
Hartebeest, red	1	0	0	0	0	1
Hippo	1	5	58	74	82	220
Impala	1101	1780	2169	2247	5457	12754
Jackal, black-backed	0	2	4	1	0	7
Kudu	0	1	0	5	5	11
Lion	22	45	31	25	22	145
Monkey, vervet	0	0	3	5	6	14
Nyala	492	558	474	570	3782	5876
Reedbuck, southern	0	40	40	0	154	234
Warthog	394	289	535	508	1043	2769
Wildebeest, black	2	9	4	0	0	15
Wildebeest, blue	36	29	209	177	253	704
Zebra, Burchell's	0	0	0	2	0	2
						23183

8 species. As with Table IV, it is unlikely that the total was reached in each case.

(d) Cape Department of Nature and Environmental Conservation: During the five years covered by this review, which in the case of the Cape was from 1976 to 1980, no culling was done in the seven provincial nature reserves listed

TABLE III. Number of Each Species Culled from the Kruger National Park from 1975 to 1979

Species	1975	1976	1977	1978	1979	Total
Buffalo	807	1439	1926	1149	3066	8387
Elephant	764	567	285	570	341	2527
Hippo	0	126	116	73	77	392
						11306

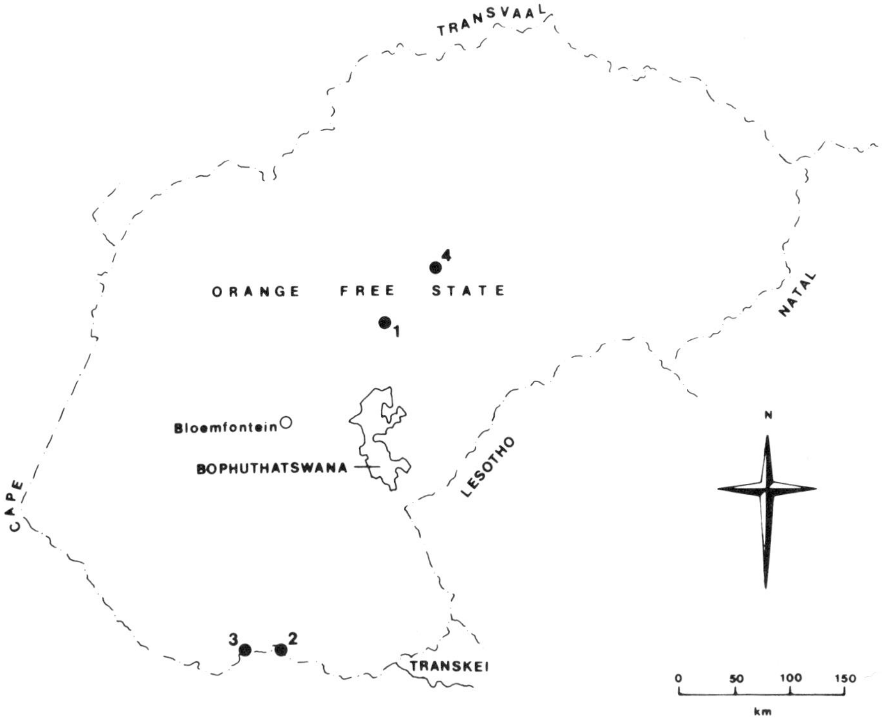

Fig. 4. The location of the four designated areas in the Orange Free State in which management of locally abundant mammals occurred. For identification of numbers, see Table I.

TABLE IV. Number of Each Species for which Hunting Permits Were Issued to Private Nature Reserves in Transvaal in 1977

Species	*1977*	*Species*	*1977*
Blesbok	*554*	*Reedbuck, mountain*	*55*
Bushbuck	*477*	*Sable*	*96*
Duiker, grey	*277*	*Springbok*	*443*
Eland	*72*	*Steenbok*	*28*
Gemsbok	*40*	*Wildebeest, black*	*70*
Hare	*1300*	*Wildebeest, blue*	*72*
Impala	*2865*	*Zebra, Burchell's*	*45*
Kudu	*1590*	*Total*	*7984*

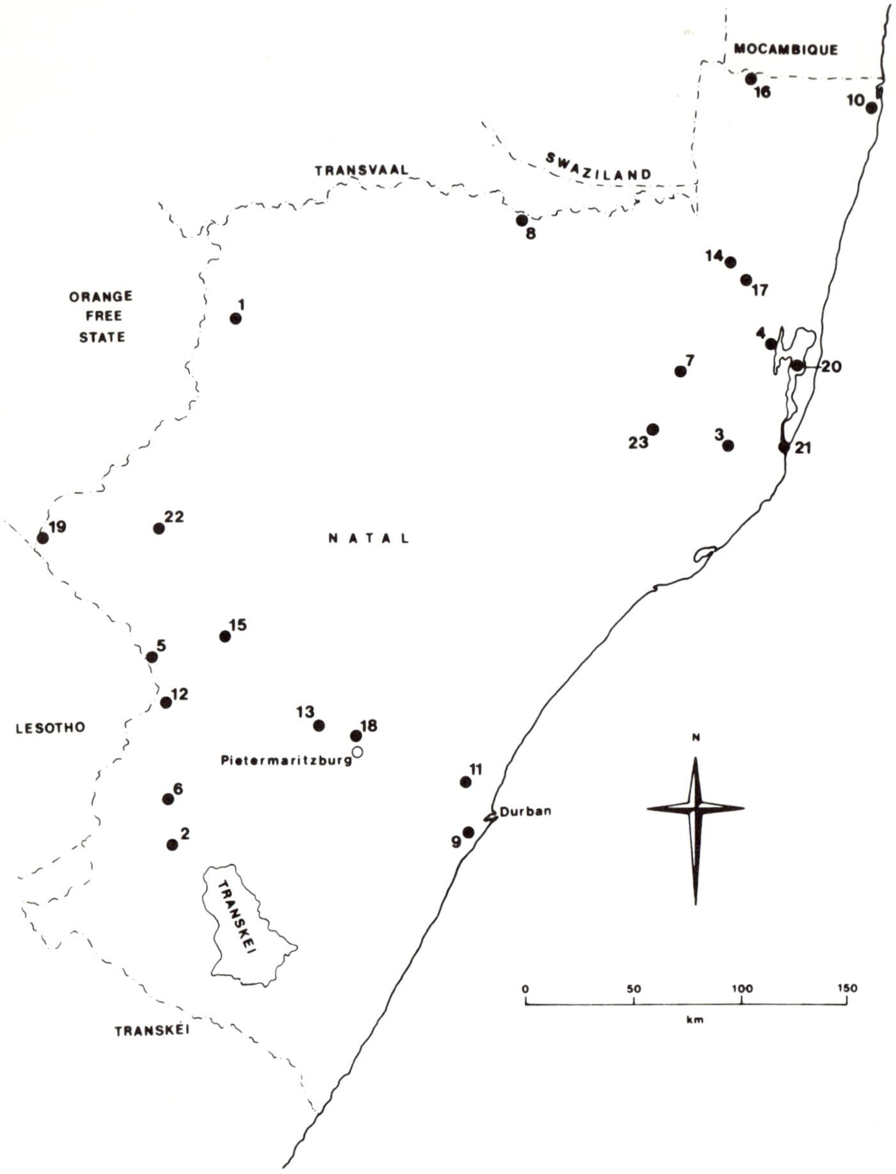

Fig. 5. The location of the 23 designated areas in Natal in which management of locally abundant mammals occurred. For identification of numbers, see Table I.

TABLE V. Number of Each Species for which Helicopter Cropping Permits (a), and Non-Helicopter Cropping Permits (b), Were Issued in Transvaal from 1976 to 1979.

Species	*1976*	*1977*	*1978*		*1979*		*Total*
			(a)	*(b)*	*(a)*	*(b)*	
Blesbok	*0*	*50*	*2642*	*0*	*2864*	*0*	*5556*
Bushbuck	*0*	*0*	*60*	*1*	*1*	*10*	*72*
Impala	*475*	*2575*	*2000*	*3180*	*5200*	*6170*	*19600*
Kudu	*0*	*42*	*135*	*15*	*310*	*0*	*502*
Springbok	*0*	*0*	*300*	*0*	*0*	*0*	*300*
Waterbuck	*10*	*0*	*80*	*0*	*1*	*0*	*91*
Wildebeest, black	*0*	*0*	*0*	*0*	*4*	*0*	*4*
Wildebeest, blue	*0*	*0*	*60*	*1*	*0*	*0*	*61*
							26186

in Table I, but there was an active program of live capture which is described later. Culling did take place in four subsidized nature reserves -- Postberg, Spitkop, Leon Taljaard, and Queenstown -- involving 26 blesbok, 11 eland, 59 gemsbok, 3 red hartebeest, 22 impala, 3 kudu, 150 mountain reedbuck, 332 springbok, 2 black wildebeest, and 2 blue wildebeest in the five-year period.

(e) Orange Free State Division of Nature Conservation: Table VI shows that from 1975 to 1979 a total of 3,439 individual animals from 11 species were culled in the four designated areas in the Orange Free State listed in Table I.

(2) Number of each species live captured in designated areas.

(a) Natal Parks Board: From 1975 to 1979, the Parks Board carried out a very successful live capture program involving 15,057 individuals from 21 species (Table VII), which must be one of the most efficient and comprehensive capture programs in the African continent. Requests or pressure for more animals from live animal purchasers is not permitted to influence removal quotas, which are set

TABLE VI. Number of Each Species Culled from Designated Areas in the Orange Free State from 1975 to 1979

Species	*1975*	*1976*	*1977*	*1978*	*1979*	*Total*
Blesbok	*0*	*51*	*42*	*85*	*71*	*249*
Eland	*0*	*35*	*39*	*10*	*18*	*102*
Gemsbok	*0*	*19*	*11*	*10*	*9*	*49*
Hartebeest, red	*0*	*42*	*19*	*15*	*17*	*93*
Impala	*0*	*11*	*18*	*0*	*0*	*29*
Kudu	*0*	*10*	*11*	*4*	*8*	*33*
Reedbuck, mountain	*114*	*43*	*9*	*6*	*24*	*196*
Springbok	*359*	*450*	*428*	*488*	*618*	*2343*
Wildebeest, black	*0*	*51*	*29*	*90*	*129*	*299*
Wildebeest, blue	*0*	*6*	*2*	*5*	*0*	*13*
Zebra, Burchell's	*0*	*10*	*8*	*8*	*7*	*33*
						3439

solely for conservation management reasons. However, in the case of healthy populations of rare or endangered species, such as black rhino, special requests will be considered for translocation to other reserves. This is classified as removal for conservation reasons.

(b) Kruger National Park: The live capture program in Kruger National Park was of a very limited nature, involving 42 elephant in 1979 and 119 Burchell's zebra in 1975.

(c) Transvaal Nature Conservation Division: From 1975 to 1979 the Transvaal Nature Conservation Division captured 3,096 individuals from 17 species (Table VIII), many of these animals coming from private nature reserves. The Division has a long waiting list for live game of most species. A minimal herd size of three males and five females is adhered to for capture and translocation of rarer species -- such as giraffe, roan, sable, and tsessebe -- as practical experience has shown this to be the most successful minimum herd size. Permits were also issued to the private sector to capture animals, but accurate totals of numbers captured are not available.

(d) Cape Department of Nature and Environmental Conservation: Live capture from the seven provincial nature reserves in the Cape totaled 3,178 individual animals from 9 species (Table IX). An additional 453 individuals from

TABLE VII. Number of Each Species Live Captured from the Natal Parks System from 1975 to 1979

Species	1975	1976	1977	1978	1979	Total
Baboon	0	0	0	0	86	86
Blesbok	93	96	86	39	93	407
Buffalo	0	0	22	0	0	22
Bushbuck	0	0	0	10	0	10
Duiker, red	0	0	10	10	0	20
Eland	13	34	6	9	17	79
Giraffe	0	0	0	0	9	9
Hartebeest, red	17	0	20	12	8	57
Hippo	0	0	0	0	2	2
Impala	2019	1362	1756	989	2433	8559
Kudu	0	26	46	33	47	152
Nyala	21	63	305	584	1285	2258
Reedbuck, southern	0	22	28	208	349	607
Rhino, black	0	0	6	5	5	16
Rhino, white	89	54	24	33	156	356
Springbok	0	0	0	0	27	27
Warthog	179	122	68	0	0	369
Waterbuck	17	11	6	12	8	54
Wildebeest, black	46	37	31	31	11	156
Wildebeest, blue	149	184	219	455	738	1745
Zebra, Burchell's	0	32	12	22	0	66
						15057

9 species were captured from the subsidized nature reserves, comprising 20 blesbok, 57 bontebok, 85 eland, 19 gemsbok, 20 red hartebeest, 60 mountain reedbuck, 181 springbok, 3 black wildebeest, and 8 blue wildebeest.

(e) Orange Free State Division of Nature Conservation: Live capture from the four designated areas in the Orange Free State from 1975 to 1979 involved a total of 3,134 individual animals from 12 species (Table X).

(3) Criteria used to reach decisions to remove animals.

(a) Natal Parks Board: Recommendations for removal are based on an annual assessment of veld condition carried out by field staff, coupled with the latest estimates of animal population numbers and population vigor, as assessed by natural mortality, physiological condition,

TABLE VIII. *Number of Each Species Captured by the Transvaal Nature Conservation Division from 1975 to 1979*

Species	1975	1976	1977	1978	1979	Total
Blesbok	36	0	50	374	261	721
Eland	21	0	0	0	0	21
Gemsbok	0	20	40	0	0	60
Giraffe	8	33	13	19	22	95
Hartebeest, red	29	21	40	66	88	244
Impala	191	48	56	184	162	641
Kudu	16	31	21	38	35	141
Oribi	29	0	0	0	0	29
Reedbuck, mountain	9	0	0	0	0	9
Roan	0	0	12	10	17	49
Sable	19	38	0	29	62	148
Springbok	49	0	4	0	30	83
Tsessebe	24	19	25	65	75	208
Waterbuck	0	27	19	20	7	73
Wildebeest, black	18	0	23	14	38	93
Wildebeest, blue	29	50	21	121	24	245
Zebra, Burchell's	61	9	27	97	42	236
						3096

calving rates, calf survival, and intercalving periods. Veld assessments are undertaken annually at the end of winter and include a visual assessment of soil erosion and an assessment of the extent to which preferred grasses and browse are used. If an area is underutilized, animal removal programs cease and consideration is given to adjusting surplus grazing by either introducing herbivores, mowing, or burning, or a combination of all three. If the extent of utilization is satisfactory, removal is maintained at current rates, based on the mean number removed over the past few years. If there is evidence of overutilization, but a distinct improvement is noticeable, removal is continued at current rates. If overutilization remains a problem and no improvement is evident, current removal rates are arbitrarily increased, and the effects are assessed at the next review. If overutilization is worsening, the current removal rate is increased, and consideration is given to a "capital" reduction over and above the annual rate.

TABLE IX. Number of Each Species Live Captured from Seven Provincial Nature Reserves under the Control of the Provincial Administration of the Cape of Good Hope from 1976 to 1980

Species	*1976*	*1977*	*1978*	*1979*	*1980*	*Total*
Blesbok	1	0	0	12	0	13
Bontebok	9	25	12	15	6	67
Eland	9	0	25	0	11	45
Gemsbok	3	37	0	0	1	41
Hartebeest, red	0	0	0	6	4	10
Kudu	0	0	0	0	39	39
Reedbuck, mountain	0	15	0	17	0	32
Springbok	4	579	491	306	1450	2830
Wildebeest, black	0	0	0	66	35	101
						3178

The NPB research effort is directed toward the development and improvement of population census techniques, detailed long-term vegetation monitoring programs, veld utilization assessment, and assessment of physiological condition. Detailed studies have been completed recently on the biology of several of the mammals affected by removal programs, and some of these investigations have formed the basis of M.Sc. and Ph.D. theses (Anderson, 1978; Attwell, 1977; Hancock, 1978; Jeffery, 1978; Melton, 1978; Owen-Smith, 1973; Venter, 1979; Vincent, 1980). Particular emphasis is given in these and other studies to obtaining information on habitat preference, population dynamics, and interrelationships between various species.

Present policy in NPB areas favors regular annual removals in contrast to irregular operations in response to extreme conditions. The removal rates are reviewed at three to four-year intervals giving time for both the veld and the populations themselves to respond to removal programs.

(b) Kruger National Park: In the Kruger National Park decisions to remove animals are based on three major considerations. These are the impact the overabundant species have on the composition and structure (physiognomy) of the vegetation, food and water resources -- especially during the periods of prolonged drought to which the Transvaal Lowveld is subjected, and associated animal species frequenting the same habitat.

TABLE X. Number of Each Species Live Captured from Designated Areas in the Orange Free State from 1975 to 1979

Species	*1975*	*1976*	*1977*	*1978*	*1979*	*Total*
Blesbok	*162*	*140*	*0*	*7*	*10*	*319*
Eland	*25*	*8*	*0*	*19*	*74*	*126*
Gemsbok	*0*	*8*	*0*	*6*	*19*	*33*
Giraffe	*0*	*0*	*0*	*0*	*7*	*7*
Hartebeest, red	*22*	*23*	*0*	*48*	*38*	*131*
Impala	*26*	*53*	*2*	*7*	*2*	*90*
Reedbuck, mountain	*108*	*35*	*49*	*23*	*23*	*238*
Rhino, white	*0*	*0*	*6*	*0*	*0*	*6*
Springbok	*408*	*286*	*329*	*245*	*574*	*1842*
Wildebeest, black	*108*	*80*	*24*	*15*	*96*	*323*
Zebra, Burchell's	*0*	*3*	*0*	*0*	*0*	*3*
Zebra, mountain	*0*	*0*	*7*	*9*	*0*	*16*
						3134

Primary management objectives in the Kruger National Park are to maintain the diversity of life forms in a naturally interacting and functioning ecosystem. Reduction in the numbers of overabundant species, is, therefore, considered necessary where these species attain densities which could disrupt (or endanger) the natural habitats in which they occur and/or populations of associated animal species. Manipulation of animal populations is considered justifiable only where the numbers (densities) deviate from natural fluctuations due to unnatural pressures.

(c) Transvaal Nature Conservation Division: Removal programs under the direct control of the Transvaal Nature Conservation Division are based on the need to reduce populations well before the carrying capacity of the specific reserve is exceeded; this is weighed against the desirability of taking off surplus animals to form new breeding nuclei elsewhere. (The big program of commercial sustained cropping that takes place in Transvaal falls outside the scope of this chapter and will not be considered further.)

(d) Cape Department of Nature and Environmental Conservation: The decision to remove animals from designated areas is made annually based on the veld condition, number of animals, and in some cases on the need for restocking other designated areas. In those nature reserves where management plans exist, the optimum numbers of the different species have been based on the carrying capacities of the different plant communities and the known feeding preferences of the species involved. In these cases game removal programs are based on the numbers proposed in the management plans and the veld condition.

For most of the larger nature reserves the main management objective is to conserve a representative example of the veld type or ecosystem, with secondary aims of establishing breeding nuclei of indigenous species for restocking purposes and providing facilities for outdoor recreation and nature education.

(e) Orange Free State Division of Nature Conservation: In the Orange Free State a decision is made each year to remove animals from designated areas based on the condition of the animals, the availability of food, and the carrying capacity of specific veld types. Annual population censuses are conducted, using drive counts, strip censuses, or helicopter total counts. Veld types are classified by means of the Braun-Blanquet method, and carrying capacities are determined for each area in relation to veld types and a classification of the grazing animals present.

(4) Subsequent monitoring programs to assess the effectiveness of animal removals in relation to specified management goals.

(a) Natal Parks Board: The vegetation monitoring programs outlined in Section 3a form a base for assessment of the effectiveness of animal removals. In addition to the visual assessment of veld condition, detailed long-term vegetation monitoring programs are being initiated in most of the larger reserves based on fixed transects, exclosure plots, aerial photographic interpretation, and photopoints in different vegetation types. In Mkuzi Game Reserve the relationship between diversity and structure of the various plant communities and the diversity and structure of the large herbivore communities utilizing them is being examined -- a study that should yield important results for future monitoring programs. In addition two important related projects are due to be initiated in late 1981 -- a study of the habitat utilization and the grazing impact of the white rhino in Umfolozi Game Reserve, and a study of the impact of browsing ungulates on the woody plant dynamics of Umfolozi Game Reserve.

(b) Kruger National Park: In the Kruger National Park extensive aerial surveys are conducted annually during which the 15 major herbivore species are censused and their distribution, sex, and age structure are recorded. Surface water is mapped, and a number of environmental variables are recorded -- including the height, cover, growth stage, greenness, litter, utilization, and burning record of the field layer, and the phenology, burning record, and elephant damage in the woody vegetation. The interpretation of this information is related to the overall research program which has been devised to study and interpret the ecology of the Kruger National Park in a holistic, ecosystem approach.

(c) Transvaal Nature Conservation Division: Monitoring programs after game removal are routine in several of the larger reserves, mainly to assess the effect of burning to provide grazing without causing detrimental changes to the habitat. Translocated animals have also been monitored, and these studies have indicated that giraffe herds smaller than 3 males:5 females were seldom successful, and that black wildebeest herds should be in a group of at least 10 males:10 females. Management goals are specified as "the restocking of the full spectrum of indigenous game species into suitable habitats where they formerly occurred."

(d) Cape Department of Nature and Environmental Conservation: The only monitoring program in operation to assess the effectiveness of animal removals is a long-term monitoring of the vegetation by means of descending point methods.

(e) Orange Free State Division of Nature Conservation: Monitoring programs consist of annual game counts, while efforts are made to determine the population dynamics of the different species. The increase in population size determines the overall removal program. The behavior of the different animal species is also studied to determine their concentration in specific areas. The deterioration of veld types where animals concentrate is closely studied and efforts are made to capture these animals specifically when it is noticed that extraordinary deterioration occurs. Veld monitoring programs consist of wheelpoint analysis of the plant cover. Production of the veld is measured by means of plot clipping methods.

(5) Methods used in animal removal programs. In South Africa animal removals from designated areas were initiated in the mid-1950's when animal numbers in various reserves were building up. Early control methods included night shooting and net capture for live capture and translocation. Capture by

drugs (darting) was then developed for certain species such as white rhino and gradually extended to other species. The development of the use of an overdose of scoline administered by a dart to kill animals, particularly elephant and buffalo, came simultaneously. More recently, large-scale live capture of animals has come into widespread use in South Africa, using a helicopter and/or vehicles to drive animals into enclosures (bomas) of plastic sheeting and nets, a method pioneered by the staff of NPB. A detailed description of the techniques in use is beyond the scope of this chapter and the aims of this volume; furthermore the more important techniques have already been described adequately elsewhere. (See the reference list for more information.)

(6) Destination and/or use of removed animals.

(a) Natal Parks Board: Live animals are translocated to other reserves in Natal and other provinces. The majority (80.5%) of animals live captured are sold to private landowners in South Africa (at a subsidized fee in Natal) to promote interest in wildlife. In the five-year period of this review, 201 individuals (1.7% of the total live captured) were sent out of South Africa; 119 of these were white rhino (Table XI).

The animals shot in the Natal parks system from 1975 to 1979 were used as indicated in Table XII. Use by NPB staff includes staff rations, research, and predator bait for lion control programs. In addition, 18 hippo were donated to local black residents.

(b) Kruger National Park: In and immediately around the area of the Kruger National Park, veterinary restrictions, particularly those related to the spread of foot-and-mouth disease, have placed severe limitations on the live sale of animals, which in the last five years has been limited to elephants and zebra. Of the 42 elephant calves live captured, 32 were exported to the United States, 5 died in captivity, and the remainder were translocated locally. The majority of animals removed from the park are used for fresh meat production, biltong, and canned meat; the hides and ivory are also used.

(c) Transvaal Nature Conservation Division: Animals removed by live capture are moved to ecologically suitable areas for restocking. Culled animals are sold as venison.

(d) Cape Department of Nature and Environmental Conservation: Animals removed by live capture are used for stocking other reserves or sold to farmers.

(e) Orange Free State Division of Nature Conservation: Captured animals are used for reintroductions in new game reserves and for stocking farms. Culled animals are used for local meat production.

TABLE XI. Destination of Animals Live Captured from the Natal Parks System from 1975 to 1979

Species	Natal Parks Board	South African conservation authorities	South African private farms	Bophuthatswana and Transkei
Blesbok	61	0	208	6
Buffalo	22	0	0	0
Baboon	86	0	0	0
Duiker, red	0	0	10	0
Eland	44	0	0	0
Giraffe	4	0	0	0
Hartebeest, red	50	0	4	0
Hippo	0	2	0	0
Impala	746	0	6300	0
Kudu	115	0	21	0
Nyala	34	0	1219	4
Reedbuck, southern	176	70	350	10
Rhino, black	10	5	0	0
Rhino, white	32	14	147	12
Springbok	0	0	27	0
Warthog	165	0	197	0
Waterbuck	49	0	6	0
Wildebeest, black	66	0	46	0
Wildebeest, blue	248	0	1135	11
Zebra, Burchell's	69	0	2	0
Total	1977	91	9672	43

South-West Africa/ Namibia	Botswana	Zimbabwe	Kenya	South African zoos	Countries outside of Africa
0	0	0	0	1	0
0	0	0	0	0	0
0	0	0	0	0	0
0	0	0	0	0	0
0	0	0	0	0	0
0	0	0	0	0	0
0	0	0	0	0	0
0	0	0	0	0	0
0	0	0	0	20	21
0	0	0	0	10	0
0	0	0	0	0	31
0	0	0	0	0	0
0	0	0	0	0	0
34	12	7	9	7	57
0	0	0	0	0	0
0	0	0	0	0	0
0	0	0	0	0	0
0	0	0	0	0	0
0	0	0	0	0	30
0	0	0	0	0	0
34	12	7	9	38	139

(7) Revenue generated from animal removal programs.

(a) Natal Parks Board: From 1975 to 1979, R1,108,078 was generated from the sale of live animals, meat, and by-products (Table XIII). Revenue from live sales is set aside in a special Land Acquisition Fund to help the NPB purchase new land for conservation. During the period under review, this amounted to R729,785 (Table XIII). All other revenue received goes into the NPB's general revenue account.

(b) Kruger National Park: The revenue received from the products derived from removal operations includes the sale of fresh meat, canned meat, biltong, hides, and ivory, and this totaled R4,123,820 from 1975 to 1979 (Table XIV). However, the expenditure and depreciation associated with these operations totaled R2,804,820, giving an overall net income for the period of R1,319,000.

TABLE XII. Use of Animals Shot in the Natal Parks System from 1975 to 1979

Species	*Used by Natal Parks Board staff*	*Sale of meat*
Blesbok	*81*	*69*
Buffalo	*177*	*78*
Eland	*7*	*0*
Hartebeest, red	*1*	*0*
Hippo	*274*	*7*
Impala	*4670*	*6697*
Kudu	*0*	*0*
Nyala	*2933*	*2912*
Reedbuck, southern	*80*	*154*
Warthog	*2751*	*0*
Wildebeest, black	*15*	*0*
Wildebeest, blue	*347*	*224*
Zebra, Burchell's	*2*	*0*
	11338	*10151*

TABLE XIII. Revenue Generated from Animals Removed in the Natal Parks System from 1975 to 1979

Source of revenue	1975 R	1976 R	1977 R	1978 R	1979 R	Total R
Live sales	116,972	62,406	159,799	175,358	331,242	845,777
Carcass sales	0	1,429	2,328	47,722	209,398	260,877
By-product sales	405	347	672	0	0	1,424
	117,377	64,182	162,127	223,080	540,640	
						R 1,108,078
Amount paid to Land Acquisition Fund	4,541	58,845	159,799	175,358	331,242	
						R 729,785

TABLE XIV. Revenue Generated from Animals Removed in the Kruger National Park from 1975 to 1979.

Year	Income	Expenditure and depreciation	Net income
	R	R	R
1975	636,744	501,903	134,841
1976	621,121	453,564	167,557
1977	840,200	528,067	312,133
1978	1,100,401	618,093	482,308
1979	925,354	703,193	222,161
Total	4,123,820	2,804,820	1,319,000

(c) Transvaal Nature Conservation Division: Revenue generated by the Division goes into the provincial account. From 1975 to 1979 this totaled R116,229 (but this includes revenue from birds and fish). Additional funds are derived from the hiring out of capture apparatus and crates.

(d) Cape Department of Nature and Environmental Conservation: The revenue obtained from the sale of live captured animals from the seven provincial nature reserves (Table IX) amounted to R24,726, and this was paid into the provincial account.

(e) Orange Free State Division of Nature Conservation: During the period of this review, revenue generated from animal removal programs totaled R203,700; this was paid into the provincial account.

(8) Suggestions on alternatives to animal removal as a method of management of locally abundant mammals. No suggestions were received on realistic alternatives to the present programs of culling and/or live capture.

IV. DISCUSSION

In South Africa management of locally abundant mammals has become a regular annual event in the designated wildlife areas. During the period of this review, approximately 104,000 individuals from 26 species were culled. However, of equal importance and interest is the fact that 25,079 individuals from 30 species were live captured and translocated to other areas, the majority of these animals (80.5% in the case of Natal) being sold to private landowners in South Africa. In that the state and provincially-administered designated wildlife areas in South Africa occupy a mere 3% of the land surface of the Republic compared with the 75% covered by privately-owned rural land (Mentis, 1980), coupled with the fact that many of these designated areas are small (Table I) and isolated from one another, private landowners in South Africa are assuming increasing importance in the conservation of mammals and their habitats. The recent publication of the *World Conservation Strategy* (IUCN, 1980), which has placed great emphasis on a positive linking of conservation and development and which defined conservation as "the management of human use of the biosphere so that it may yield the greatest sustainable benefit to present generations while maintaining its potential to meet the needs and aspirations of future generations," should encourage private landowners to introduce species that are at present being removed from designated areas. An inevitable accompaniment of these introductions is the retention of adequate areas of indigenous vegetation as suitable habitats, although priority is nearly always given by farmers to those species which have their living requirements met by the prevailing land use practices and which do not conflict with the farmer's aspirations (Mentis, 1980). One of the best examples of a suitable species for introduction is the impala, which increasingly is being used as a secondary crop under extensive beef ranching in the savanna areas. Recent estimates indicate that under such conditions, with one impala per steer stocked on the land, it is possible for a farmer to double his net income compared with pure beef ranching (Collinson, 1979). Springbok and blesbok are two other species which have already proved their value and suitability for introduction into farming areas and subsequent utilization. Such introductions have great potential for expansion elsewhere in the continent, but the relevant research and advisory services should first of all give consideration to the following (Mentis, 1980):

(1) Identify which species are compatible with what types of land use.
(2) Accumulate data, by literature survey and field study, on the life histories of each species.
(3) Develop, test, and refine strategies for efficiently providing appropriate farm game habitat and for managing and utilizing the animals.
(4) Prepare simple how-to-do-it literature, understandable to the farmer, on farm game management and utilization.

In addition further consideration should be given to the problem of the minimum group size and composition that is necessary for successful introductions.

The revenue generated from animal removal programs is usually lost to the conservation authorities concerned. The approach of the Natal Parks Board, where revenue generated from live capture (R729,785 during the five years covered by this review) is paid into a Land Acquisition Fund, deserves consideration by other conservation authorities in Africa, although this will probably not be possible within a state financing system.

From the brief discussion on the theory of island biogeography in the introduction to this chapter, the importance of long-term genetic conservation must receive further attention in the management planning of small designated areas. Few, if any, of the wildlife sanctuaries of Africa have placed emphasis on genetic conservation, and if the speed at which biotas in small "islands" collapse to lower equilibria is historical in scale rather than geological, then the maintenance of an effective *minimum* population size that guarantees long-term survival in nature, and the determination of that minimum population, assumes great significance. Present culling and translocation programs are invariably dictated solely by habitat considerations, but an ignorance of genetic principles could cause irreversible genetic damage. For example, culling of complete herds of animals in one area of a reserve could erode genetic variation faster than is necessary since the genetic relatedness of individuals within social groups will usually be greater than the relatedness of individuals between groups. These considerations are particularly relevant to the small designated areas in South Africa where the culling of complete groups could seriously deplete genetic variation. As an alternative, the culling of selected individuals from all of the groups is less likely to eliminate rare alleles or sets of coadapted alleles (Soule' and Frankel, 1980). In summary, genetic considerations *must* become an integral part of the management planning of wildlife sanctuaries.

Local overabundance of mammals will continue to be a problem in South Africa, and it will almost certainly become more of a problem in the rest of Africa as the designated areas become more isolated and smaller. It is important to recognize the development of overabundance before serious environmental degradation has occurred. The approach of the Natal Parks Board in this regard has considerable merit, and this aspect will not be considered further in this chapter. However, in addition to veld assessment and monitoring, consideration should be given to characterizing mammal population condition in terms of both physiological condition and demographic vigor (Hanks, 1981) as indicators of overabundance. Deposited fat reserves, adrenal cortical hypertrophy, blood chemistry and hematology, urinary excretion of hydroxyproline, and body growth have all been used and considered as indices of physiological condition, but deposited fat reserves as indexed by perinephric and bone marrow fat appear to be the most useful and least contentious methods for assessing physiological condition. The use of a single statistic (such as r_s) to describe demographic vigor presents many problems (Hanks, 1981); ideally, demographic vigor should be assessed together with physiological condition.

If there is a problem of overabundance, realistic options for management action appear to be limited to culling and/or translocations. The conservation authorities in this review did not have any alternative suggestions.

With the increase in sophistication and specialization of wildlife research, there is a danger that the professional wildlife biologist could become alienated from his nonprofessional field counterpart, and consequently the results of research programs that were initiated to provide answers to specific management problems will remain as academic exercises and will not be translated into action on the ground. Both the wildlife managers *and* the biologists have a duty to try to overcome this rapidly developing communications gap, especially in the sphere of the management of locally abundant mammals.

ACKNOWLEDGMENTS

This chapter would not have been possible without the help of the various South African conservation authorities, and we would like to acknowledge all those who assisted with data gathering and gave permission for these data to be used. Mrs. J. Hancock typed the manuscript.

REFERENCES[1]

Anderson, J. L. (1978). Aspects of the ecology of the nyala *(Tragelaphus angasi* Gray, 1849) in Zululand. Ph.D. Thesis, University of London.

Anderson, J. L. (1979). The social organization and aspects of behaviour of the nyala *Tragelaphus angasi* Gray, 1849. *Z. Säugetierkunde 45 (1980)*, 90-123.

Attwell, C. A. M. (1977). Reproduction and population ecology of the blue wildebeest *Connochaetes t. taurinus* in Zululand. Ph.D. Thesis, University of Natal.

Bourquin, O. (1973). Utilization and aspects of management of the Willem Pretorius Game Reserve. *J. So. Afr. Wildl. Manage. Assoc. 4(1)*, 51-58.

Brooks, P. M. (1975). The sexual structure of an impala population and its relationship to an intensive game removal programme. *Lammergeyer 22*, 1-8.

Brooks, P. M., Hanks, J., and Ludbrook, J. V. (1977). Bone marrow as an index of condition in African ungulates. *So. Afr. J. Wildl. Res. 7(2)*, 61-66.

[1] *In addition to references referred to in the text, this list includes other studies related to the management of locally abundant mammals in South Africa.*

Collinson, R. F. H. (1979). Production economics of impala. *Hlabisa Soil Conserv. Comm. Symp. 3.* (In press).

Densham, W. D. (1974). A method of capture and translocation of wild herbivores using opaque plastic material and a helicopter. *Lammergeyer 21,* 1-25.

Densham, W. D. and Tomkinson, A. J. (1979). Economics of game cropping. *Lammergeyer 27,* 18-24.

Diamond, J. M. and May, R. M. (1976). Island biogeography and the design of natural reserves. *In* "Theoretical Ecology. Principles and Applications" (R. M. May, ed.) Blackwell, Oxford.

Dodds, D. G. and Patton, D. R. (1968). "Wildlife and Land-use Survey of the Luangwa Valley." U. N. Food and Agriculture Organization, Rome.

Hancock, P. G. (1978). Dynamics of an overutilized area in the Ndumu Game Reserve Tongaland. M.Sc. Thesis, University of Natal.

Hanks, J. (1979). "A Struggle for Survival. The Elephant Problem." C. Struik, Cape Town.

Hanks, J. (1981). Characterization of population condition. *In* "Population Dynamics of Large Mammals" (C. W. Fowler, ed.) Wiley, New York.

Hanks, J., Cumming, D. H. M., Orpen, J. L., Parry D. F., and Warren, H. B. (1976). Growth, condition and reproduction in the impala ram (*Aepyceros melampus*) *J. Zool., London 179,* 421-435.

Hey, D. (1977). The history and status of nature conservation in South Africa. *In* "The History of Scientific Endeavour in South Africa" (A. C. Brown, ed.) Royal Society of South Africa, Cape Town.

Holt, S., Laws, R. M., Beddington, J., Harwood, J., and Jewell, P. (1979). The dilemma of local overabundance of protected mammals and the possible means to resolve it: a preliminary statement. Unpublished manuscript. LAM Workshop Steering Committee, Cambridge.

International Union for the Conservation of Nature and Natural Resources (1980). "World Conservation Strategy." IUCN, Gland, Switzerland.

Jeffery, R. C. V. (1978). Age determination, growth and condition of a population of eland *Taurotragus oryx* (Pallas, 1766) under semi-intensive management at Coleford Nature Reserve. M.Sc. Thesis, University of Natal.

Laws, R. M. (1969). The Tsavo research project. *J. Reprod. Fertil. Suppl. 6,* 495-531.

Laws, R. M. (1970). Elephants as agents of habitat and landscape change in East Africa. *Oikos 21,* 1-15.

Laws, R. M., Parker, I. S. C., and Johnstone, R. C. B. (1975). "Elephants and Their Habitats. The Ecology of Elephants in North Bunyoro, Uganda." Oxford Univ. Press (Clarendon), London and New York.

Meester, J. and Setzer, H. W. (1971). "The Mammals of Africa. An Identification Manual." Random House (Smithsonian Inst. Press), New York.

Melton, D. A. (1978). Ecology of waterbuck *Kobus ellipsiprymnus* (Ogilby, 1833) in the Umfolozi Game Reserve. D.Sc. Thesis, University of Pretoria.

Mentis, M. T. (1970). Estimates of natural biomasses of large herbivores in the Umfolozi Game Reserve Area. *Mammalia 34(3)*, 363-393.

Mentis, M. T. (1972). A review of some life history features of the large herbivores of Africa. *Lammergeyer 16*, 1-89.

Mentis, M. T. (1974). Distribution of some wild animals in Natal. *Lammergeyer 20*, 1-68.

Mentis, M. T. (1977). Stocking rates and carrying capacities for ungulates on African rangelands. *So. Afr. J. Wildl. Res. 7(2)*, 89-98.

Mentis, M. T. (1978). Economically optimal species-- mixes and stocking rates for ungulates in South Africa. *Proceedings of the First International Rangeland Congress 1978*, 146-149.

Mentis, M. T. (1980). Game conservation on privately owned rural land. Paper presented at the meeting of the Zimbabwe Grassland Group, Zimbabwe, August 1980.

Mentis, M. T., Collinson, R. F. H., and Wright, M. G. (1978). Legal harvest of protected game in Natal, 1973-1976. *Lammergeyer 24*, 13-18.

Mentis, M. T. and Duke, R. R. (1976). Carrying capacities of natural veld in Natal for large wild herbivores. *So. Afr. J. Wildl. Res. 6(2)*, 65-74.

Miller, R. I. (1978). Applying island biogeography theory to an East African reserve. *Environ. Conserv. 5(3)*, 191-195.

Miller, R. I. (1979). Conserving the genetic integrity of faunal populations and communities. *Environ. Conserv. 6(4)*, 297-304.

Miller, R. I. and Harris, L. D. (1977). Isolation and extirpations in wildlife reserves. *Biol. Conserv. 12*, 311-315.

Naylor, J. N., Caughley, G. J., Abel, N. D. J. and Liberg, O. (1973). "Luangwa Valley Conservation and Development Project, Zambia. Game Management and Habitat Manipulation." U. N. Food and Agriculture Organization, Rome.

Owen-Smith, R. N. (1973). The behavioural ecology of the white rhinoceros. Ph.D. Thesis, University of Wisconsin.

Pienaar, U. De V. (1969). Why elephant culling is necessary. *Afr. Wildl. 23(3)*, 180-194.

Pienaar, U. De V., Van Niekerk, J. W., Van Wyk, P., and Fairall, N. (1966). An experimental cropping scheme of hippopotami in the Letaba River of the Kruger National Park. *Koedoe 9*, 1-33.

Pienaar, U. De V., Van Wyk, P., and Fairall, N. (1966). An aerial census of elephant and buffalo in the Kruger National Park, and the implications thereof on intended management schemes. *Koedoe 9*, 40-107.

Polunin, N. and Eidsvik, H. K. (1979). Ecology principles for the establishment and management of national parks and equivalent reserves. *Environ. Conserv. 6(1)*, 21-26.

Reilly, G. (1980). Population jumps 27% in 10 years. Rand Daily Mail. September 12, 1980.

Smuts, G. L. (1974). Growth, reproduction and population characteristics of Burchell's Zebra (*Equus burchelli antiquorum,* H. Smith, 1841). D.Sc. Thesis, University of Pretoria.

Smuts, G. L. (1974). Game movements in the Kruger National Park and their relationship to the segregation of subpopulations and the allocation of culling compartments. *J. So. Afr. Manage. Assoc. 4(1)*, 51-58.

Smuts, G. L. (1975). Home range sizes for Burchell's zebra *Equus burchelli antiquorum* from the Kruger National Park. *Koedoe 18*, 139-146.

Smuts, G. L. (1978). Interrelations between predators, prey, and their environment. *BioScience 28(5)*, 316-320.

Soulé, M. and Frankel, O. (1980). Nature reserves. *In* "Conservation Genetics." (In press).

Starfield, A. M., Smuts, G. L., and Shiell, J. D. (1976). A simple wildebeest population model and its application. *So. Afr. J. Wildl, Res. 6(2)*, 95-98.

Terborgh, J. (1974). Preservation of natural diversity: the problem of extinction prone species. *BioScience 24(12)*, 715-722.

Thomson, P. J. (1975). The role of elephants, fire and other agents in the decline of a *Brachystegia boehmii* woodland. *J. So. Afr. Wildl. Manage. Assoc. 5(1)*, 11-18.

Venter, J. (1979). Ecology of the southern reedbuck *Redunca arundinum* Boddaert 1785 on the eastern shores of Lake St. Lucia, Zululand. M.Sc. Thesis, University of Natal.

Vincent, J. (1980). The population dynamics of impala *(Aepyceros melampus)* in Mkuzi Game Reserve. Ph.D. Thesis, University of Natal.

Vuilleumier, F. (1970). Insular biogeography in continental regions; 1, the northern Andes of South America. *Am. Nat. 104(938)*, 373-388.
Young, E., ed. (1973). "The Capture and Care of Wild Animals." Human and Rousseau, Cape Town.

STABILITY PROPERTIES OF SEMIARID SAVANNAS IN SOUTHERN AFRICAN GAME RESERVES

Brian H. Walker

Center for Resource Ecology
University of the Witwatersrand
Johannesburg
South Africa

I. INTRODUCTION

The question which we are being asked to consider in this volume is essentially this: With regard to large mammal species that have become locally very abundant, can we avoid having to kill them, and if not, how few can we kill in order to avoid some unacceptable change? In most instances the unacceptable change will apply not to the abundant mammal itself, but to some associated species, such as a competitor or one of its favored food species (i.e., one of Caughley's [this volume] Class 2 reasons for culling). In other cases, however, the decision may be to allow any change short of complete extinction of the mammal itself.

As a result of these two views there has developed a dichotomy into the two extremes of strict controlled culling to keep things constant, and laissez faire. The dichotomy is, however, unnecessary, since there are various degrees of control. The essence of the argument which follows is that it is necessary to try to determine the limits of change in each of the key variables of the system, beyond which irreversible and unacceptable changes occur, and then to allow all changes up to these limits to take place.

II. EFFECTS OF CULLING ON ECOSYSTEM RESILIENCE

This section begins with a series of admittedly untested hypotheses, and follows with two examples to back up these statements. Restraining the normal dynamics of a population

ISBN 0-12-385280-3

or some community of populations, attempting to keep it constant near some defined (or supposed) equilibrium -- often by culling -- is usually counterproductive with respect to what is hoped will be achieved. It leads to a decline in the resilience of the system concerned, and it is therefore important to let the natural variation in numbers take place.

A stable equilibrium, by definition, means that if the system (or population) is moved away from the equilibrium value it will tend to return to that equilibrium. However, unless it is globally stable (which is rare) the equilibrium has some bounds, or threshold, and if the variability in the system allows it to exceed this threshold, then the system will not return to its original state. In general terms the amount of change from the equilibrium to the threshold is a measure of the system's resilience. Once the threshold has been exceeded, the system will then either tend to some new equilibrium density (in the case of a single species) or it may change completely to something else -- e.g., the change from grassland to chaparall in the southwestern United States, and from grassland to Karoo shrubland in South Africa.

Under normal circumstances the systems concerned have evolved a variety of mechanisms which ensure that the variability does not exceed, or very seldom exceeds, the bounds. Most of the culling problems in African savannas arise from the fact that through previous interference (reduction of area, fencing, introduced water supplies, predator control, artificial burning, elimination of fire) we have altered the system so that it can now exceed the bounds. The art of management is to know when the system is approaching these bounds, and when it is safe to let the populations vary.

With this introductory statement in mind I would like to make five sequential points. None of them is new and they have been stated more eloquently by others, but they are sufficiently important to bring together at this stage.

(1) With respect to any problem species, the analysis and understanding of the problem, and therefore its solution, first requires the identification of the essential subsystem of which that species is a part -- in particular, which variables must be included, and which are the processes that essentially govern their dynamics.

(2) With regard to this minimum set of variables, is there a potential for multiple equilibria? If so, the thresholds between these equilibria are the key to the management of the system. In particular we need to know what equilibria are (i.e., between which variables?), where they are (i.e., the relative amounts of the variables concerned), where the stability boundaries are (or thresholds?), and finally, what determines the existence and positions of these boundaries.

(3) The boundaries around equilibria are not constant. They change in response to external or environmental change. In addition, as a result of parameter evolution in the system, they may change intrinsically, especially if the normal dynamics of the system are altered.

(4) These internal parameter changes often lead to a contraction of the boundaries around an equilibrium if the system is always prevented from approaching these bounds. To keep the system in a resilient state, such that it can undergo considerable change and yet remain within its bounds, it must be allowed to change.

(5) Very often, many of our management actions which are aimed at conserving ecosystems and species are the opposite of what we should do. We cull animals when they increase, and we save them or introduce more when they decrease, to keep things constant. We spread water supplies evenly all over the reserves to remove spatial variation, and we use patch burning for the same reason. Most of our actions are in fact aimed at preventing the variability in time and space which in all probability is essential for the maintenance of ecosystem resilience.

In support of these statements, two examples follow, which together illustrate each of the five points.

III. THE TREE/GRASS BALANCE IN SEMIARID SAVANNAS

In an earlier paper (Walker and Noy-Meir, in press) the dynamics of grass *(G)* and woody vegetation *(W)* have been analyzed in terms of Walker's (1971) hypothesis that semiarid savannas depend on grass roots being more efficient than trees in taking up water from the upper layer of the soil, but that trees have exclusive access to some deeper layer of soil water. Provided there is sufficient rain for some water always to get through to the lower layer, but insufficient for a woodland to develop, a single stable equilibrium of *G* and *W* can in fact occur.

This model was then extended (Walker *et al.*, in press) to consider the effects of grazing at various stocking rates. The analysis was aimed at the changes from open grassland savanna to thicket, which have occurred on cattle ranches, so browsing effects were omitted. Four important assumptions underlie the results of this analysis. First, the vegetation consists of only two variables, grass and woody vegetation. (Later, an important subdivision of grass becomes apparent.) Second, the competition between *G* and *W* is determined only by available water, divided into topsoil *(T)* and subsoil *(S)* water. Third, soils are medium to heavy textured and as *G* declines, soil

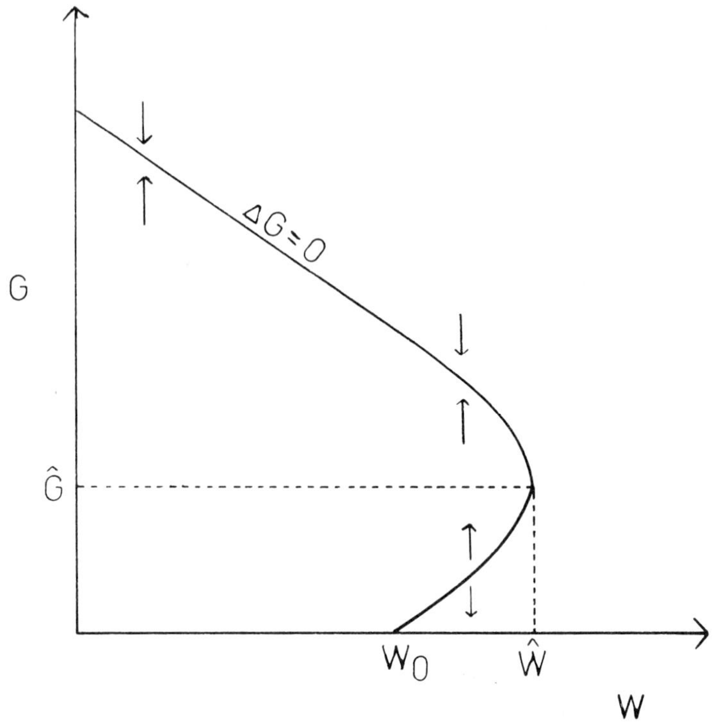

Fig. 1. The grass zero-isocline in the grass (G)-woody vegetation (W) plane. The lower, recurved section of the isocline, below $\hat{G}$, represents the set of unstable equilibria, or threshold. (From Walker et al., *in press, by permission of the Journal of Ecology.)*

capping occurs, resulting in a reduction in water infiltration down to around one tenth of the maximum when $G = 0$. Fourth, per capita consumption of G is zero at some positive amount of G (i.e., not all G can be eaten) and increases to satiation at imtermediate levels.

The details of the analysis are given in the publication referred to above. In essence, the results are as follows. The isocline for zero change in grass (i.e., the amounts of G at fixed levels of W where G is in equilibrium) has the general shape as shown in Fig. 1. The unstable, recurved section of the isocline results from the reduction in infiltration (and therefore reduced growth) at low G. The corresponding isocline for zero change in W appears as in Fig. 2. The lower asymptote in W is determined by its exclusive access to subsoil water,

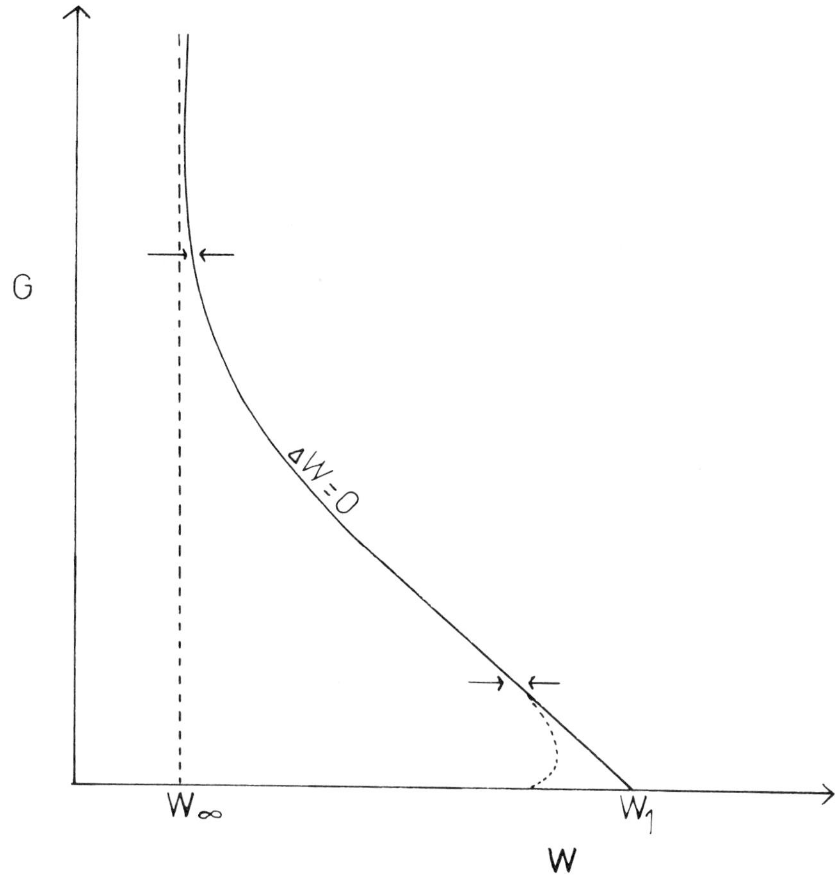

Fig. 2. The zero-isocline for W in the G-W plane. The value of W_∞ depends largely on the amount of subsoil water. (From Walker et al., *in press, by permission of the Journal of Ecology.)*

The combined dynamics of *G* and *W* can lead to various configurations of the two isoclines, depending on the parameters of the model (relative uptake rates for water, ratio of *T*:*S*, the minimum infiltration value, and others). Fig. 3 represents the most likely combination under semiarid conditions on medium textured soils which are over one meter deep. The two intersections of the isoclines give rise to two stable equilibria (one with high *G* and low *W*, the other with *W* alone) separated by an unstable equilibrium. The *G*-*W* space is thus divided into two stable regions by the separatrix which passes through the unstable equilibrium. (This separatrix represents one of the thresholds referred to earlier in the paper.)

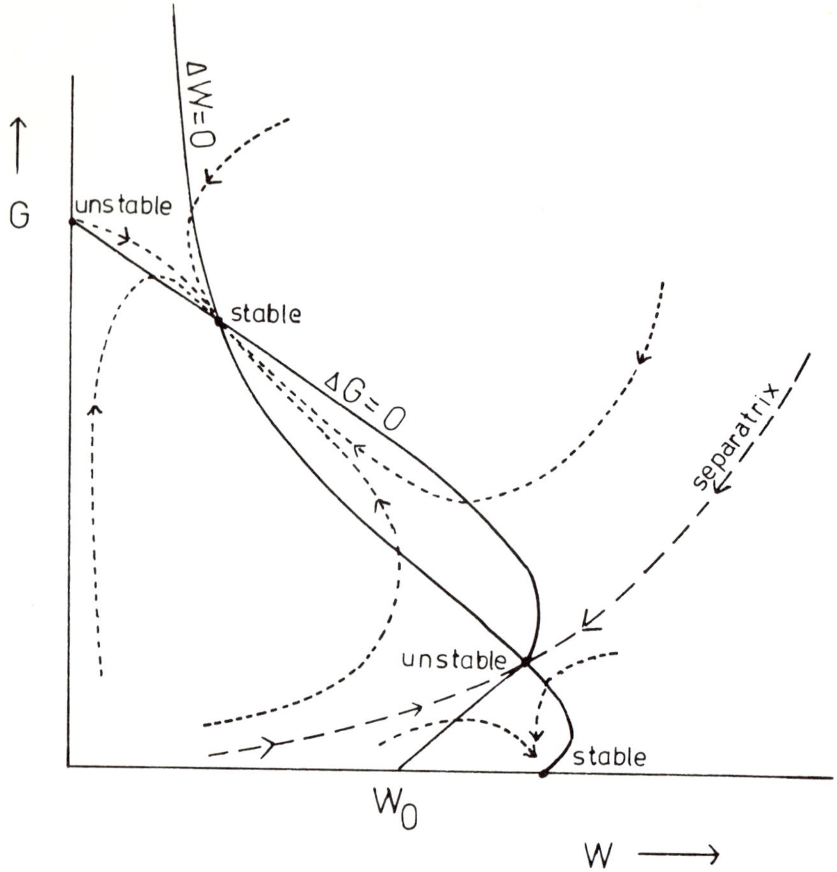

Fig. 3. Likely relative position of the G and W zero-isoclines, resulting in two stable and two unstable equilibria. The G-W plane is divided into two stable regions by the separatrix which passes through to joint, unstable equilibrium. The dynamics of the G-W system are indicated by the fine dashed lines. (From Walker et al., *in press, by permission of the Journal of Ecology.)*

The effects of grazing on the grass zero-isocline are shown in Fig. 4. The isocline is depressed as stocking rate increases. Of considerable importance in this regard is the effect of the level of G_0, the amount of grass below which no grazing can take place. Fig. 5 presents the results of two runs of the model, using data for an "average" savanna (derived from various savanna studies in southern Africa). In Fig. 5a, the value of G_0 was kept high (at 500 kg/ha), and at all stocking rates the joint, stable equilibrium was maintained --

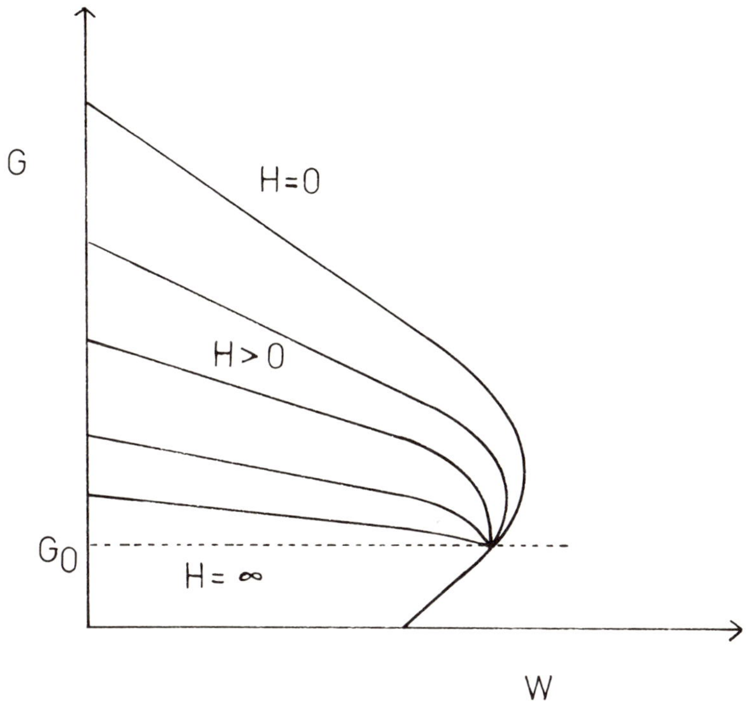

Fig. 4. The effects of increasing stocking rate (H) on the position of the grass zero-isocline. Below G_0 no grazing is possible. (From Walker et al.*, in press, by permission of the Journal of Ecology.)*

although at the highest stocking rate the equilibrium is very close to the threshold, and the system is not very resilient under these conditions. In Fig. 5b, G_0 is low (10 kg/ha) and at a stocking rate of just less than 100 kg/ha the joint, stable equilibrium is lost, and the system can progress only to the thicket condition of W alone.

The significance of these results lies in the determinants of G_0. A high level of unavailable (or at least inedible) grass -- thereby ensuring a high minimum grass cover on the soil -- is achieved by the presence of species which are conventionally regarded as undesirable. Such species are those which either taste bad (*Cymbopogon*, *Bothriochloa*), are very fibrous and tough (*Eragrostis plana*), are stoloniferous and rhizomatous, or in general have a large root and crown-to-shoot biomass. Under light grazing such species are inferior competitors to the leafy, erect grasses which channel their photosynthate into new leaf, and they gradually die out. It is only

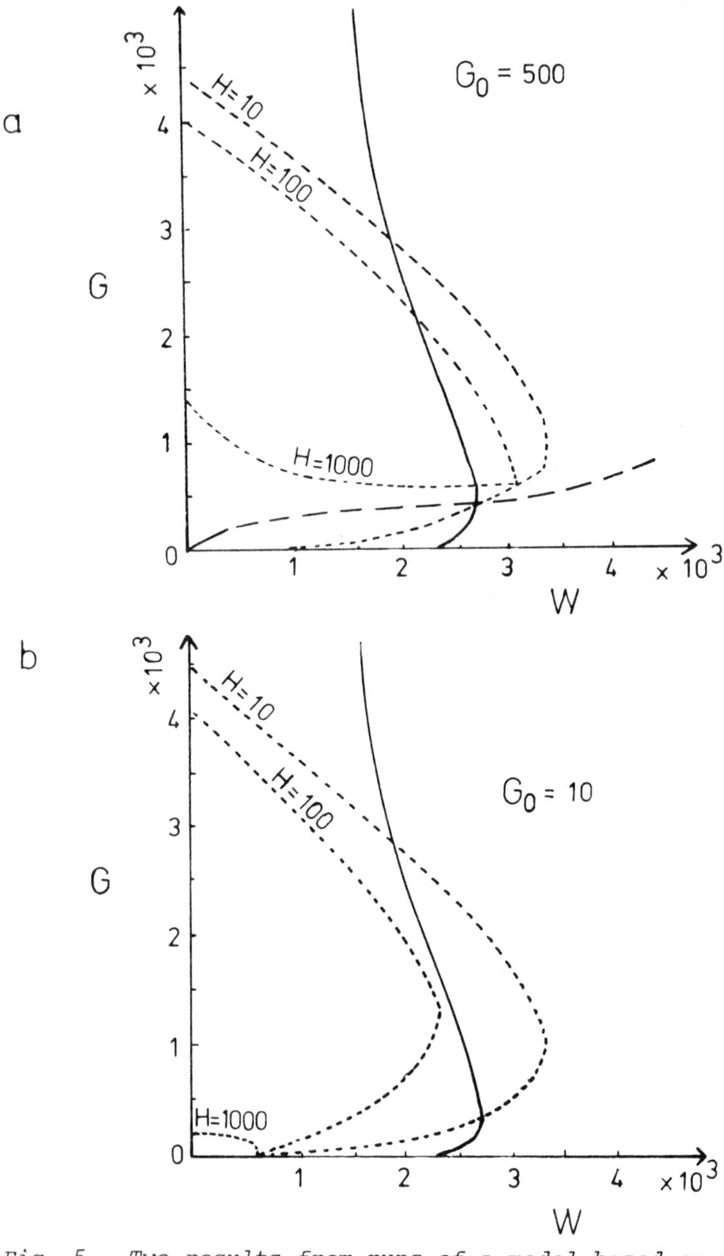

Fig. 5. Two results from runs of a model based on data combined from various southern African savannas. All values are in kg/ha. In (a) the grazing cutoff G_0 is high, and the joint, stable equilibrium exists for all stocking rates (H). In (b) G_0 is small, and G is eliminated when H is slightly less than 100. (From Walker et al., *in press, by permission of the Journal of Ecology.)*

by periodic heavy or overgrazing that these resistant species can persist, and thereby maintain a high level of G_0. If herbivore numbers are always controlled and prevented from reaching high levels, the resilience of the system declines (as G_0 decreases) in the sense that the number of herbivores required to push the system beyond the threshold for a joint, stable equilibrium of G and W becomes progressively less and less.

IV. GRASS, WILDEBEEST, AND LIONS IN THE KRUGER NATIONAL PARK

The following account is based on the excellent set of data summarized by Smuts (1978) and any misinterpretations of the data are mine. A more detailed description has been presented elsewhere (Walker and Noy-Meir, in press).

In the Kruger National Park wildebeest thrive best during dry periods, when grass is short and of high quality. In wet periods, with plentiful dispersed water, there is a fourfold increase in subadult male lions, which rove independently of prides and survive due to the dispersed groups of wildebeest. The recruitment curve for wildebeest during dry periods is assumed to be something like the curve *d* in Fig. 6. Whether

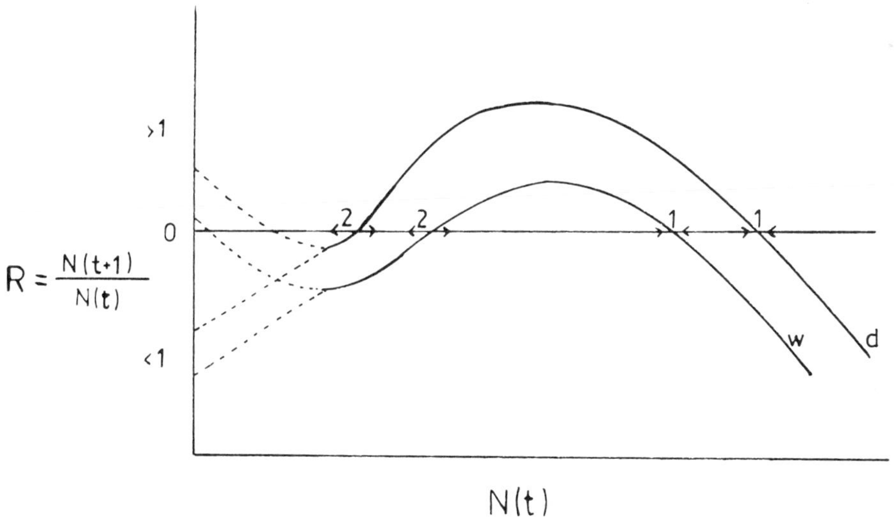

Fig. 6. Suggested recruitment curves (R) for wildebeest population in Kruger National Park, during wet (w) and dry (d) periods. The upper equilibrium (1) is stable, the lower equilibrium (2) is unstable. (After Walker and Noy-Meir, in press.)

or not the depression in the curve at low densities, due to lion predation, drops below 1.0 is not known. In wet periods the "predator pit" in the curve is deeper and extends to higher densities, and the upper food equilibrium declines somewhat, with the result that the curve drops to position *w*. The shape of the curve at very low densities is also unknown. If refugia exist or if prey switching occurs, it would rise again. If not, the curve would continue down, leading to extinction. The important point to note is that while the upper intersection *1* of the curve with the $R = 1.0$ line represents a stable equilibrium, the intersection at *2* is an unstable equilibrium, or threshold.

The mid-1960's was a dry period in the Kruger National Park, and wildebeest numbers increased to high levels. Their habitat appeared to be degrading and conventional range management wisdom dictated that numbers should be reduced. Culling was introduced in 1965. In 1971 a wet period set in. By 1973 culling was stopped since wildebeest numbers were very low. However, after culling ceased, the wildebeest did not recover, but their numbers continued to drop until 1979, at which stage culling of lions and burning of the breeding areas (to recreate short grass conditions) halted the decline.

The suggested explanation for this event is that culling reduced the wildebeest to below the threshold, *2*, which had risen during the wet period, so that even after culling was stopped the numbers continued to decline. Its significance in the context of this chapter is that it highlights the third point in Section II above -- the positions of the thresholds change in response to changing external conditions. If this explanation of the dynamics is correct, it means that wildebeest populations are less resilient in wet periods and should not be interfered with at such times.

V. CONCLUSIONS

As outlined in the introduction, the essence of this chapter is that wildlife systems should be allowed to change and vary naturally with minimum interference in order that they maintain a high resilience. However, owing to previous interference, the preexisting checks and mechanisms which prevented these systems from exceeding their stable bounds have been altered. It is, therefore, necessary to intervene (usually by culling) when these bounds are very closely approached. Since we know so little about the existence of these thresholds or the factors that determine them, there is a tendency to be conservative, usually too much so.

REFERENCES

Smuts, G. L. (1978). Interrelationships between predators, prey, and their environment. *BioScience 28*, 316-320.

Walker, B. H. and Noy-Meir, I. (In press). Aspects of the stability and resilience of savanna ecosystems. *In* "Ecology of Tropical Savannas" (B. J. Huntley and B. H. Walker, eds.)

Walker, B. H., Ludwig, D., Holling, C. S., and Peterman, R. M. (In press). Stability of semi-arid savanna grazing systems. *J. Ecol.*

Walker, H. (1971). "Ecology of Tropical and Subtropical Vegetation" (J. H. Burnett, ed.) Oliver and Boyd, Edinburgh. (Translated by D. Mueller-Dombois)

AN OUTLINE OF A MANAGEMENT PLAN FOR KASUNGU NATIONAL PARK, MALAWI

R. H. V. Bell

Department of National Parks and Wildlife
Kasungu
Malawi

I. INTRODUCTION

There are two aspects to the problem of overabundance of large mammals in African national parks. The first aspect concerns the technical questions: What are the facts of the ecological situation, what changes are taking place, what management techniques are practicable, and what will happen if we use them -- in short what management options are open to us? The second aspect concerns the questions of decision making -- the choice between options. I have argued at length elsewhere (Bell, 1978) that this aspect is an aesthetic procedure, in that a decision is a choice between options on the basis, not of the facts themselves, but of a value judgment concerning the facts.

In African wildlife management these two aspects have had different relative prominence in different situations. The great East African controversies over cropping in the 1960's and early 1970's concerned the aesthetic, decision making aspect almost exclusively and had very little to do with the technical aspects of overpopulation. To date, the East African conservation establishment has tended to adopt a no-cropping stance, while the de facto managers of the wildlife population (illegal hunters) have taken a heavy crop in response to external, economic factors.

In Southern Africa, by contrast, management decisions have been related to the technical aspects of overpopulation, and the region now has considerable experience with cropping in the interests of habitat control. In some cases the results have not corresponded with expectations, resulting in reassessments of both technical and decision making aspects.

ISBN 0-12-385280-3

Bell (1978) has argued that wildlife management should be based on management plans that are continuously reviewable, integrated feedback processes between policy making, research, management, and monitoring. The Southern African experience to date appears to approach this ideal more nearly than the East African.

It is also argued that the question of local overabundance cannot be viewed in isolation from the overall requirements of the area concerned. For this reason, our approach to the problem is offered as an integral component of an overall management plan for Kasungu National Park, Malawi. This management plan has not yet been put into effect, and thus may be regarded as still being in the decision making phase.

II. KASUNGU NATIONAL PARK: THE PROBLEM

Kasungu National Park (Fig. 1) is an area of 2,000 km^2 situated in the central region of Malawi, Central Africa. The landform, soils, and vegetation are typical of the Central African plateau with its basement geology, ancient, gently undulating land surface of poor sandy soils supporting closed canopy *Brachystegia/Julbernadia* woodland, interspersed with the seasonally flooded grassland of the "dambo" drainage system. Mean annual rainfall is about 31 inches.

The Central African plateau generally supports low densities of humans and wildlife as a result of its low soil nutrient status (Bell, in press). However, human densities in Malawi are high by African standards -- averaging over 100 people/km^2. The area now occupied by Kasungu National Park was settled until 1923, when the villages were removed as a result of sleeping sickness. Most villagers settled in the Chulu area on the park's northeast boundary. Since then human densities on the park boundaries have steadily increased as a result of population growth and immigration to agricultural schemes and estates. Today the park is bounded on the east (Malawi) side by solid settlement and on the west (Zambia) side by ribbon settlement separating it from the Lwangwa escarpment.

Following the evacuation of 1923, no serious enforcement of wildlife legislation took place until about 1965 when national park developments were initiated in the form of game scout camps, roads, and a small tourist lodge. Since then, management has been aimed primarily at encouraging visitor use and law enforcement.

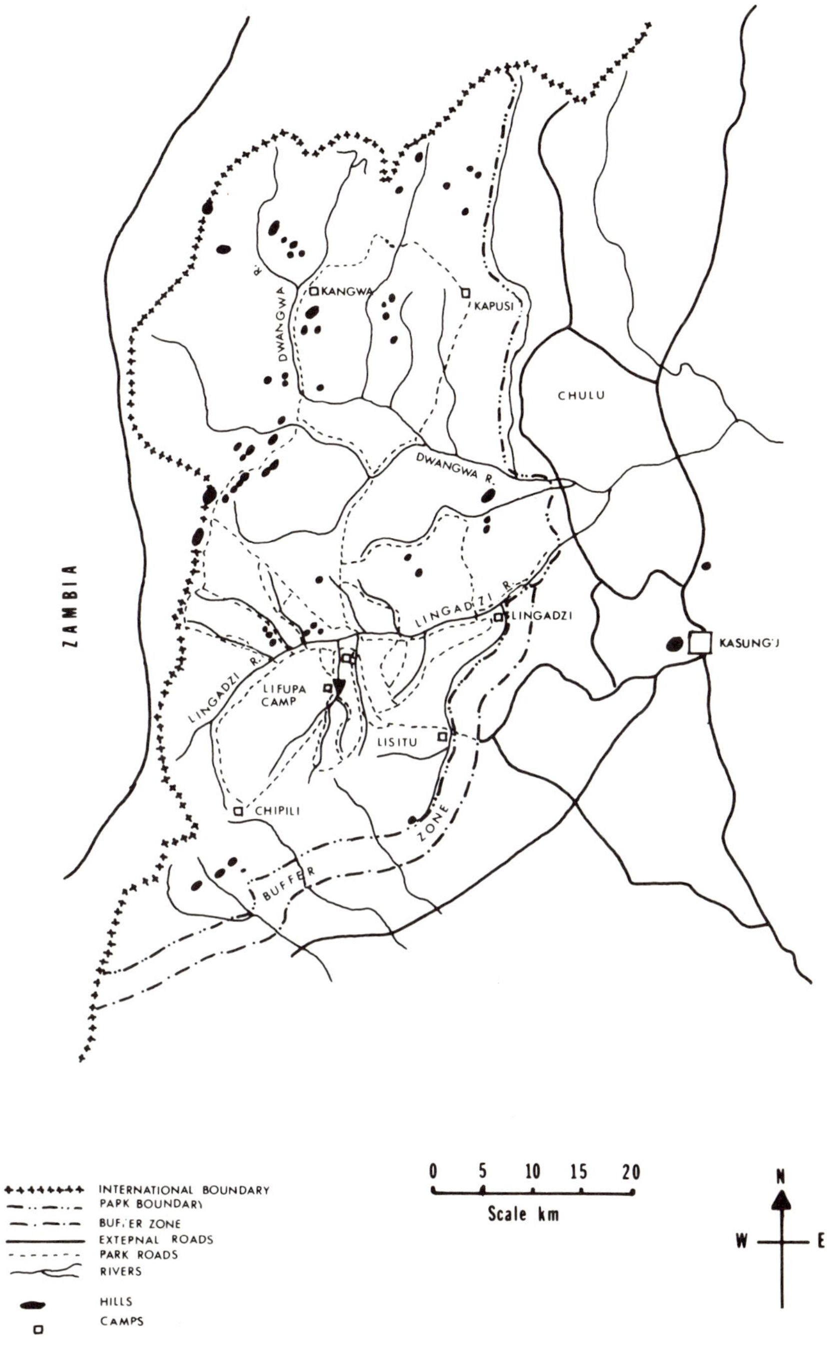

Fig. 1. Kasungu National Park.

It seems clear from the subjective accounts of former inhabitants, poachers, and park wardens, that wildlife densities have gradually increased since 1923, but the extent of the increase is difficult to assess. Current estimates are shown in Table I, but these are tentative and probably inaccurate.

In the case of elephants, estimates by various workers have been made by aerial transect sampling since 1970. Confidence limits are wide, but the total does not appear to have varied much from a figure of about 1,000. It will be argued that further increase is being prevented by illegal hunting and shooting to protect crops.

However, in spite of the fact that the elephant population does not appear to have increased substantially within the last ten years, it is clear -- both from subjective accounts and monitoring data -- that elephant are contributing to a change in the structure of the vegetation in parts of the park. This change affects the aesthetic value of the woodland itself, the visibility of animals by tourists, and the food supplies and habitat requirements of other wildlife species. In these terms the situation may be characterized as an "elephant problem."

III. ECOLOGICAL FACTORS

A. *Landscape Classification*

All aspects of ecology and management are related to landscape diversity. In Kasungu National Park landscape classification is based on differences of parent geology and on the phase of landscape evolution. The most important division is between "plateau" areas, making up about 80% of the park, and "valley" areas, making up most of the remainder (Fig. 2).

"Plateau" areas are very flat, with deep sandy soils of low nutrient status and high water infiltration rates. The vegetation consists of closed canopy *Julbernadia/Brachystegia* woodland with a short to medium length herb layer beneath it. Dambos tend to be wide, flat, and sandy, supporting medium height grass of poor quality.

"Valley" areas consist of the slightly incised valley slopes of the two major dambo systems of the park, the Dwangwa and the Lingadzi, and are mainly confined to the central part of the park. Because of downcutting of the valleys into the zone of rock weathering, valley soils have higher clay contents

TABLE I. *Common Wildlife Species of Kasungu National Park*[a]

Species	Estimated numbers	Average weight (kg)[b]	Biomass (kg)	Percentage of total biomass
Elephant	1000	1725	1,725,000	65.0% } 77%
Buffalo	700	450	315,000	12.0% }
Waterbuck	100	160	16,000	0.6%
Eland	300	340	102,000	4.0%
Reedbuck	1000	40	40,000	1.5%
Warthog	1000	45	45,000	1.7%
Roan	500	220	110,000	4.0%
Zebra	500	200	100,000	4.0%
Sable	500	185	92,500	3.5%
Hartebeeste	500	125	62,500	2.4%
Hippo	10	1000	10,000	0.4%
Rhino	20?	816	16,320	0.6%
Lion	40?	-	-	-
Leopard	common			
Hyena	scarce			
Cheetah	scarce			
Wild dog	relatively common			
		Total	2,634,320	100.0%

[a]All estimates are tentative and subject to correction.

[b]The average weight for a population includes an estimate of the weights and relative numbers of all age classes and is, therefore, lower than the mean adult weight. (Figures from Coe et al., 1976.)

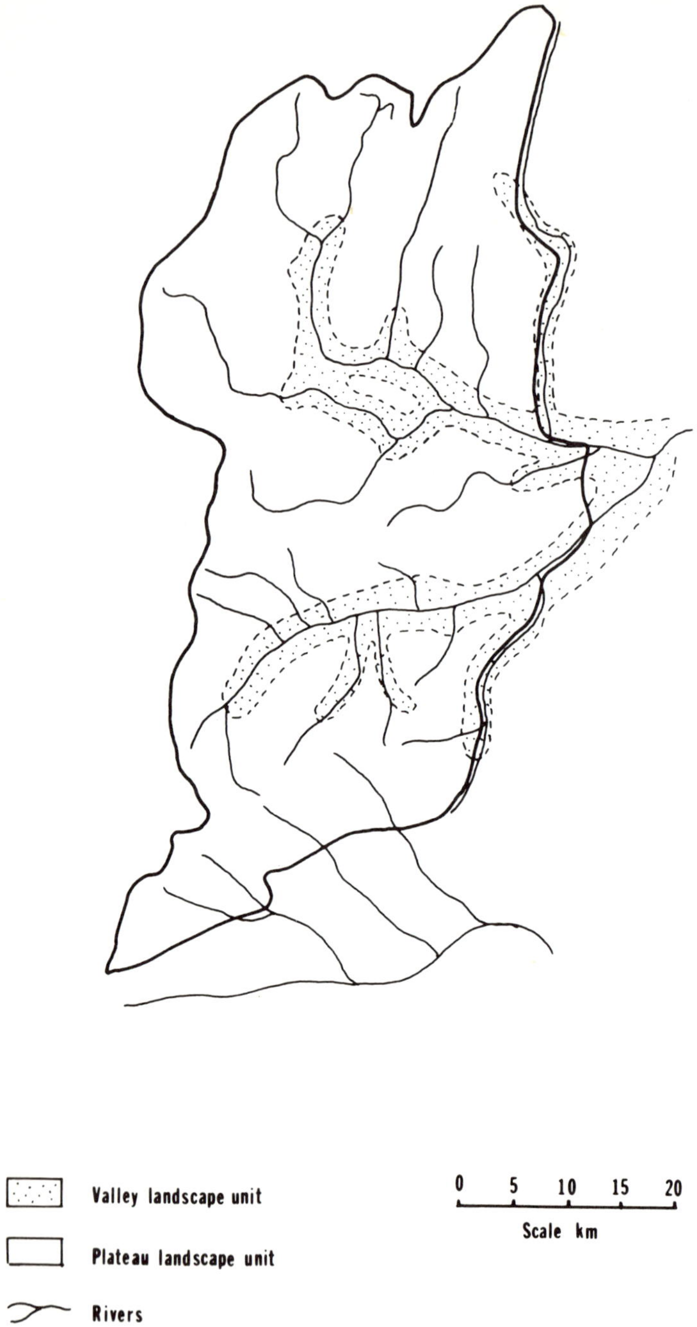

Fig. 2. Kasungu National Park landscape classification.

than plateau soils, and hence higher soil nutrient status and lower water infiltration rates. The vegetation is distinctive, consisting of open woodland including *Pericopsis, Combretum, Terminatia,* etc. and tall *Hyparrhenia* grassland. Dambo floors have high clay contents and support medium to tall grasses of high quality.

B. *Distribution of Settlement and Wildlife*

Settlement and wildlife areas are highly correlated with landscape diversity; settlement density and wildlife biomasses are disproportionately high in "valley" areas and disproportionately low in "plateau" areas. Water is freely available at all times and is not a constraint on distributions.

Briefly, the large herbivores fall into two groups with regard to distribution in the park:

(1) Large animals -- elephant, buffalo, eland, and waterbuck make up over 75% of herbivore biomass. This group is heavily concentrated in "valley" areas, to the extent that over 50% of large herbivore occupancy is estimated to occur within this 20% of the park area (see Fig. 3).

(2) Smaller animals -- including zebra, roan, sable, and Lichtenstein's hartebeeste (see Fig 4.) -- make up about 20% of total mammalian herbivore biomass. Of the total occupancy of this group, less than 10% occurs within the "valley" areas, that is they select against this landscape unit. Selection in favor of plateau areas is most marked in the case of hartebeeste and sable. It is noted that the three antelopes in this group are characteristic of the broad-leaved woodlands on the basement geology of Central Africa, and in the case of roan, the West African Guinea belt.

These differences in distribution are reflected in differences of sighting rates by tourists. (Tourist routes follow dambos and most sightings are in dambos.) "Valley" dambos show consistently higher biomass sighting rates in which large animals predominate than "plateau" dambos, in which smaller animals predominate.

Differences in seasonal sighting rates reflect changes in the proportion of time spent by animals in dambos as opposed to woodland. Most species, particularly the smaller herbivores, follow a straightforward pattern of moving up into the interfluves during the rains (low sighting rates) followed by progressively heavier use of dambos (higher sighting rates) during the dry season, particularly on green flushes following burns. This pattern is reversed in the case of large animals, particularly elephant, where sighting rate peaks in the middle of the rains.

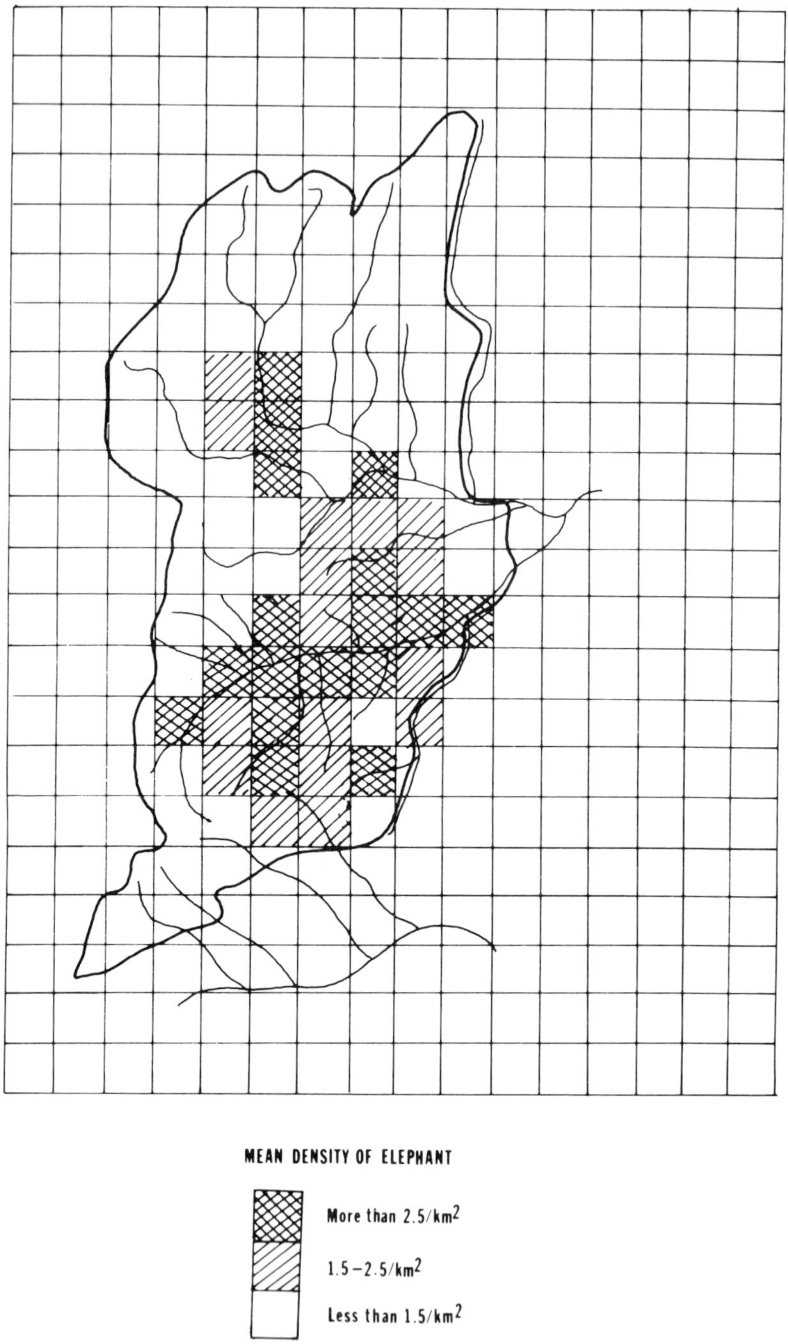

Fig. 3. Distribution of elephant from aerial surveys, 1977 to 1979.

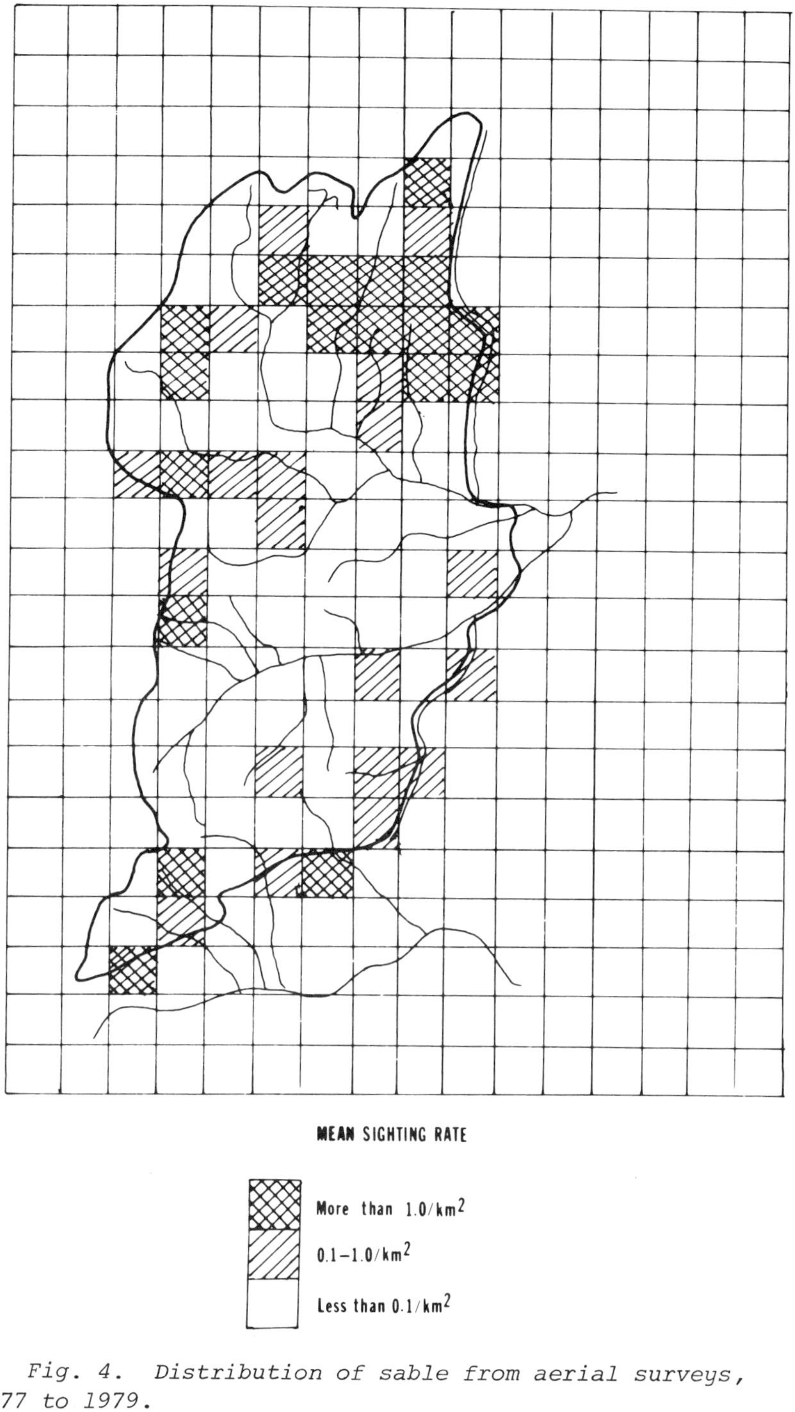

Fig. 4. Distribution of sable from aerial surveys, 1977 to 1979.

C. *Herbivore Distribution in Relation to Food Supplies*

It is currently presumed that the distributions of herbivores are determined by the production and quality of their food supplies.

The availability of food to each species of herbivore depends both on the total production of usable components (i.e., protein, soluble carbohydrates) and their ratio to the production of unusable/inhibitory components (i.e., structural carbohydrates, secondary chemicals, etc.); this ratio determines food quality. Production rates vary independently, resulting in differences of usable production and quality. As a simple generalization larger herbivores, and particularly some nonruminants such as elephant, are capable of using lower quality foods than smaller ruminants (Bell, 1971).

The hypothesis now being tested is that, while the production of both usable and unusable/inhibitory components is highly correlated with water availability, the gradient of the relationship with soil nutrient availability is steeper in the case of usable components than that of unusable/inhibitory components. Thus increase of water availability may increase total production but reduce quality, while increase of soil nutrient availability may increase total production and also increase quality.

In Kasungu National Park the sandy "plateau" soils have low nutrient status but high water infiltration rates so that water passes rapidly through the shallow grass rooting zone to the deeper woody root zone. Total woody production is high and total grass production is low. However, because of very low fiber production in this situation, grass quality is high. This is the condition required for a low biomass of smaller, more selective herbivores.

In the more clayey "valley" soils, by contrast, soil nutrient status is high, but water infiltration rates are low, so that water is retained longer in the grass rooting zone. Woody production is low, but total grass production is high; although usable component (protein) production is high, fiber production is very high, so that grass quality is low. This is the condition required for a higher biomass of larger, more tolerant herbivores. Relevant data are summarized in Figs. 5 and 6.

Emphasis has been placed on grass quality as opposed to browse since browser biomass is very low apart from elephant which appear to graze whenever possible, perhaps as a result of the secondary chemical content of *Brachystegia* woodland trees (Bell, in press). However, elephant make considerable use

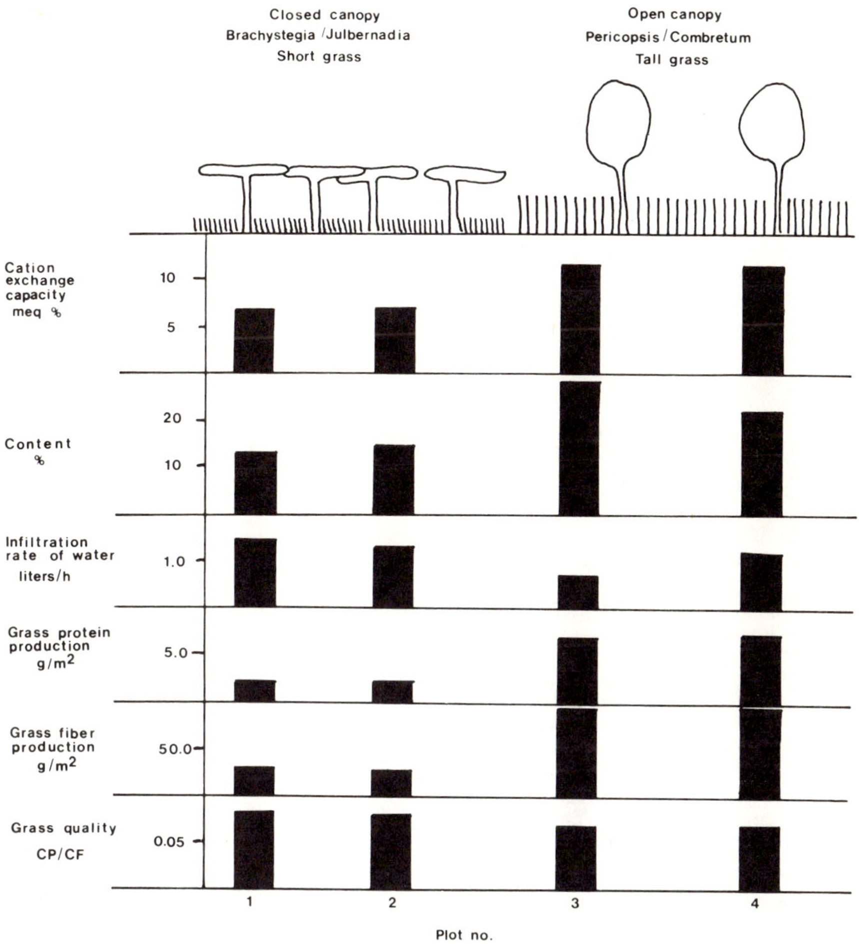

Fig. 5. This figure summarizes data from a series of sampling points at 200 m intervals along a transect crossing a boundary between the "plateau" landscape unit (points 1 and 2) and the "valley" landscape unit (points 3 and 4). In the top row, the vegetation type is indicated schematically. The next three rows give data on soils -- i.e., the cation exchange capacity, the clay content, and the infiltration rate of water in liters/h measured with infiltration rings (three replicates at each point). The bottom three rows give data on grass production and quality measured towards the end of the growing season (March) following a burn in the previous October. The figures for protein and fiber production are the total production in g/m^2 from October to March -- i.e., the majority of the growing season. It is assumed that there were no losses to grazing at this site. The bottom row gives the protein fiber ratio from the data shown in the previous two rows.

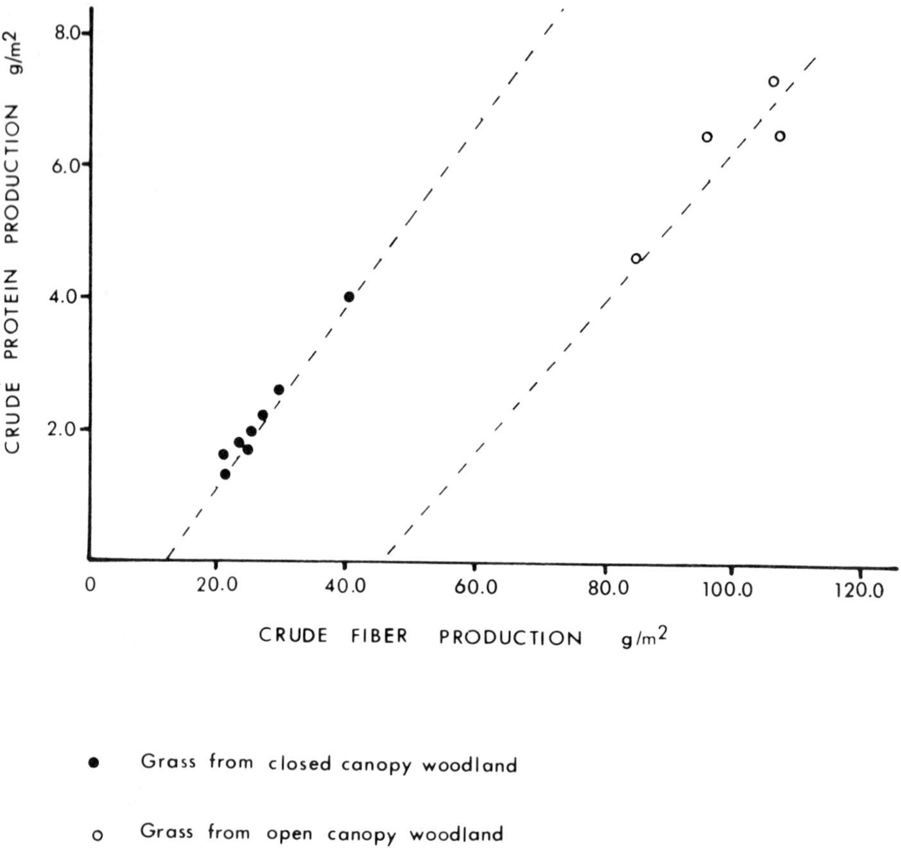

Fig. 6. This figure shows data from a series of sampling points astride the plateau/valley landscape unit boundary, including those points shown in Fig. 5. Data are again from the late growing season (March) and show total production since the previous October. Each point is the mean of 10 quadrats of 1 m^2. This method of presenting the data clarifies the point that, although protein production is higher in the open canopy woodland, the quality is lower because of the higher fiber production. In such a diagram, lines of equal quality are represented by radials from the origin.

of browse during the dry season when grass quality falls below their thresholds of quality tolerance as a result of curing and of quantity requirement as a result of burning. This is presumed to be the cause of the reverse pattern of seasonal movement of this species.

D. Vegetation Dynamics

This section considers the effects on food supplies of defoliation of woodlands by settlement, herbivores, and fire and relies heavily on the studies of Fay Robertson now in progress in Kasungu National Park.

1. Regrowth Patterns. If an area of *Brachystegia* woodland is cleared, there follows a period of 5 to 20 years of rapid regrowth by multistemmed coppicing, resulting in a dense thicket of up to 3.5 m in height. Regrowth then slows down and by a process of self thinning, most of the stems are suppressed and a closed canopy woodland 5 to 10 m high emerges after about 25 years. Trees continue to grow slowly, and the number of stems are reduced until a stable mature canopy 10 to 20 m high results with relatively little regeneration beneath it.

The regrowth pattern of the woody canopy affects grass growth in important ways by mechanisms that have not been determined, but probably involve root competition and shading. The main effects are as follows (see Fig. 7):

(1) Following clearance of woodland, grass growth rapidly increases and medium woodland grasses are replaced by tall grasses. On plateau soils, most of the extra production is in the form of fiber so that quality is very low.

(2) During the phase of rapid, dense, coppice regrowth, grass production is strongly suppressed, resulting in a short sparse grass layer. Fiber production is suppressed preferentially so that quality is fairly high although total biomass is low.

(3) During the mature canopy phase, the grass returns to the medium height, density, and quality described in the last section.

The details of these responses are related to the landscape unit -- increase in grass growth following canopy removal being greatest in "valley" situations.

2. Fire. The whole park is liable to burning either by fires originating outside the park or set illegally by poachers. The effects of fire are being investigated in a series of monitoring plots.

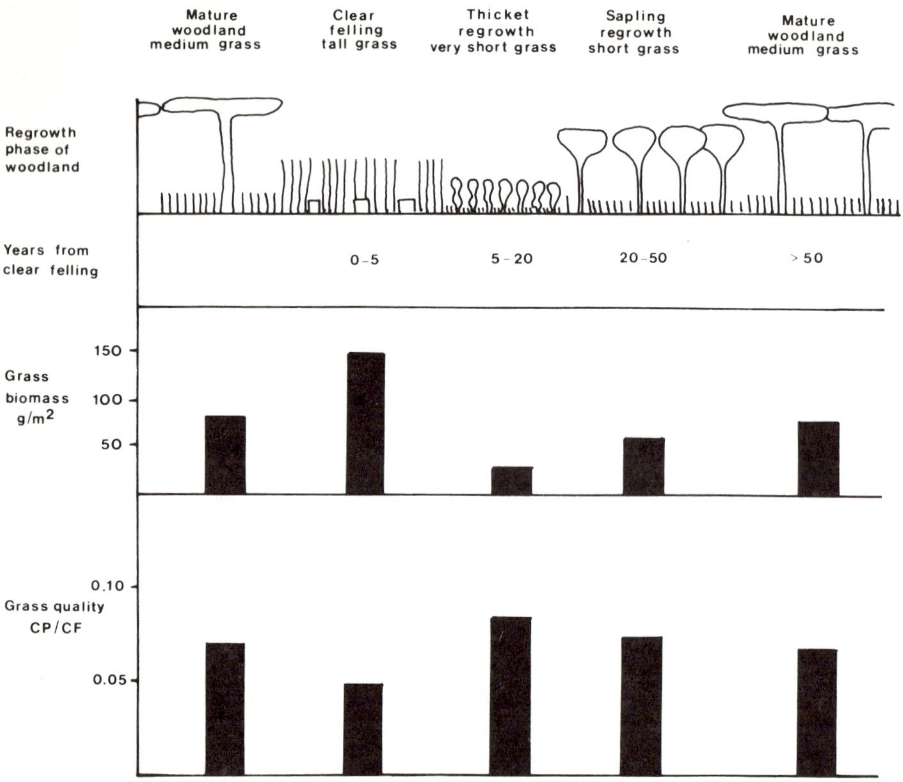

Fig. 7. The effect of woodland regrowth on grass biomass and quality. This figure refers to Brachystegia *woodland on "plateau" sites. The time scale of woodland regrowth is derived from F. Robertson (personal communication). Data on grass biomass and quality are from samples taken from sites of comparable woodland structure and soil type. However, since the history of the woodland in these cases was not known, the relationship between woodland regrowth and grass production is still hypothetical.*

Fire temperature is mainly determined by grass biomass, grass water content, and wind speed. Temperatures, therefore, vary between seasons, landscape units, and phase of woodland regeneration. "Valleys" experience the most intense fires, mainly because of high grass biomass, while mature plateau woodlands experience moderate fires, and dense coppice areas experience cool patchy burns because of low grass biomass.

The effects of fire on woodland are fairly simply related to temperature -- cool burns killing stems up to 0.5 cm in diameter and hot fires killing stems up to 5 cm.

A key question is whether fire can prevent woody vegetation from escaping from the cleared, long grass phase into the dense coppice phase. Present indications are that in "plateau" sites it cannot, but that may be capable of this in "valley" sites.

3. *Elephants*. The current estimate of elephant numbers is about 1,000, giving a low overall density of 0.5/km^2. However, elephant density is very uneven, the majority of elephants are concentrated in the Lingadzi drainage, giving high local densities of 5/km^2. The heaviest occupancy occurs in the tall grass, open woodland "valley" areas but, particularly during the dry season, elephant browse the "plateau" woodlands adjacent to the "valley" areas. In consequence much of the central part of the park, including most of the tourist area, is subjected to relatively heavy use by elephant.

The effect of elephant on woodland is monitored in permanent sites with large samples of marked trees and subjected to various fire regimes. Present data indicate that following the onset of elephant activity, the vegetation eventually takes up a new equilibrium structure, dependent on the rate and form of elephant damage, the frequency and intensity of burning, and the growth rates of trees and grasses. Clearly, under given levels of elephant and fire, the equilibrium will vary with soil factors and landscape unit. However it seems likely that there are two relatively stable equilibrium structures of *Brachystegia* woodland under heavy elephant use.

The first is found in the central part of Kasungu National Park where large areas are reaching the short dense coppice phase of growth. This is thought to be an equilibrium structure, first because its tree productivity is much higher than that of mature woodland, and secondly because its growth form suppresses grass growth and hence fire intensity. Both these factors would tend to buffer the woody vegetation against further reduction by elephant offtake within a fairly wide range of elephant densities and soil conditions.

However, in certain conditions, it appears that elephant can reach sufficient densities for their offtake to exceed the production of the coppice phase, thereby opening up the canopy to allow accelerated grass growth and increased fire intensity. This combination produces a second equilibrium of open grassland with low, scattered woody regeneration, maintained as such by intense fires. This can be seen in the peripheral parts of the Lwangwa valley, and possibly in Tsavo National Park, but is not at present seen in Kasungu, unless the open woodland of the "valley" areas is an equilibrium of this type.

The secondary effects of the elephant-induced vegetation change in the central part of Kasungu Park are as follows:

(1) The aesthetic appeal of the coppice phase woodland is itself less than that of mature *Brachystegia* woodland which has fine trees and open understory.

(2) The visibility of wildlife is less in the coppice phase than in either the mature woodland phase or the open "valley" areas.

(3) The coppice phase and the open "valley" areas provide unsuitable habitat for sable and Lichtenstein's hartebeeste and, to a lesser extent, for roan and zebra.

A key question is, why is the density of elephant so uneven? Is it due to ecological factors, so that left to themselves elephant will not spread into and modify the vegetation of the areas now unaffected, or is it due to some extraneous factor such as illegal hunting, so that effective law enforement could lead to more extensive habitat modification? The question is examined below, but our present conclusion is that illegal activity is an important factor in determining elephant density patterns.

E. Elephant Recruitment and Mortality

1. Recruitment. A brief preliminary study (Jachmann, 1979) indicates that elephant recruitment is around 5%/annum; this is close to the maximum for an even sex-ratio population and implies that the population is not close to carrying capacity. From a population of 1,000, it could be expected that 50 calves would be born.

2. Mortality. If the census data is accurate and the population is in fact roughly stable, there must be a density independent mortality of approximately 50 elephant per year. Elephant mortality in Kasungu National Park has three components: natural mortality, illegal hunting, and crop protection shooting. The figures for 1977, 1978 and 1979 are summarized in Table II.

TABLE II. Known Kills of Elephant[a]

Year	*Illegal hunting*	*Crop protection*	*Natural (lions)*	*Unknown*	*Total*
1977	*12*	*19*	*3*	*10*	*44*
1978	*15*	*22*	*4*	*5*	*46*
1979	*25*	*5*	*4*	*3*	*37*

[a]*From Bell (1979a, 1980).*

Of these figures, all may be taken as underestimates except the figure for crop protection. The combined mortality seems well able to account for the fact that the population is not increasing. Catch per effort analysis of illegal activity (Bell, 1979a, 1980) indicates that mortality due to illegal hunting is increasing rapidly, while mortality due to crop protection is being reduced as a matter of policy.

3. *Mortality Distribution.* The distribution of mortality is very uneven (Fig. 8); 90% of all crop protection shooting occurs in a 30 km sector of boundary astride the Lingadzi River where high human and elephant densities meet. Most shooting is of young males.

Illegal activity is also very uneven; from 1977 to 1979, 80% of all illegal activity was encountered in the northern half of the park, while 75% of all elephant known to have been killed illegally in the same period were killed in this same area. However, the northern half of the park receives only about 20% of elephant occupancy; that is, it usually contains only about 200 elephant which are clearly being cropped at a rate well in excess of their sustainable yield. Mortality due to illegal activity appears to be biased somewhat in favor of males.

4. *Conclusion, Present Elephant Situation.* The picture that emerges at present is that elephant distribution is primarily related to landform and vegetation but that there are two elephant "sinks" -- one in the eastern boundary due to crop protection and one in the north due to illegal hunting. The stability of the density pattern depends on the rate of diffusion towards these sinks. The relevant data are not at present available; in addition the intensity of the two sinks is changing. This situation is obviously relevant to the "vacuum zones" discussed by Owen-Smith (this volume). Finally, it appears that Kasungu National Park suffers simultaneously from both heavy illegal offtake and an elephant problem. My guess is that this is a common situation in African conservation areas, and I suggest the solutions to both problems are closely interrelated.

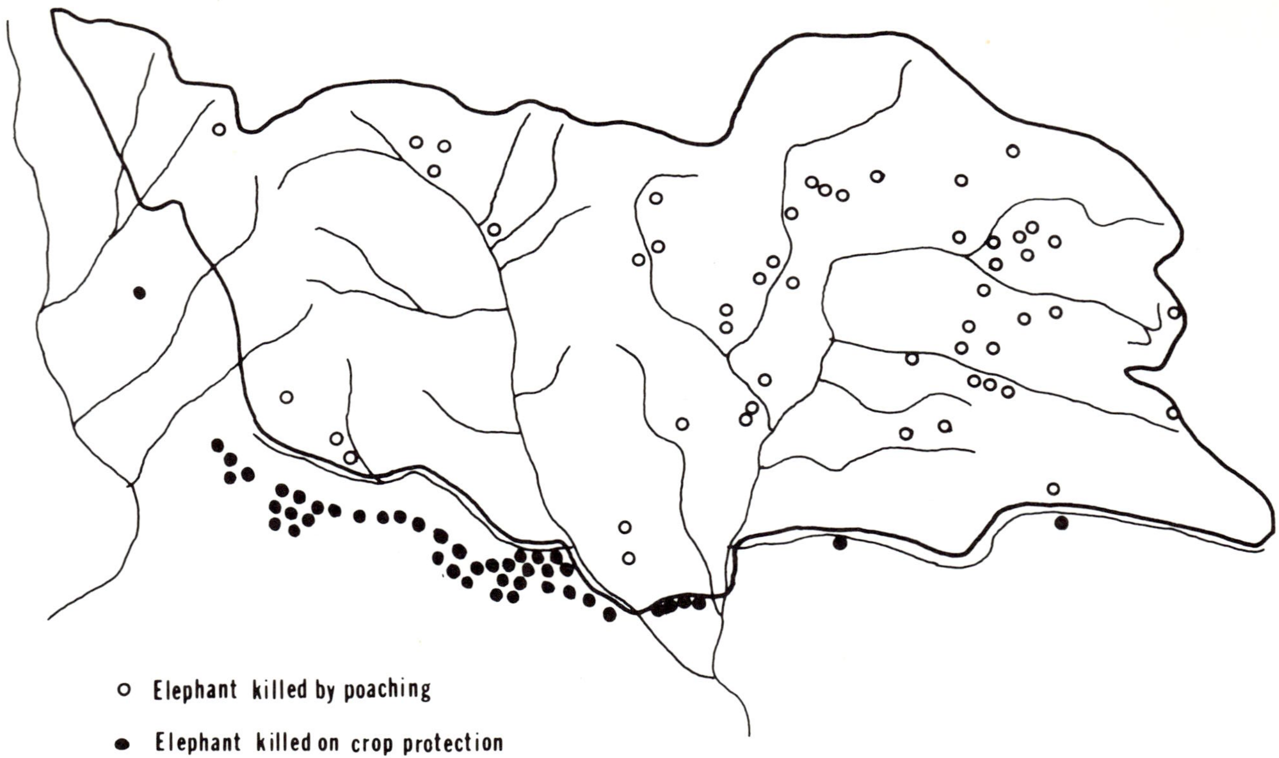

Fig. 8. Elephant deaths due to poaching and crop protection, 1977 to 1979.

IV. THE MANAGEMENT PLAN

A. *Introduction*

The main ecological facts concerning Kasungu National Park have been briefly summarized. A further important fact is that much information remains unknown, and many of the "facts" may be wrong (see Bell, 1978). The facts as presented may be judged in many different ways and an equal number of alternative management regimes may be advocated.

The management regime outlined here is put forward by the Wildlife Research Unit on the basis of our assessment of the overall objectives of the Department of National Parks in relation to Kasungu National Park. The plan has not yet been assessed by the Department as a whole and should not be regarded as official Malawi government policy.

B. *Requirements*

This section outlines the policy requirements of the Department of National Parks as seen by the Wildlife Research Unit, which are seen as determining the outline of the management plan. It is emphasized that these requirements are not direct consequences of the ecological facts, but are aesthetic judgments related to them (see Bell, 1978).

(1) The purpose of the Department of National Parks and Wildlife is to preserve in perpetuity representative examples of the plant and animal communities of Malawi. This requires that the northern half of Kasungu National Park, with its mature *Brachystegia* woodlands and relatively high densities of sable, hartebeeste and roan, must be kept at its present very low elephant density to prevent the onset of woodland coppicing. This area should also be subject to regular late hot burning.

(2) The low density in the north is at present maintained by illegal hunting. Illegal activity is not considered to be an acceptable form of management (although the possibility has been considered by Bell [1979b]) as a matter of principle and because of losses of revenue. Therefore, illegal activity must be reduced by intensification of law enforcement (Bell, 1979a) and substituted by cropping of elephant in the northern half of the park. The rate of cropping will depend on the diffusion rate of elephant from the high density area of the Lingadzi Valley.

(3) A further objective of the Department of National Parks and Wildlife is catchment conservation within its areas of jurisdiction. Habitat change must not be allowed to reach a point

at which gully erosion is initiated. This requirement sets an upper limit on permissible wildlife densities in the "valley" areas. At present there is no indication that this limit has been reached, but the few known gullies are being monitored.

(4) Tourist use, which depends on high biomass sighting rates, is to be encouraged. Therefore, elephant offtake from the "valley" areas is not indicated unless the vegetation structure passes out of the coppice phase into the open phase in the adjacent "plateau" areas. In this event the intensity pattern of cropping will be related to the diffusion rate. The burning regime in the high density elephant areas should be designed to encourage woody vegetation while assisting tourist viewing. A mosaic of time-staggered patch burning is indicated.

(5) The Department has an obligation to protect life and property from damage by wildlife, but there is little indication that crop protection shooting is effective in this respect. Efforts should be made to install effective barriers in the high risk areas, bearing in mind the reduction of this component of elephant mortality on the other aspects of management.

(6) Research and monitoring of ecological and management factors must be continued and intensified to assess progress in relation to objectives and to allow continuous reassessment of ecological interpretations, management options, management techniques, and conservation objectives.

ACKNOWLEDGMENTS

The production of this paper has depended on the active cooperation of the officers and staff of the Department of National Parks and Wildlife, Malawi, particularly the Chief Game Warden, Mr. David Anstey and the wardens of Kasungu National Park, Mr. Roy Miller and Mr. Peter Ingham. I am also grateful for discussions with and advice from Nigel Hunter and Fay Robertson, and to my wife Kathy for putting up with it all.

REFERENCES

Barnes, R. F. W. (1979). Elephant ecology in Ruaha National Park, Tanzania. Ph.D. Thesis, Cambridge University.

Bell, R. H. V. (1971). A grazing system in the Serengeti. *Sci. Am. 225*, 86-93.

Bell, R. H. V. (1978). The function of research in wildlife management. Proceedings of the 5th Regional Wildlife Conference in East and Central Africa, Gaborone, Botswana, July, 1976.

Bell, R. H. V. (1979a). Report and recommendations of law enforcement in Kasungu National Park. Report WRU/AP/8, March 20, 1979, to the Malawi Government, pp. 1-36.

Bell, R. H. V. (1979b). Letter to the editor -- "Iain Douglas-Hamilton isn't so great either." *Africana 7(1)*, 6-7.

Bell, R. H. V. (In press). The effect of soil nutrient availability on community structure in African ecosystems. *In* "Ecology of Tropical Savannas" (B. J. Huntley and B. H. Walker, eds.)

Bell, R. H. V. (1980). Elephant poaching, Kasungu National Park, 1979. Report WRU/AP/36, March 24, 1980, to the Malawi Government, pp 1-4.

Coe, M. J., Cumming, D. M., and Phillipson, J. (1976). Biomass and production of large African herbivores in relation to rainfall and primary production. *Oecologia (Berlin) 22*, 341-354.

Jachmann, H. (1979). The ecology of the Kasungu elephant. Report to the Malawi Government, pp. 1-49.

THE MANAGEMENT OF ELEPHANT AND OTHER LARGE MAMMALS IN ZIMBABWE

David H. M. Cumming

Department of National Parks
and Wildlife Management
Salisbury
Zimbabwe

I. INTRODUCTION

Many large mammal species are now more numerous in Zimbabwe than they were at the turn of the century. Human populations have increased fourteenfold since then and domestic livestock more than tenfold. The range freely available to large mammals is now less than one sixth of the country. This chapter examines aspects of the historical influences on some large mammal populations and the management applied to them over the last 25 years. Some options for future management are discussed.

The early promulgation and enforcement of game laws provided for the recovery of populations decimated by ivory hunters during the last century and the rinderpest outbreaks of the 1890's. The first game laws came into force in 1902 and the first game reserve, now Wankie National Park, was proclaimed in 1930. A further series of parks and game reserves was established during the 1960's with some changes occurring during the 1970's. Nearly 12% of the area of Zimbabwe is now national parks and wildlife land (Fig. 1) and a further 4% is state forestry land. Nevertheless, the conservation of large mammals in Zimbabwe increasingly has had to compete economically with other forms of land use. This competition is exemplified by an early commitment to game eradication when an expanding cattle industry was threatened by trypanosomiasis.

Legislation was radically altered in 1975 with the promulgation of the National Parks and Wildlife Act. This Act placed the responsibility for the conservation and use of wildlife firmly on the landowner. It has further served to enhance the value and encourage the sustained use of wildlife resources.

PROBLEMS IN MANAGEMENT OF LOCALLY
ABUNDANT WILD MAMMALS

ISBN 0-12-385280-3

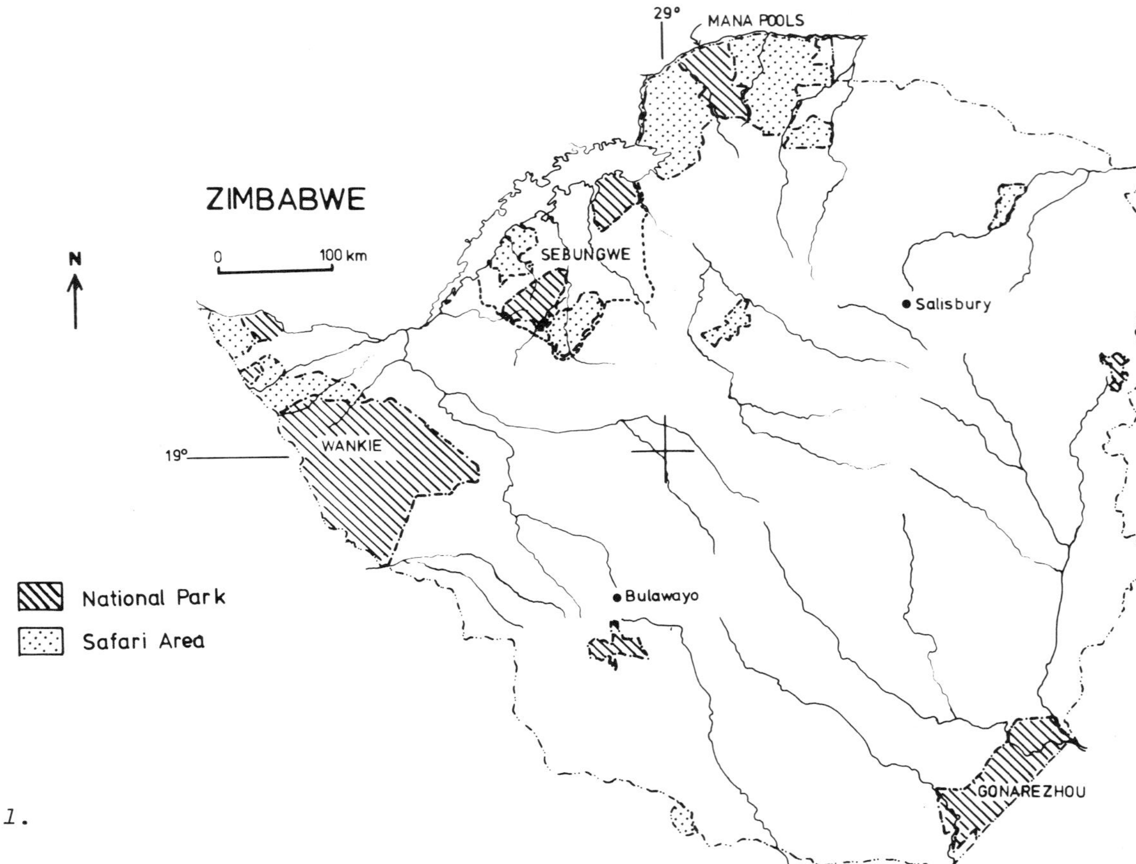
ZIMBABWE
N
0
100 km
29°
MANA POOLS
SEBUNGWE
Salisbury
WANKIE
19°
Bulawayo
GONAREZHOU
National Park
Safari Area

Figure 1.

II. THE ELEPHANT PROBLEM IN ZIMBABWE

Annual exports of ivory and records of the number of elephant killed each year bear out Spinage's (1973) contention that at the turn of the century few animals were killed and little ivory exported because there were very few elephant. Ivory exports and the number of elephant killed in Zimbabwe between 1905 and 1979 (Fig. 2) show an exponential increase of about 15%/annum up to 1965. The further increase in numbers killed after 1965 is a reflection of culling operations and tsetse control measures. Available evidence suggests that a 5% annual growth rate for elephant populations in Zimbabwe is not unreasonable.

The "elephant problem" seems to have "surfaced" in Zimbabwe in the early 1960's. Not only were elephant seen to be overeating their resource in protected areas but there was also the problem of excluding the animals from expanding areas of subsistence agriculture and from tsetse control corridors. The first culling operation took place in Wankie National Park in 1966 and 1967 when 500 elephant were killed from an estimated population of 5,000 animals. Culling then took place in the Zambezi Valley in Mana Pools National Park in 1968 and 1969, but it was not until 1971 that large-scale culling programs were executed in national parks in Zimbabwe (Table I).

Tsetse control operations over the whole country accounted for 679 elephants between 1919 and 1960 (Anon., 1961). The number killed between 1961 and 1970 was 4,066, with the bulk of the animals being shot between 1964 and 1968. Most of the control hunting during this period was carried out by a few highly skilled hunters.

When culling began in national parks, the influence of the elephant on its environment was viewed in much the same light as overgrazing by domestic livestock. The appropriate solution was seen as destocking, ameliorated by the provision of more watering points. Pressure was brought to bear on the appropriate government minister in 1964 by an executive officer of World Wildlife Fund (WWF), members of the public, and by the country's Natural Resources Board. No research or scientific surveys preceded the decision by the government to cull. The controversy has continued since, both in public and within the responsible agency. The management techniques adopted over the last 15 years have included culling, disturbance hunting,

Fig. 1. National parks and safari areas in Zimbabwe. Recreational parks, sanctuaries, and botanical reserves are not shown.

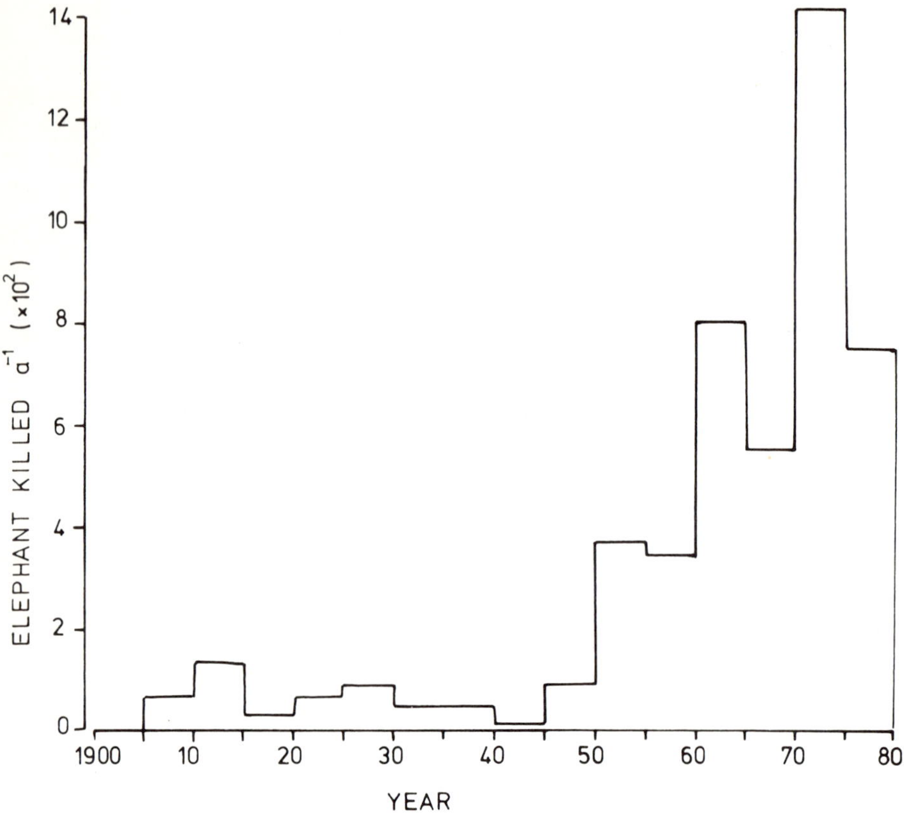

Fig. 2. Number of elephant killed per annum in Zimbabwe since 1906. The histogram shows the mean number per year in each five-year period. Data based on ivory exports until 1965 after which kills were used.

elephant drives, provision of additional watering points, and the erection of fences. Plans to manipulate elephant densities by manipulating artificial watering points in Wankie were frustrated by the recession and the war.

Three areas are examined here. Wankie National Park illustrates the dilemma of a continuing culling program if accelerated change in the flora and fauna of the park is to be avoided. The Sebungwe provides an example of foci of local overabundance resulting from elephant population growth and compression together with conflicts in land use, tsetse control, and a very rapidly expanding human population. The southern section of the Gonarezhou was subjected to a pulse of heavy culling, which was followed by effective woodland regeneration.

TABLE I. *Summary of the Numbers of Elephants Killed in National Parks on Culling Operations and in Control Operations Elsewhere in the Country*[a]
Note: This Table is Incomplete and Many Figures Still Require Further Cross Checking

Year	National parks					Safari areas, tsetse control areas, and tribal lands				Total
	Wankie	Gonarezhou	Chizarira	Matusadona	Mana Pools	Sebungwe Region	Zambezi Valley	Southeast Lowveld	Other	
1960	85					69	?	?	?	439[b]
1961	145					117	85	?	?	495[b]
1962	130					84	74	-	-	288
1963	70					132	134	?	?	336
1964	62					760	167	?	?	989
1965	204					558	605	377	?	1744
1966	114					618	262	40	?	1034
1967	433					276	92	179	?	980
1968	255					204	23	61	69	612
1969	125				150	169	86	3	27	560
1970	132				94	35	82	1	138	482
1971	1308	665			-	18	77	1	?	2069
1972	1036	1129	176	85	93	4	130	?	?	2653
1973	53	166	122	169	-	?	98	?	?	608
1974	768	541	104	104	-	?	c. 60	-	?	1671
1975	433	175	-	43	-	?	c. 80	-	?	731
1976	569	4	-	36	-	c. 35	c.100	-	?	744
1977	620	-	?	106	-	c. 40	16+	-	?	782
1978	410	-	?	100	-	c. 40	29+	68	?	647
1979	296	-	?	-	-	c. 40	10+	c.100	?	446
	7248	2680	402	643	337	3199	2210	830	234	18216 (17783)

[a]Figures for Wankie before 1966 reflect numbers killed on control operations outside the park.
[b]From annual reports without regional breakdown.

A. Wankie National Park

The ivory hunters of the last century greatly reduced the elephant population in the Wankie area and the 1896 rinderpest outbreak decimated other large mammal species. The area has since been free of tsetse fly (*Glossina morsitans* Westw.) During the 1930's and 1940's large ungulate populations were low and predators were treated as vermin. The number of head of game seen per mile traveled over the years 1937 through 1947 (E. Davison, unpublished graphs) show a rate of increase of about 12%/annum. Davison (1950) estimated the population of elephant in Wankie to number 1,000 in 1930 and 2,000 in 1944. Current estimates (Williamson, 1979) place the population at 15,564$\pm$14%. The area of the park is 14,620 km^2.

1. Environmental Features. A key feature of Wankie National Park is that present game populations depend very largely on artificial water supplies. The park is cut off from the two major rivers in the region -- the Zambezi in the north and the Gwaai in the east. Elephant are also cut off now from the Nata River in the south (Fig. 3). The annual rainfall of 630 mm in the east and of about 400 mm in the west falls during a single rainy season from November through March or April. Rainfall for October averages less than 6 mm and for April less than 25 mm, so that the dry season is effectively six to seven months. The southern and western parts of the park fall within the frost belt. The Kalahari Sands, which cover much of the park (Fig. 3) carry woody vegetation ranging from *Baikaiae plurijuga* woodlands through woodland thickets to open *Terminalis* scrub in the more arid west, with narrow vleis in shallow drainage depressions. These communities have been influenced by past logging, frost, fire, and elephant (Rushworth, 1975; B. H. Walker, unpublished manuscript). The northern part of the park is characterized by shallow basalt soils and more dissected topography. The vegetation is predominantly open *Colophospermum mopane* and *Combretum* woodlands interspersed with dry and moist vleis. Sodic soils are common in the Karroo sediments in the northeast of the park. Basalt soils carrying *C. mopane* woodlands occur in the southwest of the park.

Human influences on the ecosystem other than early logging, burning, and culling, have taken the form of farms and fences in the Gwaai Valley and north of the Deka. More recently, fences have been erected on the southern and western boundary of the park to restrict game movement and control the spread of foot and mouth disease. The development of cattle ranching schemes in Botswana during the 1950's resulted in an immigration of

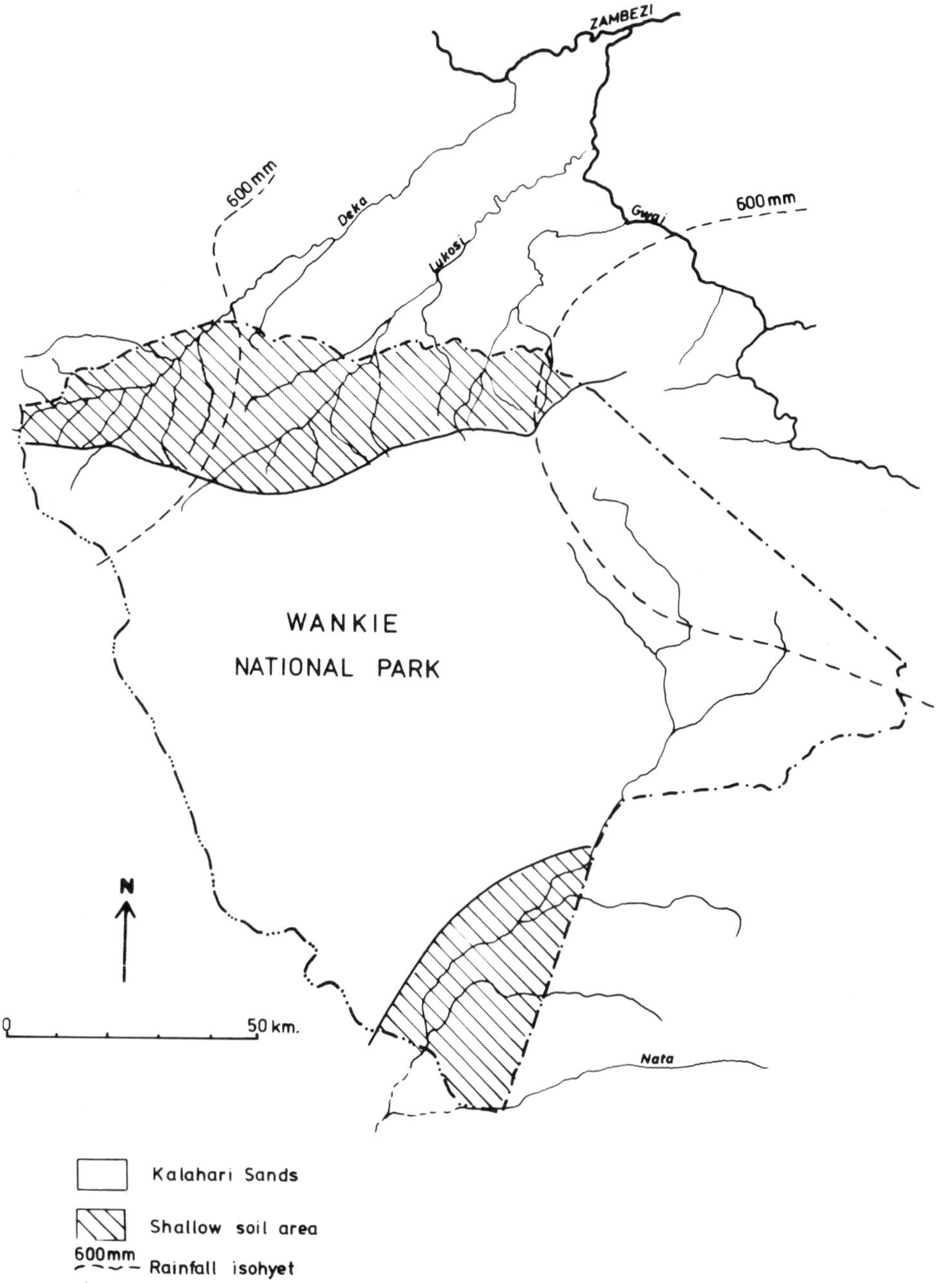

Fig. 3. Wankie National Park; showing major drainages, soil/vegetation zones, and rainfall isohyets.

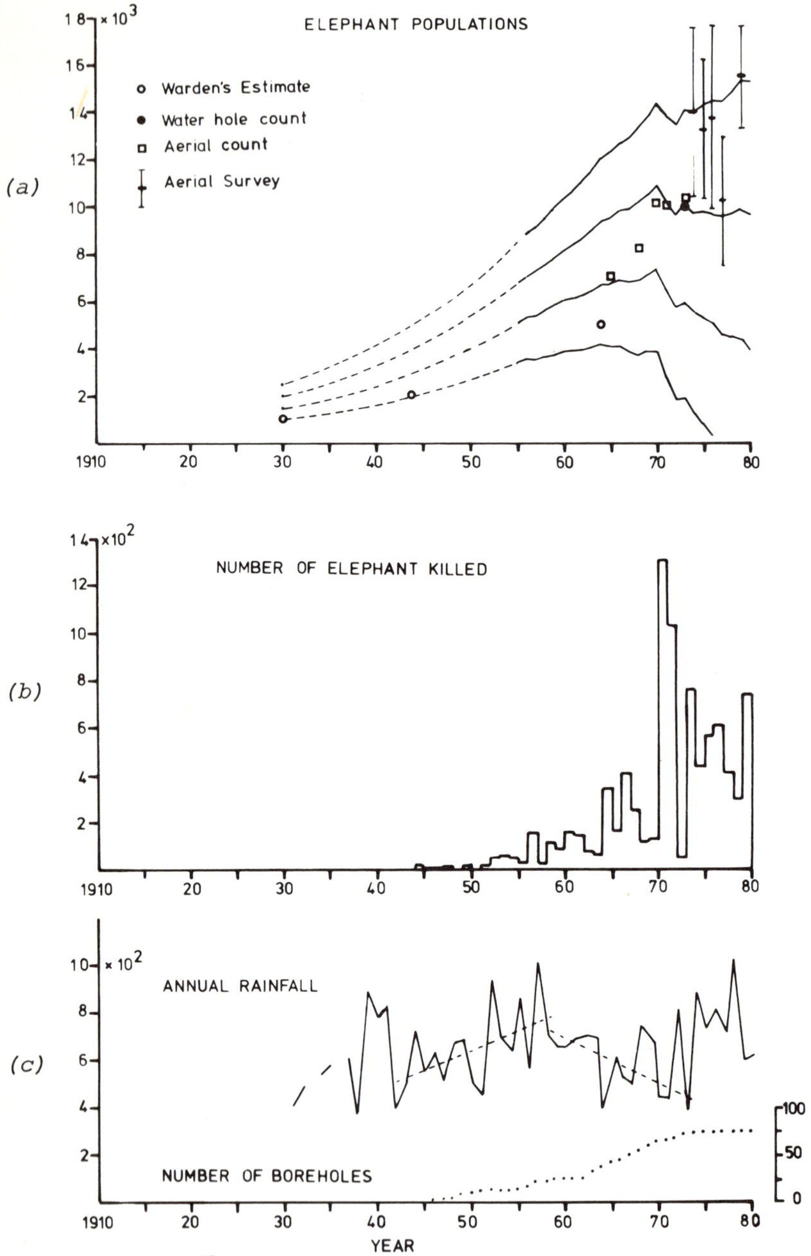
(a)
ELEPHANT POPULATIONS
Warden's Estimate
Water hole count
Aerial count
Aerial Survey
(b)
NUMBER OF ELEPHANT KILLED
(c)
ANNUAL RAINFALL
NUMBER OF BOREHOLES
YEAR

Figure 4.

elephant into Wankie National Park (E. Davison, personal communication 1980). Increasing settlement of the adjacent and previously unoccupied tribal area to the south during the 1950's will also have affected the range available to elephant. Disturbance in Botswana over the last two years may have caused further immigration into the park, but direct evidence is lacking.

2. Elephant Numbers. The estimates made by wardens between 1930 and 1964 fit a model of 5%/annum growth in the eleplant population (Fig. 4a). However, assuming an annual growth pulse of 5% and subtracting the known, usually dry season, kills of elephant, it is clear that this population could not have sustained the subsequent levels of offtake. Anderson (1965) in the first aerial count estimated a population of 6,981 eleplant in the park. Subsequent aerial counts, using a variety of aircraft with volunteer pilots and observers, indicated populations of about 10,000 elephant. A waterhole count at the end of the dry season provided a similar figure (Fig. 4a). The first rigorous sample survey carried out in 1974 provided an estimate of 14,000 elephant. Apart from the low count in 1977, subsequent surveys have confirmed the higher estimate (Fig. 4a).

A 5% growth model beginning with 2,500 elephant in 1930, or of 2,000 in 1930 with immigration during the 1950's, together with the removal of known mortality each year from 1956, fits the more recent census data well. There are, of course, many alternative models which could fit these meager data. The models nevertheless serve to show that as long as the Wankie elephant population exceeds about 9,000 and has a growth rate of 5%, current levels of offtake are unlikely to do more than stabilize the population. Accurate measures of population size and particularly of population growth rate are crucial. Evidence presently available indicates a reproductively vigorous population (Williamson, 1976) with a mean age

Fig. 4. (a) Estimates of elephant numbers in Wankie National Park and models of elephant population growth giving an annual growth rate of 5%, less known kills each year from 1956. (b) Number of elephant killed in Wankie or on control work on park boundaries. (c) Annual rainfall recorded at Wankie Main Camp, 1931 to 1980. Least square regression lines indicate period of increasing or decreasing rainfall. The cumulative number of boreholes (artificial waterholes) in the park is shown in the lower right of the figure.

of puberty at 11 years, a calving interval of 4 years and 44% of the female population less than 11 years old. The lower levels of fecundity recorded by Laws *et al.* (1975) have not yet been found in southern African populations (Kerr, 1978). The models of Hanks and Mackintosh (1973) indicated a maximum population growth rate for elephant of 4.7%. However, subsequent data from the Addo elephant (A. Hall-Martin, personal communication) and a possibly younger age at puberty (H. Croze, A. Hillman, and Lang, unpublished manuscript) indicate that higher rates are feasible.

3. Elephant Culling. The decision to cull in Wankie was made by the Minister for Lands toward the end of 1964. The first operation began nearly two years later and continued through into 1967. Five hundred elephant were removed from a population thought to number 5,000. This decision came at the end of a drought year (Fig. 4c). No culling took place from 1968 to 1970, although a number of animals were shot on control operations on the borders of the park. In 1971 the decision was taken to cull 3,000 elephant from a population estimated to be about 10,000 animals. This decision follows two years of low rainfall and, although it was not appreciated at the time, it followed a fourteen year period of declining rainfall (Fig. 4c). In the event 1,308 elephant were killed in 1971 and 971 in 1972. Despite the poor rains of 1972 and 1973, no elephant were culled. Aerial counts in October 1973 indicated a population of 10,500 elephant, and recommendations were made for this population to be reduced over a period of three years to 7,000 elephant. This recommendation was not followed, and 762 elephant were culled in 1974. After this, the Director of National Parks adopted a policy of maintaining the population at the then established level of 13,000 elephant until such time as a clear case emerged for either reducing the population or allowing it to expand. Subsequent culling has reflected this policy, i.e., the removal of 3% to 4% of the population estimated in the previous dry season. The 1980 reduction at the level of 5%, or 750 elephant, was based on a revised view of the population size and growth rate.

The distribution of culling within the park (Fig. 5) has been such as to provide protection to the most susceptible habitats in the park. Culling has mostly occurred on the shallow soil areas on the fringes of the more robust Kalahari sandveld. Culling seldom occurred in the same locality in two successive years and in some instances the gap has been up to five or more years. Air survey results indicate that densities have remained low in localized areas for two to three years after a cull.

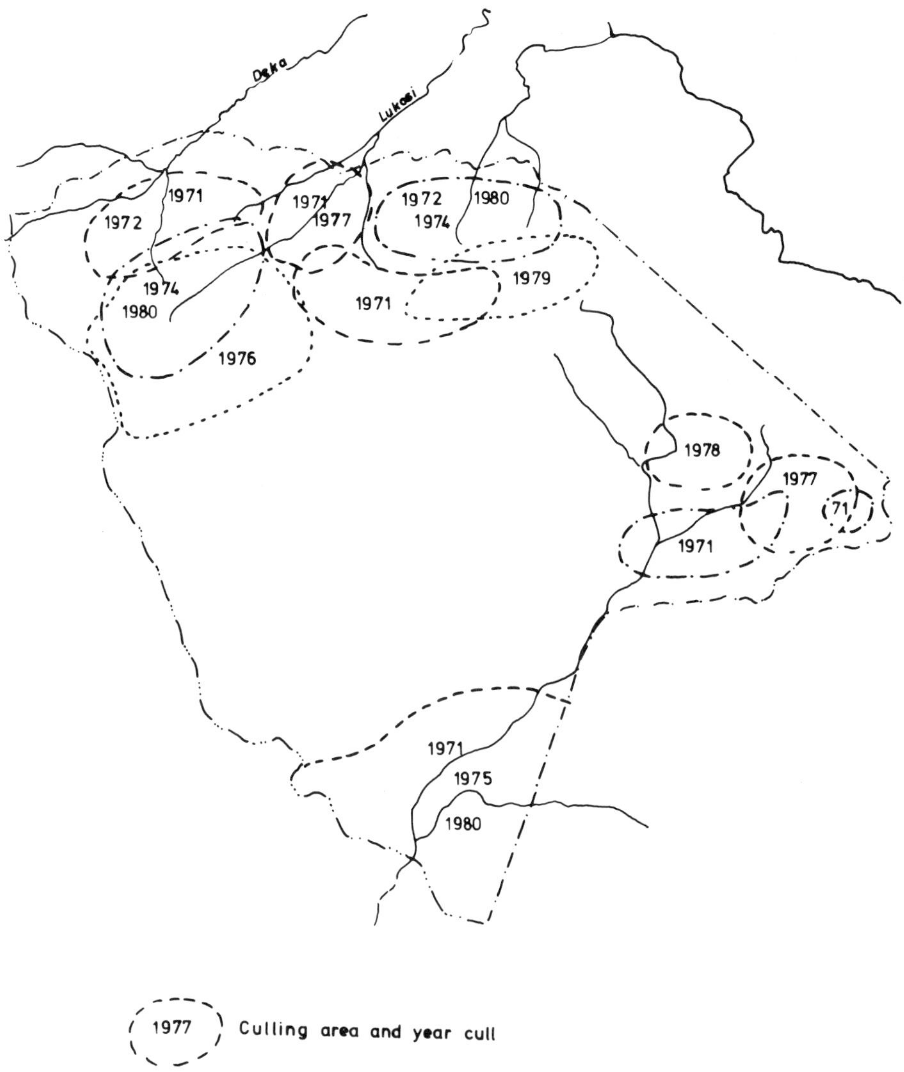

Fig. 5. Distribution of elephant culling areas in Wankie National Park between 1971 and 1980.

This, together with a few years of high rainfall, has allowed vegetation in these areas to recover from earlier heavy utilization.

4. *Elephants, Water and Vegetation.* A critical feature of the influence of elephant on vegetation in Wankie is the seasonal extremes of elephant distribution (Williamson, 1976). In 1970 Williamson found 55% of the elephant counted during an October aerial survey in only 7% of the park. Densities ranged from 2.4 to 16 elephant/km^2. The provision of additional waterholes was a second major thrust to management (Fig. 4c) and culling was centered on concentration areas, particularly on the shallow basalt soils. Elephant in Wankie now depend during the dry season on some 70 perennial water points in the park whereas before 1936 there were as few as 15 points. This has increased the range available to elephant during the peak of the dry season from c. 35% to c. 75% of the park.

Obvious though the effects of elephant on woodland were in Wankie, quantitative measures of these influences are generally lacking. Counts of particular key species in a few large plots scattered through the park indicated a mortality of c. 30% due to elephant (M. Kerr, unpublished data). As Wier (1971) has shown, the provision of artificial waterholes in Kalahari Sands leads to ecological changes about these points. These changes in the Kalahari Sands areas have not generally been viewed with alarm and have not prompted the culling of elephant in the park.

5. *Policy and Options.* Present policy is to maintain the elephant population and the woodlands in Wankie at their present level until there is good evidence to suggest we should alter this policy. This means that culling will almost certainly proceed annually or at least every second or third year. This is viewed as a conservative approach to the management of the park which serves to keep options open. Two major political forces in the country reinforce this approach as opposed to one in favor of "allowing nature to take its course." The first is the environmental conservation movement, acting through the Minister and the Wankie Park Policy Document and further backed by the Natural Resources Act, which would not tolerate range deterioration in the park. The second is that the risk of a massive die-off of elephant and the consequent "waste" of animal products cannot be tolerated in a country where the vast majority of its people are short of protein.

Future developments involve a reexamination of the manipulation of artificial watering points and the possibility of stabilizing the population by offtake in the peripheral areas of the park or preferably outside the park boundaries. Means of stimulating emigration would need to be sought. The build up of elephant on the shallow soil areas would probably still have to be dealt with by low level annual or periodic culling.

B. The Sebungwe Region

The Sebungwe has a healthy expanding elephant population, but in contrast to Wankie the problem is set in smaller wildlife areas surrounded by subsistence agriculture. The region is still poorly developed and the parks are less than 20 years old. Nevertheless, in the wake of expanding human populations, settlement schemes and tsetse control hunting, there has been a dramatic reduction in the land available to elephant. The problem is aggravated by a predominance in the national parks of woodlands sensitive to elephant depredations. These areas are also upland watersheds susceptible to erosion. The problem of deforestation by elephant became apparent in the late 1960's and the affected areas are clearly visible on satellite imagery. Culling has been and is still being used in management, and alternative strategies such as disturbance hunting and driving are to be tried. A regional plan with emphasis on the development of wildlife resource use and tourism is presently being developed for the Sebungwe by government agencies in Zimbabwe.

The old Sebungwe district (Fig. 6a), presently the Gokwe and Binga districts, was until 1956 largely unallocated land (Fig. 6b). It now comprises communal (subsistence) agricultural land and national land which is made up of national parks, safari areas, and forest areas. Commercial farming, presently at a low level, is restricted to the east of the region. Eleplant are effectively limited in distribution to the area north of the northern Sebungwe game fence (Fig. 6d). Elephant were able to wander over most of the Sebungwe 30 years ago. Although most of the changes in the region have occurred in the last 20 years, it is desirable to outline the considerable human influences which have impinged on the region and its elephant populations over the past century.

1. Historical Background. By the 1870's elephant populations were low and Selous (1881) records that the local people hunted elephant by slashing the Achilles tendon with a broad-bladed axe. The great rinderpest epizootic reduced tsetse to a few isolated pockets (Jack, 1914 and Fig. 6a) and the fly only approached its former limits in the mid-1960's. The central and southern parts of the region were inhabited by the Tonga people who were centered on the Zambezi and by the Shangwe. Human populations were low (c. 20,000 in 1904) and scattered on small shifting settlements. Upheavals occurred in 1916 when sleeping sickness resulted in the evacuation of the Busi and Lubu Valley settlements. A second major upheaval occurred when the Zambezi Valley was flooded to form Lake Kariba, and the Tonga were moved in 1957 and 1958 from the Zambezi Valley to alluvial areas further inland.

(a)

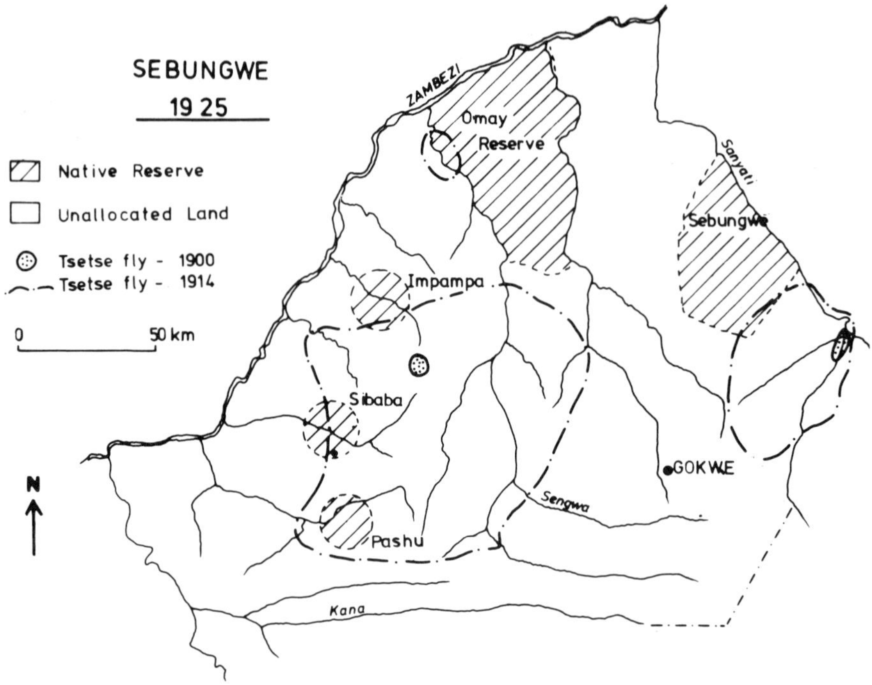

(b)

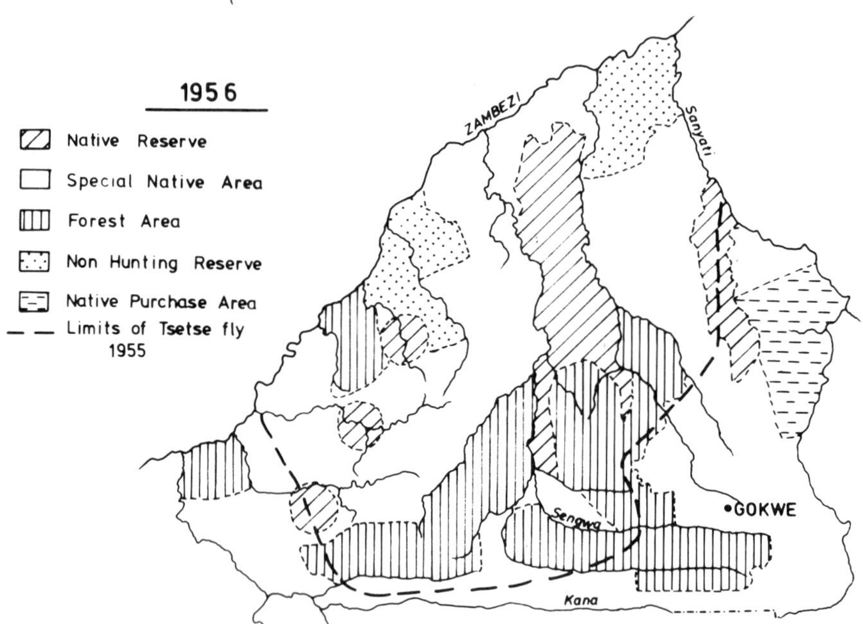

Fig. 6. Summary of the apportionment of land in the Sebungwe region, 1925 to 1980. Game reserves and forest areas were generally devoid of human settlement when proclaimed.

(c)

(d)

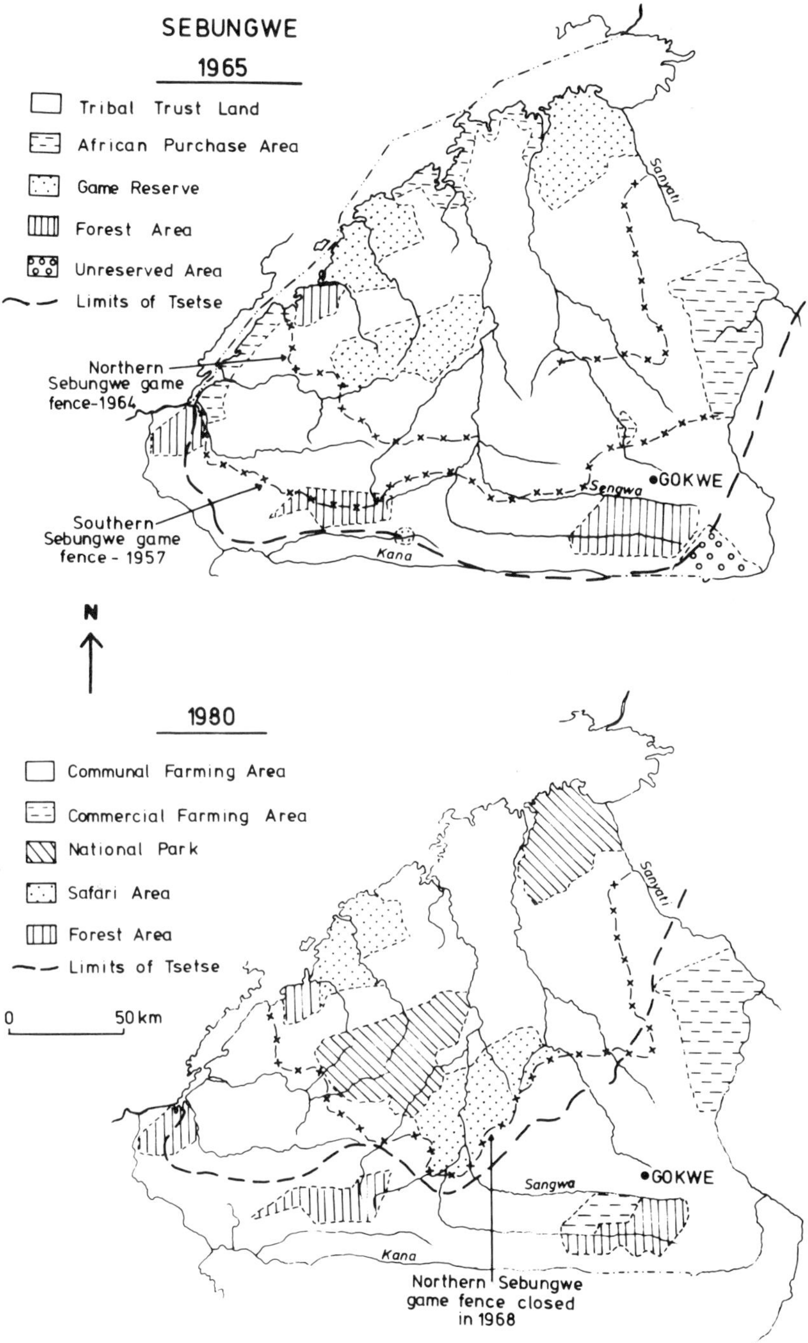

Fig. 6c and d

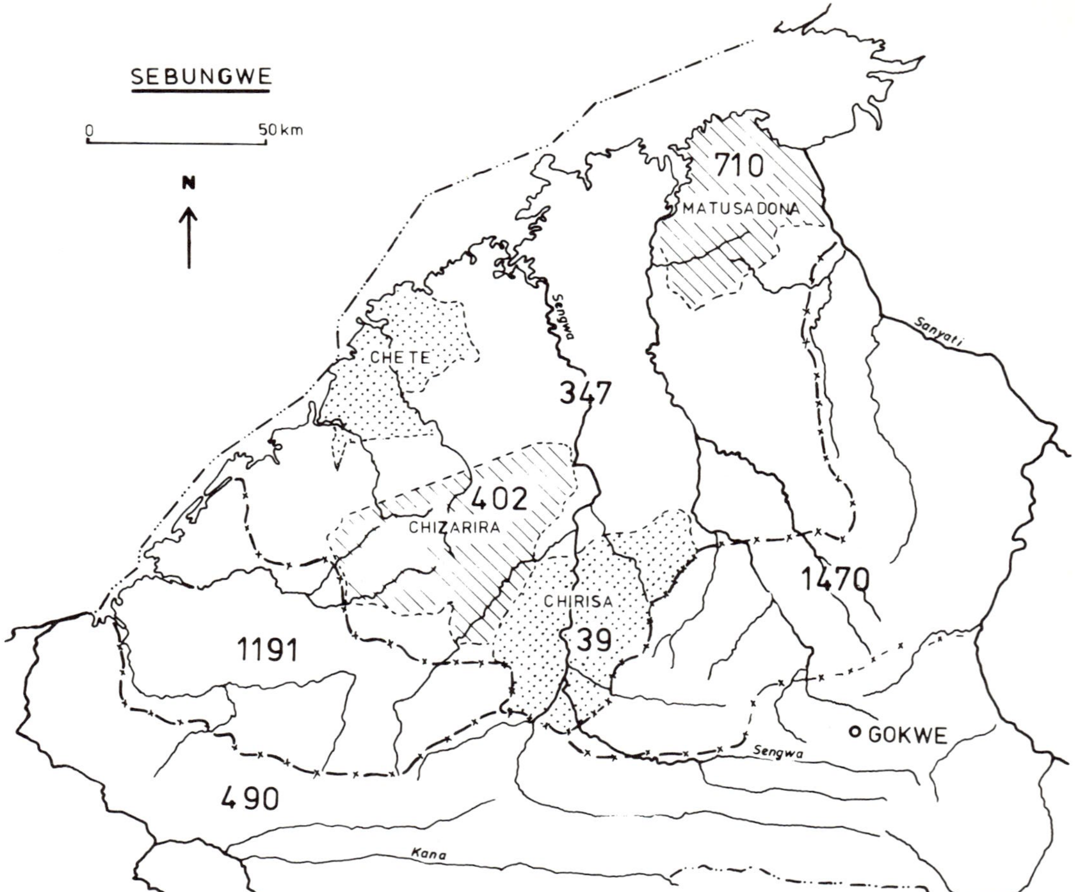
SEBUNGWE
0
50 km
N
710
MATUSADONA
Sengwa
Sanyati
CHETE
347
402
CHIZARIRA
1470
CHIRISA
39
1191
GOKWE
Sengwa
490
Kana

Figure 7.

The Sebungwe appears to have been adequately protected from illegal hunting, and poaching for ivory has not been a problem in the district. Elephant were hunted in the area before World War I, but numbers were low. In nearly a year of hunting Paxton (1914) killed only three elephant. Jack (1919) seldom mentions the spoor of elephant and does not record seeing them during his foot patrols through the Sebungwe. Tsetse control hunting did occur along the advancing fly front but remarkably few elephant were shot and G. Child (unpublished manuscript) records only thirteen killed between 1919 and 1953 in the Sebungwe operation. The Native Commissioner for the Sebungwe District in 1924 remarked that "the great proportion of the district is, to all intents and purposes, a game sanctuary" and that game populations were increasing. Changes in land apportionment in the Sebungwe are summarized in Fig. 6. The special native areas designated in 1956 were largely unoccupied land destined for planned settlement. Forest areas were also unoccupied areas of indigenous woodlands. Those in the Sebungwe have not been worked.

The situation for elephant changed after 1955. The southern Sebungwe game fence (Fig. 6c) was erected in 1956 and 1957, and elephant to the south and west of the fence were either driven north or eliminated. This resulted in a reduction of elephant range by about 30%. Meanwhile Lake Kariba began to fill and although settlement had denied much of the Zambezi alluvial area to elephant, the resettlement of people inland denied elephants some of their former haunts as well as the flooded range (c. 5% of the Sebungwe). Although protection of the fences continued, it was not until 1964 that full-scale selective game elimination started in the controlled hunting area north of the southern Sebungwe game fence. The northern Sebungwe game fence was erected during 1964 and 1965 but some sections of it were only completed over the period of 1968 to 1969. At least 2,661 elephant were eliminated in the controlled hunting areas to the south of the northern Sebungwe game fence and, between 1964 and 1968, about 40% of the area available to elephant after 1957 was removed (Fig. 7). This was followed by immigration of people from overcrowded areas elsewhere in the country and a great increase in the human population of the region from 21,215 in 1904 to 222,000 in 1979.

Fig. 7. The number of elephant killed in various parts of Sebungwe between 1955 and 1979. Animals killed north of the northern Sebungwe game fence were killed as part of culling programs (Chizarira, Matusadona) or on crop protection exercises.

2. *Elephant Numbers.* In a report to the Chief Native Commissioner in 1924 the Sebungwe Native Commissioner gave as his conservative estimate 750 elephant in the Sebungwe. I have used this estimate as a starting point from which to examine elephant populations in the Sebungwe. The known mortality of elephants in the region since 1955 could only have been sustained by a population of 5,000 or more elephant with an annual growth rate of 5% (Fig. 8). A starting population of 1,250 to 1,500 elephant in 1925 and a present population of between 8,000 and 12,000 provides a reasonable fit to the early estimate and to the results of the 1979 sample survey which covered 57% of the Sebungwe north of the fence. Not all of the area could be flown in 1979 but those parts not covered were mostly settled

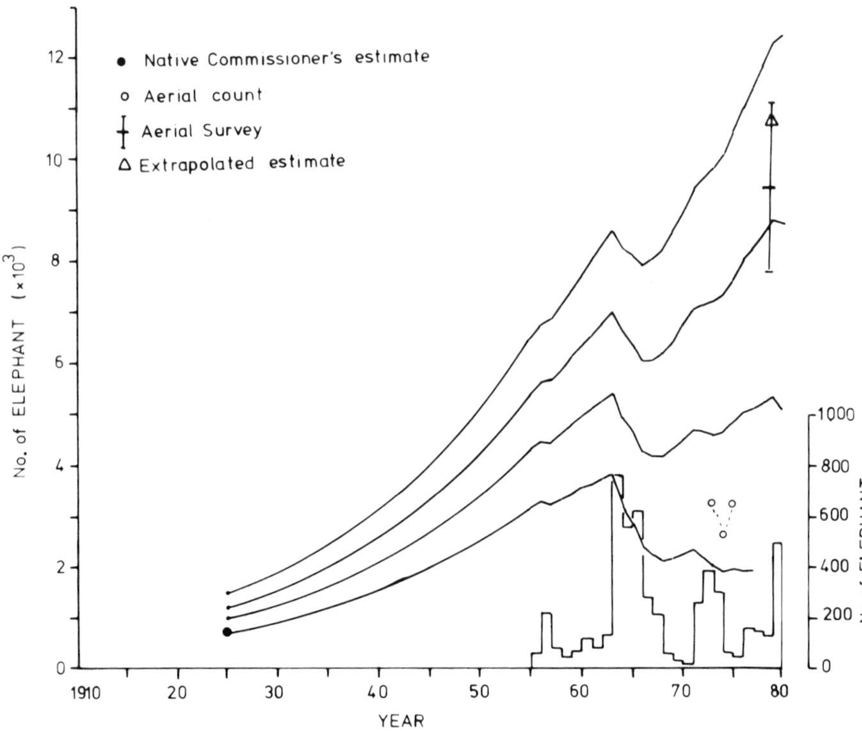

Fig. 8. Estimates of elephant numbers and models of population increase for the Sebungwe region, based on an annual growth rate of 5%, less known kills each year from 1956. The histogram shows the number of elephant killed in the region between 1956 and 1980. (Projected culling of 500 animls in 1980.)

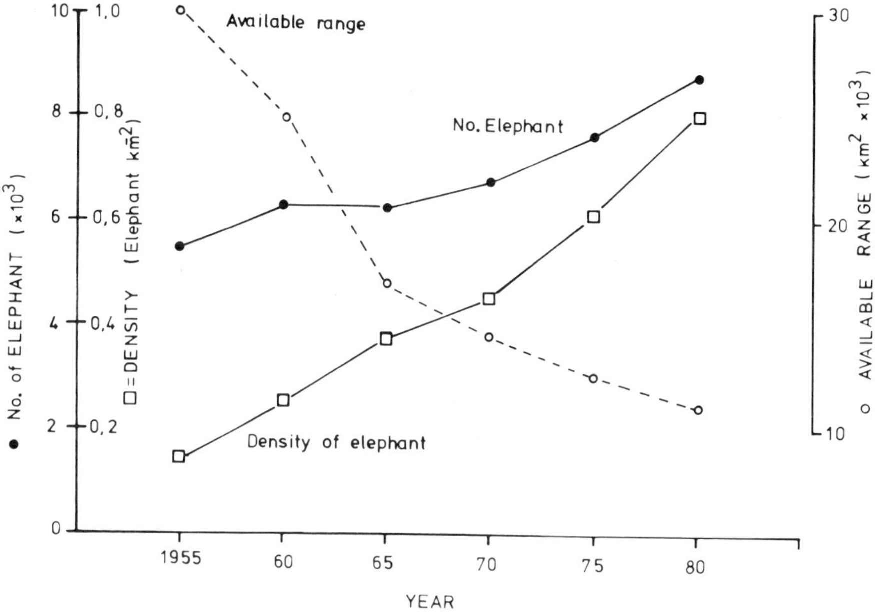

Fig. 9. Changes in elephant numbers in the Sebungwe between 1955 and 1980 and the accompanying decline in range available to elephant. The associated changes in crude, elephant density in the Sebungwe are also shown. If the present population were to be forced into existing parks and wildlife land, then the density would rise to about 1.2 elephant/km^2.

areas with few elephant. My extrapolations from the census results indicate a population of about 11,000 elephant in the Sebungwe. Earlier counts of the Sebungwe (1973, 1974, and 1976) were "minimum counts" of about 3,000 elephant thought to represent a population of some 5,000 animals. Similar "minimum counts" had been flown from a Super Cub in each of the game areas from 1968. These results were, however, very variable and generally provided totals which were 30% to 40% of those estimates obtained from ground transects in the Sengwa Wildlife Research Area. Between 1965 and 1978 these transects revealed elephant densities of about 1 elephant/km^2. The 1979 aerial census agreed with these data, and it is reasonable to suggest on the basis of the 1979 census and earlier counts that the wildlife areas of the Sebungwe, particularly the Chizarira, Chirisa, and Matusadona, have been supporting a density of 1 to 2 elephant/km^2 for the last decade. With the decrease in available range since 1955 and an increasing elephant population, the crude density of elephant in the region has increased some fourfold in 25 years (Fig. 9).

3. *Elephant Control and Culling.* Of the 4,649 elephant killed in the Sebungwe since 1955, most were killed in tsetse control operations during the mid-1960's. These animals were killed in the controlled hunting area south of the northern Sebungwe game fence. This hunting almost certainly served to drive many herds north of the fence. In 1968 some 250 elephant were driven out of the controlled hunting area across the newly erected fence into Chirisa. Culling, until this year, has been limited almost entirely to Chizarira and Matusadona National Parks. In each area culling has been carried out during the dry season and has been aimed at providing protection to specific areas of woodland. The numbers of elephant removed in various parts of the Sebungwe are summarized in Fig. 7.

The culling operations in Matusadona National Park have been successful in arresting destruction of the woodland in the valley floor but not in the escarpment. The Chizarira culling did not afford other than very short-term protection to woodlands. However, the headquarters of the park was destroyed in December 1977, and management of the park has been difficult since then.

The first recommendation for culling in Chirisa and the Sengwa Wildlife Research Area was made in 1973 following Anderson and Walker's (1974) work. They suggested that a 60% reduction in the elephant would be necessary to reverse trends in woodland deterioration in the Sengwa area. Culling was, however, delayed until such time as adequate monitoring of the vegetation had been established and until radio tracking studies on the elephant (Martin, 1978) had progressed sufficiently to support an experimental culling operation. This stage was reached in 1978 and two family units were culled in July and August before guerrilla activities brought the operation to a halt. The program resumed in June 1980. The plan is to cull only from the central clan in Chirisa and to take animals emigrating south of the fence. Up to 500 may be culled during 1980. The experimental cull is intimately linked to a project to promote wildlife utilization, Operation Windfall (Wildlife Industry for All), to benefit tribesmen in the Sebungwe, and particularly those surrounding Chirisa (Martin *et al.*, 1978).

4. *Elephants, Water and Vegetation.* The Sebungwe has an abundance of water supplies, and there are few parts more than 10 km from perennial surface water. The influence of elephant on woodlands has been studied in the Sengwa Wildlife Research Area since 1968 and was also examined in the Chizarira. The following is a resumé of the main findings:

Aerial photographs of the Sengwa Area reveal very little change in woodlands between 1951 and 1965 but a marked change after that. Good quality aerial photography is available for 1951, 1959, 1965, 1968, 1971, 1976 -- and color photography at 1:10,000 is available for 1977, 1978, and 1980.

In 1968 Jacobsen and Cumming examined damage to trees and shrubs by elephant in three major vegetation types in the Sengwa area. The percentage of trees dead and uprooted in *Brachystegia* (miombo) woodland, *Acacia tortilis/Grewia* woodland, and *Colophospermum mopane* woodland was, respectively, 11%, 10%, and 6%.

Anderson (1973) and Anderson and Walker (1974) found the following:

	Woodland type		
	Miombo	*Acacia*	*C. mopane*
Total dead tree population (%)	27	48	22
Converted to shrubs (%)	33	12	45
Intact (%)	40	40	33

They considered that the loss over the period 1965 to 1973 was on the order of 9% annually and several times greater than the replacement rate.

Martin (1974) in a detailed study of miombo and *C. mopane* woodlands calculated that some 39% or 38,500 kg/ha, in *C. mopane* woodland and 45%, or 10,000 kg/ha, in miombo woodland has been removed, or was in the form of dead wood over the previous few years.

P. R. Guy (unpublished data) followed up Anderson's (1973) work and also examined elephant feeding (Guy 1974). He found that a single animal, in the Sengwa area, may push over c. 1,500 trees/year. He found marked temporal and spatial variations in the rate of woodland damage with an overall decline of 4%/annum in the Sengwa woodlands between 1974 and 1978. *Brachystegia boehmii* (miombo) woodland was most affected with a nearly 50% decline in woody biomass (trees and shrubs) over 4 years; a decline from 23,023 kg/ha to 12,585 kg/ha.

In Chizarira, Thomson (1975) found *Brachystegia boehmmi* woodland was being knocked down at a rate of 22 trees/ha/year which was close to 20%/annum. This rate was high enough to destroy the woodland in six years, and this has occurred.

The influence of elephant in the Sebungwe woodlands is clear enough, particularly in the mountainous areas which are in Chizarira National Park and Matusadona National Park. What is not yet satisfactorily established is whether the changes they have wrought consititute damage in an ecological sense.

5. Policy and Options. Present policy is to reduce the pressure of elephant on woodlands in the Chirisa, Chizarira, and Matusadona "highland" areas. In 1980 in Chirisa this is being done by culling a single clan. Initial plans were to cull heavily in Chizarira (500 out of 2,000 animals in the peak of the dry season) and in Matusadona. The Chizarira operation was postponed for logistical reasons and the possibility that it might influence the experimental cull in Chirisa. The possibility that the high density of elephants in the Matusadona highlands in October was a seasonal phenomenon was confirmed by aerial surveys in January and again in February 1980 when the areas surrounding the park were also censused (R. Taylor, unpublished data). In the light of these results an attempt is being made to inhibit dry season movement of elephant into the park. This, we hope, will be accomplished by regular ground and aerial patrolling of the boundary, disturbance hunting, and aerial harrassment of herds approaching the boundary.

Until recently, the management of elephant in the three parks has been examined independently. The present approach is to develop an overall policy and management plan for the Sebungwe elephant population, one which will promote the long-term value and conservation of elephant throughout its present range (i.e., north of the fence) rather than just within the game parks. The possibilities include culling, safari hunting, and disturbance of animals entering fragile areas such as the upland parts of Chizarira and Matusadona. We also intend to seek ways of breaking down the "hard edges" between national parks and surrounding areas by emphasizing the development and use of wildlife resources of which elephant, because of their value, are a key species.

C. The Gonarezhou

The historical background to the elephant problem in the Gonarezhou National Park, proclaimed in 1970, has close parallels with that of the Sebungwe. The elephant population has expanded from about 1,000 animals in 1925 to some 7,000 in 1970. The population was further compressed by agricultural development, tsetse control operations, and persecution in Mozambique. The Gonarezhou is effectively divided into two regions -- the Sabi-Lundi in the north and the Nuanetsi in the south. Culling has taken place in both, but I shall deal only with that on the Nuanetsi.

The marked effect of elephant on the riverine communities of the Nuanetsi, and particularly those in the Buffalo Bend area, were first reported by Ferrar in 1966. In 1968 he estimated the elephant population in the subregion to be c. 1,400 in a dry season range of 675 km^2 with a density of c. 2 elephant/km^2. Ferrar (1969) recommended that the population should be reduced to 500 elephant. A decision to cull was delayed until 1971, and 600 elephants were removed in June 1972. An aerial census two years later provided an estimate of 1,031 elephant south of the railway line in Nuanetsi subregion, while the aerial count gave a figure of 618 elephant.

By 1971 most of the mature *Acacia tortilis* woodland in the Buffalo Bend area had been felled or ringbarked. The cull took place in June 1972. The 1971 and 1972 rains had been well above average and in 1973 the widespread regeneration of *Acacia tortilis* was apparent. Five hundred impala were culled in the riverine area during mid-1973. By 1979 dense stands of *Acacia tortilis,* some 5 to 7 m high, covered areas of former *A. tortilis* woodland. Elephant damage to woodlands in the region has not been apparent recently. Although this provides the only clear example in the country of *Acacia* woodland regeneration following an elephant (and possibly impala) cull, the influence of the last three years of the war on the elephant populations of the region has still to be assessed.

III. MANAGEMENT OF OTHER LARGE MAMMALS

Local overabundance of a variety of large mammls has occurred in national parks in Zimbabwe, and the problems have generally been managed by population reduction. This is not a euphemism. Apart from elephant and hippopotamus, capture and translocation have been used whenever practicable. The extent of these operations over the past ten years is indicated in Table II.

Pre and postreduction studies have generally been lacking, and a critical evaluation of the success or otherwise of reductions is not possible. A possible exception is the Deka catchment reclamation project in the northwest corner of Wankie National Park. Here an approximately 50% reduction in buffalo and zebra populations, together with heavy elephant culling, assisted in the reclamation of the catchment which had reached a very downgraded state by 1970 (Robinson *et al.*, 1973). The game populations have remained low since the reductions of 1974 and 1975 and, if anything, there has recently been concern at underutilization of grasslands.

TABLE II. Summary of Population Reduction Operations on Large Mammal Species Other Than Elephant in Zimbabwe between 1965 and 1980
Note: This Table is Probably Incomplete

Species	*Wankie*	*Matusadona*	*Mana Pools*	*Gonarezhou*
Hippo	-	-	-	*Cull*
Zebra	*Cull/ Translocation*	-	*Trans- location*	-
Buffalo	*Cull*[a]	*Cull*[b]	*Cull*[b]	-
Waterbuck	*Cull/ Translocation*	-	*Trans- location*	-
Wildebeeste	*Cull/ Translocation*	-	-	-
Impala	*Cull*[b]	-	*Cull/ Translocation*	*Cull*
Warthog	*Cull*	-	-	-

[a] *In addition to three culling operations removing a total of c. 2,000 buffalo, approximately 200 are killed each year for staff rations. The total population is c. 15,000 buffalo.*
[b] *Culled and also shot regularly for staff rations.*

The wildebeeste cull in Wankie was almost certainly overdone and populations of this species have shown little sign of recovery after seven years. This syndrome seems to parallel that found in Kruger and in South-West Africa. The Mana Pools populations have, superficially at least, shown no ill effects from the population reductions which occurred there five and six years ago. However, the experiences of the mid-1970's have, if anything, engendered a more cautious approach to the reduction of equid and bovid populations in national parks which have appropriate predators. Populations in the smaller game parks within recreational parks in Zimbabwe are reduced when necessary by capture and translocation and, very occasionally, by culling if live animals are not required. For the most part, captured animals are sold to farmers to stock farms.

The hippopotamus populations on the Lundi River are fully protected and can be counted accurately. Like elephant, they show a consistent growth and have been culled in 1972, 1973, and 1978.

During the last two years most game populations in the country have been subject to poaching, mainly by snaring, but the prospect of population reductions in species other than elephant and hippopotamus in the near future seems unlikely.

IV. DISCUSSION

Even a superficial examination of the historical background of protected areas in Zimbabwe shows that man for a long time has exerted, directly and indirectly, a considerable influence on large mammals and their habitats. Management of one sort or another thus becomes a necessity if we are to avoid politically unacceptable environmental change or degradation. Most of the protected areas were initially set aside as game reserves for the protection of spectacular species of large mammals. The National Parks and Wildlife Act of 1975 requires a far broader perspective and does not sanction the protection of one species or group of species at the expense of others. So far the major tool in the management of mammals judged to be too abundant has been culling or, if possible, translocation. A major and often little appreciated feature in the management of populations of large mammals in Africa, even in southern Africa, is the lack of sufficient scientific staff to fully investigate and follow up management problems. Trained and experienced field men and money are seldom available to implement more than the most rudimentary management measures. Population reduction is seen as a conservative policy which, in the conservation of ecosystems, seeks to avoid the problem of "option foreclosure" inherent in a laissez faire approach. Over the last three years, culling has only been carried out on elephant and hippopotamus -- species of which man is the only effective predator.

Wildlife conservation in Zimbabwe is regarded as a viable and legitimate form of land use. Indeed, in terms of land area, the Department of National Parks and Wildlife Management is the largest land use agency in the country. Landholders throughout the country are encouraged to conserve and use wildlife on a sustained basis. This means that many animals which would otherwise be a nuisance are nurtured by landholders. A particularly good example is the conservation of tsessebe *(Damaliscus lunatus)* on farms in Zimbabwe. The main populations of this Royal Game species occur on cattle ranches in the commercial farming area, and they have now reached a level where sustained use is both necessary and important to their increase and the further expansion of their range. The species has accordingly just been deproclaimed as Royal Game in most of the country.

A further extension of the view that wildlife is a valuable economic resource is the move toward encouraging a greater integration of wildlife and other forms of land use. The creation of peripheral hunting zones and the regulation of populations of elephant and other mobile species in national parks through emigration and harvesting of emigrants is an attractive possibility. For this concept to work it will have to be economically sustainable, and it should serve to enhance the value of protected areas to local people. The value of protected areas to those living around them is probably crucial to the long-term survival of these areas and to the plants and animals they protect.

The approach expressed here toward the problem of local overabundance has tended to emphasize the economic use of large mammals. I am sensitive, however, to the important ethical issues that are involved in the relationships between man and his fellow creatures. Perhaps the real dilemma is the extent to which we, as a species, can or should avoid being the super predator we are.

ACKNOWLEDGMENTS

I have drawn extensively on data contained in departmental files, on some documents in the National Archives, and particularly on unpublished reports by Graham Child, Tony Ferrar, Michael Kerr, Rowan Martin, Brian Sherry, Russell Taylor, and Basil Williamson.

I also thank Gerald Davison for providing me with many of the tsetse control hunting figures and Desmond Lovemore who, over many years, has provided me with much information about the Sebungwe.

REFERENCES

Anon. (1961). Annual Report of the Department of Wildlife Conservation for the year ending December 31, 1960.

Anderson, J. L. (1965). Aerial census elephant and buffalo: Wankie Game Reserve. Departmental report.

Anderson, G. D. (1973). Vegetation composition and elephant damage in the major habitat types of the Sengwa Wildlife Research Area of Rhodesia. M.Sc. Thesis, University of Rhodesia.

Anderson, G. D. and Walker, B. H. (1974). Vegetation composition and elephant damage in the Sengwa Wildlife Research Area, Rhodesia. *J. So. Afr. Wildl. Manage. Assoc. 4,* 1-14.

Davison, E. (1950). Wankie Game Reserve. Unpublished manuscript.

Ferrar, A. A. (1969). Wildlife survey: Buffalo Bend. Unpublished departmental report.

Guy, P. R. (1974). Feeding behaviour of the African elephant in the Sengwa Research Area of Rhodesia. M.Sc. Thesis, University of Rhodesia.

Hanks, J. and Mackintosh, J. E. A. (1973). Population dynamics of African elephant *(Loxodonta africana). J. Zool. London 169,* 29-38.

Jack, R. W. (1914). Tsetse fly and big game in southern Rhodesia. *Bull. Entomol. Res. 5,* 97-110.

Jack, R. W. (1919). Extracts from diaries of Rupert Jack, 1914-1919.

Kerr, M. A. (1978). Reproduction of elephant in the Mana Pools National Park, Rhodesia. *Arnoldia (Rhodesia) 8 (29),* 1-11.

Laws, R. M., Parker, I. S. C., and Johnstone, R. C. B. (1975). "Elephants and Their Habitats." Oxford Univ. Press (Clarendon), London and New York.

Martin, R. B. (1974). Structure biomass and utilization of vegetation in mopane and miombo woodlands of the Sengwa Wildlife Research Area. Certificate in Field Ecology Report, University of Rhodesia.

Martin, R. B. (1978). Aspects of elephant social organization. *Rhod. Sci. News 12 (8),* 184-187.

Martin, R. B., Conway, A., and Dix, P. (1978). Project: Windfall Sebungwe. Departmental report.

Paxton (1914). Extracts from diaries, 1913-1914.

Robinson, J. C., Hill, J. C. R., and Rushworth, J. (1973). Deka River catchment survey. Unpublished report. Department of Conservation and Extension/Department of National Parks and Wild Life Management.

Rushworth, J. (1975). The floristic, physiognomic and biomass structure of Kalahari sand shrub vegetation in relation to fire and frost in Wankie National Park. M.Sc. Thesis, University of Rhodesia.

Spinage, C. A. (1973). A review of ivory exploitation and elephant trends in Africa. *East Afr. Wildl. J. 11*, 281-289.

Thomson, P. J. (1975). The role of elephants, fire and other agents in the decline of a *Brachystegia boehmii* woodland. *J. So. Afr. Wildl. Manage. Assoc. 5*, 11-18.

Wier, J. S. (1971). The effect of creating additional water supplies in a Central African National Park. *In* "The Scientific Management of Animal and Plant Communities for Conservation" (E. Duffy and A. S. Watt, eds.), *Symp. Brit. Ecol. Soc. 11*, 367-385.

Williamson, B. R. (1975). Seasonal distribution of elephant in Wankie National Park. *Arnoldia (Rhodesia) 7(11)*, 1-16.

Williamson, B. R. (1976). Reproduction of female elephant in the Wankie National Park, Rhodesia. *So. Afr. J. Wildl. Res. 6*, 89-93.

Williamson, B. R. (1979). Elephant census Wankie National Park. Departmental report.

THE PROBLEMS OF OVERPOPULATION OF DEER IN NORTH AMERICA

David R. Klein

Alaska Cooperative Wildlife Research Unit
Fairbanks, Alaska

I. INTRODUCTION

In the North, human impact on the environment has been curtailed by unsuitable conditions for agriculture and by the harshness of the environment. This has minimized the problem of wildlife habitat destruction and overexploitation of bird and animal populations, which has been common to much of the rest of the world. Exceptions have been the overexploitation of marine mammals in the Arctic for commercial utilization and local extermination of ungulate populations to provide meat for explorers, whalers, and, increasingly, numerous and effective indigenous hunters equipped with the tools of western technology. Much of the terrestrial habitat for mammals in the Arctic is on islands and local populations are, therefore, particularly vulnerable to overexploitation and extermination. This pattern has been associated with man-caused local extinctions of caribou in Greenland (Vibe, 1967), Svalbard (Hjeljord, 1973), the Canadian Arctic archipelago (Parker *et al.*, 1975) and of muskoxen in Alaska, portions of the Canadian Barren Grounds (Tener, 1965), and presumably in Siberia in prehistoric times (Klein, 1979). Now, with the increasing focus on the Arctic as a new source for the world's energy and mineral resource needs, further threats exist to populations of large mammals in the Arctic.

Responses by ungulates to human influences have generally been more subtle in the north temperate regions than in the Arctic. Man has altered the natural environment largely through agriculture, forestry, predator removal, and in the West, grazing of livestock on open range. Ungulate populations have responded in differing ways depending upon specific situations. Variables in weather, predation, plant resources, and other factors often complicate ungulate habitat relationships causing differing ungulate responses in what may seem to be similar situations.

PROBLEMS IN MANAGEMENT OF LOCALLY
ABUNDANT WILD MAMMALS

ISBN 0-12-385280-3

As many of the eastern and north central regions of the United States were settled, conifer forests were cleared by logging and fire to be replaced by deciduous trees and shrubs of moderate to low value for deer. The more open canopy of the deciduous forests permitted increased growth of understory vegetation important as deer forage. Often several environmental factors, both man-caused and natural, simultaneously influenced deer populations, thus stimulating deer irruptions or die-offs. Historical examples are the increase in forage through logging at the same time that intensive predator removal was taking place; in addition, deer were given increased protection through the establishment of the buck law and other restrictions on hunting. This sequence, which favored deer population increase, was often followed by maturing of the forests and thus a reduction in available forage, the loss of remaining conifer stands which were important to deer as yarding sites during extreme snow conditions, the use of dogs in hunting deer, and more liberal hunting regulations. In fact, timing of changes in hunting regulations in North America has often been out of phase with the population status of deer, so that the desired effects have often not been achieved, or at worst, results opposite to those desired have been the outcome.

In any examination of the consequences of overpopulations of deer in North America it is important to bear in mind that these "imbalances" of deer with their habitat have usually been in association with habitat instability brought about by human manipulations of the environment. "Overpopulation," as used here is restricted to Caughley's (this volume) Class 3, "too many animals for their own good," and Class 4, "the system of plants and animals is off its equilibrium."

Where high deer densities have led to deterioration of the condition of the deer, changes in the vegetational composition of the habitat have often also occurred (Klein, 1970). Changes in plant composition are also accompanied by deterioration in the quality of the diet of deer as favored forage species are depressed and are replaced by less preferred plants. Although this is apparently primarily associated with changes in plant species composition in the diet, there is also evidence that plants may respond to grazing pressure through production of secondary compounds, which make the plants less palatable and reduce their value as forage (Haukioja and Niemela, 1976).

Examination of specific case histories provides insight into the complexities of deer and habitat relationships which may be of value in dealing with problems of local overabundance of threatened or rare mammals.

II. CASE HISTORIES

Deer on islands, including ecological islands of isolated habitat, may show population characteristics paralleling those of protected mammals in reserves.

In Alaska, deer *(Odocoileus hemionus sitkensis)* occur naturally on the islands of the Alexander archipelago and adjacent mainland; however, ecological conditions vary markedly among the islands (Klein, 1965a). There is a cline from moderate to more severe winter conditions from the outer islands with strong maritime influence, to those close to the mainland with continental climatic influence. A similar cline exists from south to north with respect to winter temperatures, but precipitation is highly variable. Winter snow cover may be markedly less on some islands as a result of cold dry weather while on others reduced snow accumulation may occur because much of the winter precipitation falls as rain.

Where they occur, wolves are the major predators of deer on the mainland and some of the islands, but they have not reached many islands isolated by wide water channels (Klein, 1965b). Human hunting pressure is variable in relation to proximity to settlements.

Forage available to deer during summer varies in quality from island to island but increases with the extent of alpine vegetation present and the associated topographic variation (Klein, 1965a). These factors influence forage quality through their effects on plant phenology and the extent and amount of vegetation, within reach of deer, growing away from the growth-limiting effects of the dense rain forests common at lower elevations. Deer respond to these conditions by following altitudinal and exposure clines in plant development to feed selectively on the highest quality forage available as the summer progresses.

In winter deep snows at higher elevations cover forage plants, and deer are forced to winter at lower elevations. In winters of heavy snow accumulation deer concentrate in old growth forests adjacent to the beaches where snow depths are least and irregular due to interception of the snow by the crowns of the large trees. In such winters mortality from malnutrition is high although the effect of deer on the vegetation is relatively light because much of the forage normally available to deer is covered by snow. Current forest cutting practices, which are directed toward the old growth forest stands adjacent to the beaches, appear to be causing a deterioration in the quality and quantity of suitable winter deer habitat in this region of Alaska.

The population responses of deer under these insular conditions, where dispersal as a regulator of population density is not a possible option, can be summarized as follows:

(1) Deer on islands with wide topographic variation, in contrast to deer on islands with most of the land area at low elevations, achieve more rapid growth rates and larger body size. Populations of deer on islands of varied topography, fluctuate more widely than those on the low-lying islands, with peaks following years of mild winters. After severe winters, deer are considerably less abundant on the islands with wide topographical variation. Forage plants are more directly influenced by deer grazing pressure on the low-lying islands.

(2) If one contrasts islands under strong maritime influence with those with a more continental climate, the patterns of deer response show differences similar to those of the high altitude versus low altitude islands. On islands close to the mainland with more severe winters deer populations show wider fluctuations but have less overall influence on the vegetation. Growth rates of individuals are most rapid on these islands because competition for high-quality forage is minimal. This is due to the fact that average deer densities are lower than on the maritime islands, where population numbers are in more direct relationship to the total forage present, and the influence of winter snow cover is less.

(3) Wolves complicate the response of deer to their food resources. On islands without wolves the influence of deer on the vegetation, particularly in the wintering areas, has generally been greater than on islands where wolves are present. This presumably results in the potential carrying capacity of the range for deer being higher on the wolf-occupied islands. However, the conditions under which this potential can be realized (i.e., the coincidence of low wolf numbers and mild winters) occur infrequently.

Observations during the last three decades in the Alexander archipelago suggest that a more rapid recovery of deer populations occurred on the wolf-occupied islands in the early 1950's following widespread deer die-offs associated with a series of severe winters. Wolves at that time were poisoned by government hunters and wolf populations were apparently low. The year in which deer reached high densities on the wolf-occupied islands coincided with the granting of statehood to Alaska and consequent changes in wildlife management policy resulted in the discontinuing of wolf control so that wolf populations expanded rapidly in the presence of an extremely abundant prey. For several years deer numbers remained large in the presence of large and expanding wolf populations. During the same period deer densities on the wolf-free islands reached moderate levels but their overall rate of increase appeared to be slower than among deer on islands where wolves were present. On the basis of the few years of concurrent vigorous deer and wolf populations, some wildlife biologists and members of the public assumed that wolves were an insignificant factor in the regulation of deer numbers.

In the late 1960's and early 1970's a series of severe winters depressed deer populations to very low levels. Snow depths during this period were generally less on the wolf-free islands, which lie to the north of the wolf-occupied islands, and deer populations there did not suffer such heavy winter losses. With the return of more moderate winter conditions deer numbers increased on the wolf-free islands but continued to decline to all-time lows on the wolf-occupied islands. Apparently the large wolf populations, which had not been limiting to the high deer populations during the period of mild winters, became an important factor in further suppressing the low deer populations and in holding them at extremely low levels on some islands following the heavy losses during the severe winters. A lag of approximately five years occurred before wolf densities declined to sufficiently low levels to allow the number of deer once again to rise, an increase that continues at present.

Wolves were introduced in 1961 to Coronation Island as an experiment to determine the response of deer on a wolf-free island to the effects of wolf predation. This island is under a strong maritime influence and deer had previously been controlled directly by limitations of their food resource. Forage had been greatly reduced there through the grazing pressure of the deer. The island is 78 km^2 in area and is, therefore, too small to support both deer and wolves on a sustained basis. The wolves increased from four (two pairs) to fifteen. The deer declined to extremely low levels. Forage plants showed marked recovery within a few years and the wolf population subsequently declined to extinction (H. Merriam, personal communication). The deer population subsequently recovered from near extinction and is continuing to increase.

Deer/wolf relationships on the islands of the Alexander archipelago differ markedly from Mech's (1977) experience in Minnesota. He found that although wolf packs depressed deer populations within their territories, there were "buffer zones" between pack territories which served as refuges for deer, thus preventing the wolves from forcing deer densities to such low levels that local extinctions or greatly delayed recoveries occurred. In Alaska wolf packs often occupy entire islands or large portions of islands, and they readily move between small islands separated by narrow channels. The boundaries of wolf pack territories are, therefore, most frequently the perimeters of islands so that there is little opportunity for the existence of wolf-free buffer zones.

Another example of population control on an island is the rapid expansion of reindeer *(Rangifer tarandus)* following their introduction to St. Matthew Island (330 km^2) in 1944 (Klein, 1968). On a virgin range free of predators and human harvest, they increased from 29 to 6,000 animals in 20 years and then crashed to 42 animals. Since no mature males survived, the population became extinct. The reindeer greatly overgrazed the range and with no freedom to migrate or to disperse from the island there was extreme intraspecific competition for the greatly reduced forage. Because of the rapid increase on a previously ungrazed range, there was insufficient time for such intrinsic population regulatory factors as reduced fecundity to play a significant role in controlling the rate of increase. It was coincidental, however, that a winter of such severe proportions that all records were broken occurred at the time that the reindeer had reached a density far in excess of available winter forage. Under more moderate winter conditions, presumably, the population still would have undergone a pronounced decline, but it is doubtful that extinction would have ensued.

III. DISCUSSION

If we compare deer in insular situations with those in mainland habitats, we find marked differences as well as similarities. Where predators have been removed, hunting by man may become a regulating factor, but hunting varies greatly in its effectiveness largely in relation to habitat type. Under dense forest conditions hunting rarely can be an effective population regulating mechanism. Hunting is usually restricted to the periphery of areas where habitat units are extremely large and human access limited. If the deer are nonmigratory, only a small proportion of the population will be vulnerable to hunting mortality.

Dispersal as a population regulatory mechanism is effective among deer when the most favored habitat units are small and they are surrounded by less favorable habitat where mortality risks are greatest (Hawkins *et al.*, 1971). A classic example of the extreme importance of dispersal as a population regulatory mechanism occurs among roe deer (*Capreolus capreolus)* in Denmark (Klein and Strandgaard, 1972). The scattered forest plantations with interspersed fields are optimal habitat for roe deer but hunting is not effective in controlling numbers because of the dense cover available. Roe deer are territorial and the surplus population is expelled, largely during the breeding season, to surrounding habitat with marginal cover where hunting easily controls the population. When dispersal is possible, there is little winter mortality or habitat destruction through overgrazing by roe deer in Denmark. When optimal roe deer habitat is fenced, roe deer invariably increase without the option for dispersal, and deterioration in the condition of the animals occurs with associated winter mortality and overgrazing of forage species (Klein and Strandgaard, 1972).

In North America, where deer (*Odocoileus* spp.) often occur in very large habitat units showing little diversity, such as the second growth forests of the Great Lake States, dispersal as a population regulatory mechanism is possible only at the periphery of the habitat units. Under such conditions it is common for deer to show characteristics of overpopulation, including malnutrition, reduced fecundity, distorted sex ratios, and poor over-winter survival.

IV. CONCLUSIONS

The most obvious consequences of deer overpopulations have been:

(1) Elimination of the most palatable forage species, which are usually those of highest quality as forage. Thus, the resulting altered plant species composition becomes dominated by the less desirable forage species or non-forage species. As a consequence plant succession has often been accelerated to stages less productive of deer forage.

(2) The food habits of deer change as they are forced to feed on less desirable forage. Overall quality of the diet declines and the digestible component is reduced. The consequences for the deer are reduced growth rates, smaller body size, poorer body condition, and reduced fat reserves before winter. The vulnerability of the deer to disease, parasites, starvation, and predation increases.

(3) Where food quality is limiting, delayed sexual maturity and lowered conception rates result and under such conditions fertility appears to be influenced more by rate of conception than by *in utero* mortality after conception (Cheatum and Severinghaus, 1950). Decreased post partum fawn survival has been associated with poor nutrition of the female (Verme, 1969) and heavy losses of fawns during their first winter are common on ranges where competion for available food is strong (Taber, 1956).

(4) When food becomes limiting because of high deer densities, ensuing competition, range deterioration, or for other reasons, the age of sexual maturity is delayed, fertility is lowered, fawn survival decreases, and the population becomes dominated by older animals. Such populations, although usually

stable or declining, can respond rapidly through population increase if habitat conditions are markedly improved. Under optimal range conditions deer populations are dominated by young animals, and while such populations are usually expanding, they are vulnerable to heavy mortality and population decline if habitat conditions deteriorate through overgrazing, plant successional changes, or from other factors.

(5) Under poor range conditions males suffer heavier mortality than females so that populations become skewed toward females. This factor, although often associated with the deterioration of range and population decline, nevertheless increases the reproductive potential of the population; this potential, however, is rarely realized.

(6) Wolves as predators of deer are apparently less effective in controlling deer numbers where habitat units are large and uniform than under insular conditions. Wider fluctuations or "imbalances" can be expected, therefore, among deer and wolves on islands or in disjunct habitat units, which are comparable to islands, than in uniform habitat on the mainland.

REFERENCES

Cheatum, E. L. and Severinghaus, C. W. (1950). Variations in fertility of white-tailed deer related to range conditions. *Trans. N. Am. Wildl. Conf. 15*, 170-189.

Haukioja, E. and Niemela, P. (1976). Does birch defend itself actively against herbivores? *Rep. Kevo Subarctic Res. Sta. 13*, 44-47.

Hawkins, R. E., Klimstra, W. D., and Autry, D. C. (1971). Dispersal of deer from Crab Orchard National Wildlife Refuge. *J. Wildl. Manage. 35*, 216-220.

Hjeljord, O. (1973). Studies of the Svalbard reindeer. *Norsk Polarinst. Aarbok*, 113-124.

Klein, D. R. (1965a). Ecology of deer range in Alaska. *Ecol. Monogr. 35*, 259-284.

Klein, D. R. (1965b). Post glacial distribution patterns of mammals in the southern coastal regions of Alaska. *Arctic 18*, 7-20.

Klein, D. R. (1968). The introduction, increase and crash of reindeer on St. Matthew Island. *J. Wildl. Manage. 32*, 350-367.

Klein, D. R. (1970). Food selection by North American deer and their response to overutilization of preferred plant species. *In* "Animal Populations in Relation to Their Food Resources" (A. Watson, ed.), *Symp. Brit. Ecol. Soc. 10*, 25-46. Blackwell, Oxford.

Klein, D. R. (1979). Aspects of population regulation in high arctic ungulates. Paper presented at XIV Inter. Wildlife Conf., Dublin, Ireland. 16 pp.

Klein, D. R. and Strandgaard, H. (1972). Factors affecting growth and body size in roe deer. *J. Wildl. Manage. 30,* 64-79.

Mech, L. D. (1977). Wolf-pack buffer zones as prey reservoirs. *Science 198,* 320-321.

Parker, G. R., Thomas, D. C., Broughton, E., and Gray, D. R. (1975). Crashes of muskox and Peary caribou populations in 1973-74 on the Parry Islands, Arctic Canada. *Can. Wild. Ser. Prog. Notes 56,* 10 pp.

Taber, R. D. (1956). Deer nutrition and population dynamics in the north coast range of California. *Trans. N. Am. Wildl. Conf. 21,* 159-172.

Tener, J. S. (1965). "Muskoxen in Canada: A Biological and Taxonomic Review." Queen's Printer, Ottawa.

Verme, L. J. (1969). Reproductive patterns of white-tailed deer related to nutritional plane. *J. Wildl. Manage. 33,* 881-887.

Vibe, C. (1967). Arctic animals in relation to climatic fluctuations. *Medd. Groenl. 170,* 1-227.

THE WHITE RHINO OVERPOPULATION PROBLEM AND A PROPOSED SOLUTION

Norman Owen-Smith[1]

Center for Resource Ecology
University of the Witwatersrand
Johannesburg
South Africa

I. INTRODUCTION

The white rhino problem in the Umfolozi-Corridor-Hluhluwe Game Reserve Complex in South Africa encapsulates, in perhaps more acute form, the dilemmas of overabundance that have arisen with regard to elephants and hippos within the national park remnants of terrestrial ecosystems elsewhere in Africa. The impact on the vegetation of a grazing megaherbivore, such as the white rhino, is possibly more serious than that of elephants on the woody plant stratum, since it threatens the persistence of the ground cover of grasses and hence of underlying soils. While the southern white rhino was recently removed from the International Union for the Conservation of Nature and Natural Resources (IUCN) Red Data Book, populations of rhinoceros elsewhere in Africa plummet toward extinction. White rhinos are bumbling, inoffensive giants, for which culling by bullet would seem especially repugnant. Much of the Umfolozi Reserve is proclaimed as a wilderness area, and such operations intrude upon the aesthetic qualities for which it was set aside.

In this chapter I will analyze some of the facts of the white rhino problem based on my own research carried out between 1966 and 1971 (Owen-Smith, 1981). I will show the following:

(1) The intrinsic eruption hypothesis is adequate to account for the overpopulation that has arisen.

(2) Overpopulation must result from the time lags associated with age structure effects.

[1]*Research supported by Natal Parks Board and out of an NSF grant to J. T. Emlen, University of Wisconsin.*

ISBN 0-12-385280-3

(3) In the absence of barriers dispersal would have alleviated the biomass increase.

(4) Habitat changes involve complex interactions between plant species and topographic soil/nutrient conditions.

(5) In order to simulate former conditions with the least population disruption, removals should be concentrated within "vacuum zones" serving as population sinks.

(6) For current predator/prey theory to be practically relevant to plant/large herbivore systems, competitive relations between plant types and their interaction with external factors such as climatic cycles, fire regimes, and nutrient recycling must be taken into account.

II. WHITE RHINO POPULATION EXPANSION

Regular aerial censuses of the white rhino population within the Umfolozi-Corridor-Hluhluwe Complex have been carried out since 1959. Complete counts have been carried out by helicopter since 1970, while prior to that a fixed-wing aircraft was used. Comparison of results obtained by the two methods in 1970 suggested that the fixed-wing aircraft recorded only 60% of the white rhinos seen in the helicopter censuses. This correction factor must hence be introduced for the earlier counts. Since 1961, the rate of population increase has been slowed by the expanding removal program -- white rhinos being translocated to restock other areas and zoos (Fig. 1). To

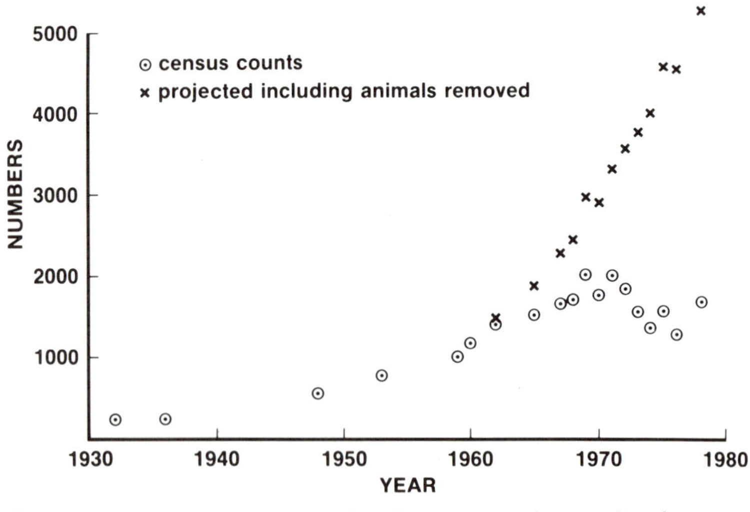

Fig. 1. White rhino population expansion, showing the effect of removals in checking the population increase in the Umfolozi-Corridor-Hluhluwe Game Reserve Complex.

estimate the true rate of population expansion, these animals have to be taken into account as well as their presumed recruitment (crudely estimated as 10%/annum). A log plot of the census data with these corrections reveals that the total white rhino population expanded at a constant rate of 9.5%/annum between 1959 and 1972 (Fig. 2). (Note that this is not the rate of density increase, since densities were controlled by the removal program.) Prior to 1959 the rate of increase was apparently somewhat slower, averaging about 6.3%/annum. It is clear that the entire population could well have arisen from fewer than 100 individuals prior to 1920, when effective conservation measures were first introduced. Rather than showing a density dependent decline, the rate of increase has apparently accelerated with increasing density. However, the high rate of increase exhibited over the period 1962 to 1971 could have been associated with the below-average rainfall that prevailed during this decade since the former distribution center of the subspecies lay in relatively arid savanna in Botswana.

This high rate of increase was the consequence of a high natality rate for so large a species and extremely low mortality rates. The mean intercalving interval averaged 2.4

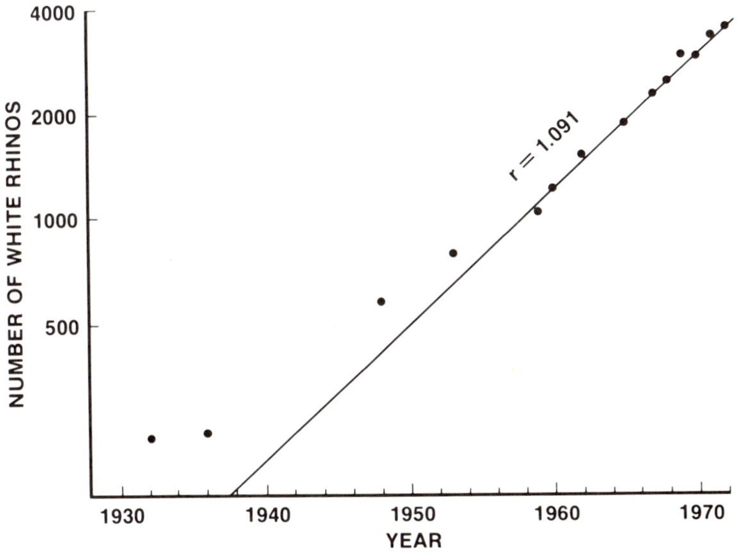

Fig. 2. Logarithmic plot of the projected expansion of the total white rhino population (including animals removed), revealing a constant rate of increase between 1960 and 1971.

years, with a shortest interval of 22 months. Animals introduced into other reserves have exhibited somewhat longer mean calving intervals. The mean annual adult mortality rates observed were 3% for males and 1.5% for females. Total infant losses, including prenatal losses and barren females as well as immediately postnatal mortality, amounted to 8%/annum. Thereafter juvenile mortality averaged 3.5%/annum. The age structure showed a marked preponderance of young animals, with 54% of the reconstructed 1969 population under breeding age (i.e., under 10 years for males and under 7 years for females). Predators, including lions and hyenas occurred in the reserve, but their increase had been recent and, with an abundance of alternative prey, their impact may have been insignficant although it could increase in the future. The population seemed to be increasing at very close to its intrinsic rate (r_{max}).

A major disturbance to the ecosystem was the virtual elimination of nearly all of the larger mammals, apart from the two rhinoceros species, during the anti-tsetse fly campaign carried out in the Umfolozi Reserve prior to 1950. However, during the 1960's large numbers of other ungulates followed by predators recolonized the area from the nearby Hluhluwe Reserve. In 1965 a rhino-proof fence was constructed around three sides of the Umfolozi Reserve, with the Black Umfolozi River providing the fourth boundary adjoining the Corridor between Umfolozi and Hluhluwe. White rhino capture and removal operations were initiated in 1961, but prior to 1967 most of the animals removed were taken from outside the boundaries of the reserve.

Regional breakdown of the rhinos counted in different censuses suggests that the population within the core area increased at a lower rate than that of the overall population. Between 1953 and 1962 the population in the section between the rivers of the Umfolozi Reserve increased by 5.2%/annum, while the population in the overall Complex increased by 7.6%/annum. Over the period 1967 to 1970 the population in the western, high density section of Umfolozi increased by 5.1%/annum, while the overall population increased by 8.6%/annum. This suggests a rate of emigration from high density areas of 3.5%/annum, despite the restrictions of the boundary fence (some of this difference, however, may have been the result of a more mature age structure in the core area). Animals colonizing new areas included subadults of both sexes plus some adult males, but no parous females.

The 1971 helicopter census counted 2,002 white rhinos within the 950-km^2 area of the Umfolozi-Corridor-Hluhluwe Complex. Of these, 1,529 occurred within the 477-km^2 area of Umfolozi Game Reserve, at a mean density of 3.2/km^2. Within the 21-km^2 area of my western study areas, the mean white rhino density was 5.6/km^2. There was no indication of any social disruption to suggest that such densities were abnormally high for the species. Subordinate bulls share the territories of some territorial bulls even in low-density introduced populations, such as that in the Kyle Game Park, Zimbabwe (Condy, 1973). Only a few aged individuals showed any notable deterioration in body conditions during the dry season. However, although annual rainfalls were below average over the study period, the seasonal distribution was altered so that the winter dry periods were shorter than usual. In 1965 a 50% increase in mortality was reported, following the failure of the late summer rains. In 1933 following a severe drought, reportedly over 100 white rhinos died out of an initial population of perhaps 300 animals (Player and Feely, 1960). The most recent census carried out in 1978 indicates a total white rhino population of 1,677, only 16% lower than that present in 1970, despite the subsequent removal of 1,529 animals.

III. HABITAT IMPACT

The Umfolozi Basin is characterized by moderately fertile soils derived from alternating shale and sandstone beds of the Ecca Series of the Karoo sediments. The mean annual rainfall (1963 to 1978) measured at the Mpila camp is 690 mm. The typical vegetation is *Acacia* savanna ("Zululand thornveld"), with medium tall *Themeda triandra* predominating in the climax grasslands.

The changes in the vegetation that were occurring in the Umfolozi Reserve included the following:

(1) The replacement of climax *Themeda* grasslands by extensive areas of short grass grassland, with creeping species such as *Panicum coloratum, Urochloa mossambicensis* and *Sporobolus smutsii* taking over.

(2) Increasing prominence of areas of denuded and eroding soil.

(3) The drying up of springs, and reduced flow of both Umfolozi rivers during the dry season.

The total large herbivore biomass within Umfolozi Game Reserve in 1970, based on the helicopter census data with corrections for undercounting of species other than white rhinos, amounted to 82 kg/ha. This is a minimum figure, since 14% more white rhinos were counted the following year (subsequently the numbers of other ungulates have been reduced considerably by culling). This is nearly twice the carrying capacity for biomass estimated on the basis of rainfall from the regression of Coe *et al.* (1976). However, it is only 40% of the biomass existing in sections of the Rwenzori National Park, Uganda, with similar rainfall, that were reportedly showing good recovery from a previously overgrazed condition (Field and Laws, 1970). White rhinos comprised about 60% of the total biomass of grazing large herbivores in the Umfolozi Reserve but contributed only about 45% to total metabolic demands (based on body mass$^{3/4}$). However, as relatively inefficient digesters of forage, offtake of herbage by white rhinos is likely to be somewhat higher than is indicated by simple metabolic relationships. White rhinos are probably responsible for at least half of the total consumption of grass attributed to large herbivores.

White rhinos are capable of using both tall and short grasses, and with their lip plucking action can graze as close as within 2.5 cm of ground level. When subjected to repeated defoliation, erect tufted grasses such as *Themeda triandra* are unable to maintain enough leaf area to replenish root reserves, and eventually plants lose vigor and die. They are replaced by recumbent species which suffer proportionately less reduction in leaf area and which are also better adapted to exploit raised soil nutrient levels. Short grass grasslands are favored by white rhinos for grazing and become closely cropped quite early in the dry season. By promoting an increase in short grasses, which are relatively protein-rich, white rhinos increase the favorability of their habitat, considered in terms of the nutrient flow to their own population. As nonruminants, they can tolerate a high content of structural tissues in their diets, thereby increasing the rates of turnover and nutrient return from fibrous plant parts into the soil for subsequent plant growth. A fiber tolerant grazer such as a white rhino could exert a less detrimental effect on grassland vigor than a species such as impala which exhibits a selective removal of green leaves alone. McNaughton (1979) has shown for the Serengeti grasslands that aboveground production under a moderate grazing regime can be double that which occurs in an ungrazed state; some stimulation of production occurred even at levels of consumption amounting to 70% of the production measured under ungrazed conditions. At Umfolozi the invasion of grasslands by unpalatable species such as *Cymbopogon plurinodis* or *Bothriochloa insculpta* was not evident, possibly because of the intolerance of these species to high soil nutrient levels.

In terms of the production of animal biomass per unit of land area, the white rhino represents the most productive single species of large herbivore occurring in the Umfolozi ecosystem. It may well be that it is only in mixed species communities, including megaherbivores, that the full productive potential of habitats in terms of animal protein can be attained because of the ability of such species to make use of a much greater fraction of the vegetation than is the case for smaller herbivores. To some degree megaherbivores could possibly replace fire in their effects on nutrient recycling, without the danger of nutrient losses that occur from the pulsed releases due to fire. Such acceleration of nutrient cycling rates could have a major effect on secondary productivity, not only of the megaherbivore population itself, but also of the many other herbivores within the ecosystem (see Botkin *et al.*, 1981).

However, white rhinos are dependent upon reserves of tall *Themeda* grassland to carry them through the late dry season. Furthermore, soils in short grass areas are exposed to the eroding influences of wind and early season thunderstorms. Pedestalling and reduced water infiltration result, which soon lead to reduced grass cover and rapid expansion of erosion gullies. At the same time, the short grass grasslands do not sustain fires and are thus readily invaded by woody plant seedlings, leading toward bush encroachment and reduced grass cover.

The rapidity with which the vegetation deteriorated over the short duration of my study was alarming. However, 1962 to 1971 was a decade of below-average rainfall, and thus some of the change was probably climatically induced. The situation was exacerbated by an influx of wildebeest and zebra into the region. Changes in the vegetation were spatially highly variable. Large areas of climax grassland on hillslopes remained relatively untouched by grazing animals. The western section, which sustained some of the highest white rhino grazing pressures, retained a mosaic of tall and short grassland areas with a good basal cover, probably aided by gentle relief and dolerite-influenced soils. Many of the areas showing the worst denudation were on deep alluvial soils which had been covered by thickets prior to the bush clearing which took place during the tsetse control operations in the late 1940's. These areas seemed to be reverting inexorably to their former vegetational state, though heavy grazing pressure was probably a factor promoting this trend. Much of the variability in consumption could be correlated with underlying soil fertility.

During the wet decade of the 1970's these trends were reversed, helped by extensive removals of grazing ungulates of all species. The white rhino removal quota was accordingly greatly reduced. However, rainfall was well below the average during the seasons of 1978-1979 and 1979-1980. Considerable alarm has once more been expressed over the rapidly spreading desertification of the area, and large-scale removals of all large herbivores, including white rhinos, are being carried out.

The margin between increased production and overutilization seems to be a delicate one. Simultaneous trends of increasing white rhino density and declining grass cover cannot persist for long. A paradoxical situation exists in that the white rhino population has clearly not yet attained the carrying capacity of the environment in terms of the logistic model of population growth, while on the basis of vegetational changes the capacity of the vegetation to support even the existing white rhino population on a long-term basis has apparently been exceeded.

IV. DEMOGRAPHIC MODELS

Riney (1964) described the pattern of the eruptive fluctuations observed for introduced populations of large herbivores in New Zealand and suggested that they could arise in any situation where a large discrepancy was created between the potential carrying capacity of the vegetation and current population levels. Basically a time lag in population response results in a period of overshoot of the carrying capacity, initiating a downward trend in the capacity of the vegetation to support the herbivores. Eventually the herbivore population stabilizes at a rather lower density in a somewhat degraded habitat.

Caughley (1976a, 1977) developed analytical models of trophic level interactions between plants and herbivores. In these models both populations undergo a series of oscillations before settling down to equilibrium biomasses which are lower than those attained in the early phase. It is important to note that the effective plant biomass is that of potentially edible plant material available for consumption. Any increase in the productivity of vegetation in terms of edible material, such as a shift from climax to short grasses, would increase the equilibrium biomass of the herbivore population, while leaving that of the plant component unchanged. If the standing crop of edible plant material is reduced by consumption to less than half of that present under ungrazed conditions, persistent oscillations develop. Such a situation could be created by the delayed action of compensatory responses within the herbivore population. As stated by May (1976), sustained oscillations also result when time delays in the regulatory mechanism of the consumer population are long compared with the characteristic response time of the system. This seems to apply to the white rhino/grasslands relationship. A low intrinsic rate of increase of the animal population, while conferring the advantage that the population can maintain steady values over short-term environmental fluctuations, presents the disadvantage of slow recovery from large disturbances (May, 1976).

Such theoretical analyses bode ill for the likely trajectories of the white rhino/grassland interaction but do not necessarily chart out the actual pathway that will be followed in a real ecosystem where relationships are more complex. Thus it will be necessary to focus more specifically on the details of the white rhino population. Table I presents the detailed age structure of a hypothetical population increasing at a constant rate of 9.5%/annum with a stable age distribution. Age-specific parameters are the same as those observed in the real white rhino population. The proportion of immature individuals (subadults plus calves) in this population is a little higher than that observed in the real population (57% versus 54%). This discrepancy could be explained on the basis that the real population has not yet attained a stable age distribution, since its rate of increase was lower prior to 1960. The model population demonstrates, in only slightly exaggerated form, the extreme preponderance of young animals that characterizes an expanding population. Even if all reproduction was to cease instantly (say by 100% infant mortality), the size of the breeding segment of the population would continue to increase at 9.5%/annum for a further seven years. It is the adult segment that makes up most of the biomass, and thus exerts the greatest pressure on food resources. There is no way in which any realistic adjustment in breeding rates or infant mortality could act quickly enough to avoid a period of overshoot of carrying capacity; the severity of this overshoot will be greater the longer the period over which the population has been increasing at its intrinsic rate and the later the age of sexual maturity. It is for this reason that a laissez faire management policy towards the white rhino population cannot be advocated.

There is only one way whereby the expansion of the breeding segment could be halted immediately: by the sudden disappearance from the population, whether by emigration or culling, of the excess number of subadults just prior to their attainment of sexual maturity. In Table II, population *A* represents an example of a population with similar attributes to those of the real population where stability has been achieved solely by the emigration of surplus subadults at six years of age. The only other change in population parameters that has been allowed is a reduction in the specific natality per adult female to 0.33 (intercalving interval = three years), which is a likely result of the shift in the age structure of breeding females toward a higher proportion of older animals. This is based on the observation that the study population at Madlozi in the western high-density section of Umfolozi exhibited a mean natality rate over the four years observation period of 0.34 calves per adult female per year, while other study

TABLE I. Composition of Hypothetical Population Increasing at a Constant Rate of 9.5%/Annum, after Attainment of Stable Age Distribution[a]

Age range	Number of individuals	Age range	Number of individuals	Age range	Number of individuals
0- 1	200	15-16	39	30-31	8
1- 2	177	16-17	35	31-32	7
2- 3	157	17-18	32	32-33	7
3- 4	140	18-19	29	33-34	6
4- 5	126	19-20	26	34-35	5
5- 6	113	20-21	23	35-36	5
6- 7	101	21-22	21	36-37	4
7- 8	91	22-23	19	37-38	4
8- 9	82	23-24	17	38-39	4
9-10	74	24-25	15	39-40	3
10-11	67	25-26	14	40-41	3
11-12	60	26-27	13	41-42	3
12-13	54	27-28	11	42-43	2
13-14	49	28-29	10	43-44	2
14-15	44	29-30	9	44-45	2

	Number	Percentage
Adult males (over 9 years)	363	19.0%
Adult females (over 7 years)	449	23.5%
Subadults (males 2.5 to 9 years; females 2.5 to 7 years)	644	33.7%
Calves (0 to 2.5 years)	455	23.8%
Total	1911	

Mortality of cohort (adults + subadults over 3 years) = 26 or 1.8%

Specific natality: Fertile cohort = $\frac{202}{556}$ = .363

Parous cohort = $\frac{202}{449}$ = .450

Postparous cohort = $\frac{110}{357}$ = .308

Proportion of individuals in age range 0 to 1 year = 10.4%

[a] Assumptions: (1) size of total population = 1918 individuals; (2) age-specific sex ratios = 100:100; (3) mortality rates as follows -- juveniles (0 to 3 years) = 3.5%/annum, subadults (3 to 7 years) = 2%/annum, adults = 1.5%/annum, plus terminal mortality of all individuals reaching 45 years of age.

TABLE II. Composition of Two Hypothetical Stable Populations[a]

Age range	Number of individuals A	Number of individuals B	Age range	Number of individuals A	Number of individuals B	Age range	Number of individuals A	Number of individuals B
0- 1	155	75	15-16	28	30	30-31	23	20
1- 2	150	67	16-17	28	30	31-32	22	19
2- 3	145	60	17-18	27	29	32-33	22	18
3- 4	141	54	18-19	27	28	33-34	22	18
4- 5	137	49	19-20	27	27	34-35	21	17
5- 6	133	44	20-21	26	26	35-36	21	17
6- 7	32	40	21-22	26	26	36-37	21	16
7- 8	32	38	22-23	26	25	37-38	21	16
8- 9	31	37	23-24	25	24	38-39	20	15
9-10	31	36	24-25	25	23	39-40	20	15
10-11	30	35	25-26	24	23	40-41	20	15
11-12	30	34	26-27	24	22	41-42	19	14
12-13	29	33	27-28	24	21	42-43	19	14
13-14	29	32	28-29	23	21	43-44	19	13
14-15	29	31	29-30	23	20	44-45	19	13

	Number A	Number B	Percentage A	Percentage B
Adult males (over 9 years)	435	431	23.6%	32.0%
Adult females (over 7 years)	466	469	25.3%	34.9%
Subadults (males 3 to 9 years, females 3 to 7 years)	495	243	26.8%	18.1%
Calves (under 3 years)	450	202	24.4%	15.0%
Total	1846	1345		
Mortality of cohort (adults + immatures over 3 years)	43	63	3.3%	5.5%
Proportion of individuals in age range (0 to 1 year)	155	75	8.4%	5.6%

[a]Assumptions: Population A -- as in Table I, with specific natality per postparous females of .33/annum, stability achieved by emigration or death of surplus subadults promptly at age 6; Population B -- specific natality = .25/postparous female/annum, first parturition at age 8, postnatal mortality (0 to 6 months) = 40%, immature mortality to age 6 = 10%/annum, subadult mortality to age 9 = 4%/annum, adult mortality = 3%/annum. Size of adult cohort in both populations adjusted to 900.

populations exhibited somewhat higher natality rates. To achieve equilibrium, the removal of 75% of the subadults reaching six years of age is required. However, if the process took the form of emigration spread evenly over the entire subadult period, then a specific emigration rate of only 20% per individual per annum would suffice. In terms of the overall population, the emigration rate amounts to 5.5% per individual per annum. This is only 1.5 times greater than the rate of dispersal that was estimated to be occurring from the population core between 1967 and 1970, despite the existence of the boundary fence at that time. (The proportion of subadults present in the model population *A* is, however, 17% lower than that estimated for the real 1969 population.) Thus it is quite feasible that the expanding population of Table I could be transformed smoothly into a stable population as in Table II entirely by a big enough increase in the dispersal rate.

Interestingly the mean adult mortality of population *A* in Table II is twice as great as that of the expanding population of Table I, though age-specific mortality rates remain the same in both. This difference is due to the much greater proportion of animals reaching the age of terminal senescence in population *A* (Table II). In the real population there should be very few individuals dying of old age, since such individuals would have been born during the 1920's, when the total population numbered only 200 to 300 individuals. Assuming that only 25% of the individuals born die of old age, it is projected that only 5 rhinos per annum would have died of old age in the entire population of 2,000 individuals in 1970 (assuming that none of these old individuals had been caught and removed elsewhere).

In the absence of opportunities for dispersal, equilibrium would have to come about through appropriate changes in fecundity and mortality rates. Population *B* in Table II exemplifies the possible characteristics of such a stable population:

(1) The mean intercalving interval is increased to 4 years.

(2) Age at first calving is retarded to 8 years.

(3) Infant mortality is increased to 40%.

(4) Juvenile and subadult mortality is increased to 10%/annum.

(5) Adult mortality is increased to 3%/annum.

In this population only 33% of the animals are under breeding age. Such attributes would be expected only for a white rhino population living in poor condition in a degraded habitat. If all opportunities for dispersal are blocked, this is the type of population that must eventually result.

It is noteworthy that the black rhino in the neighboring Hluhluwe Reserve exhibited all the characteristics of such a population. A population estimated to number about 300 animals in 1961 had declined to 199 by 1972. Compared with the smaller but expanding population of black rhinos in the Umfolozi Reserve, the mean calving interval had lengthened from 2.3 to 2.7 years, infant mortality had increased from 9% to 63% (partly due to hyena predation), age at first calving had been retarded from 6.5 to 12 years, and mean adult mortality had risen to 5%/annum (P. M. Hitchins, unpublished data). The Hluhluwe black rhinos occurred at a mean density of $0.9/km^2$ (compared with $3.2/km^2$ for the Umfolozi white rhinos). Paradoxically, while both bush encroachment and grass disappearance were the major problems of habitat management in the Complex, the browsing species of rhinoceros was declining in numbers while the grazing species was still increasing at three times the density and six times the biomass.

V. MANAGEMENT SOLUTIONS

Management steps were urgently needed in 1970 to achieve the following objectives, given in order of priority:

(1) Reduction of the overall biomass of white rhinos and other grazing ungulates to a level which would halt further deterioration of the habitat.

(2) Transformation of the characteristics of the white rhino population to that of an equilibrium rather than an expanding population.

(3) Stabilization of the white rhino population at a level in balance with habitat resources, with the least disruption of natural population processes.

Table III outlines the removal program that I proposed in 1972 aimed at reducing white rhino biomass by one third and at the same time reducing the high proportion of immature animals in the population. The starting point was the population composition estimated for 1969, with account taken of the animals that were known to have been removed between 1969 and August 1971. The program required the capture and relocation of a total of 1,258 white rhinos over the four year period 1971 to 1975.

Independently, the need for a massive removal campaign had already been accepted by the Natal Parks Board. The removal quota was stepped up from 151 animals during the calendar year of 1971 to 448 in 1972, 293 in 1973, and 243 in 1974 (Brooks, 1974, and personal communication). In all, 1,093 white rhinos were captured over the four year period between mid-1971 and mid-1975, 13% fewer than I had proposed. However, the population remaining in September 1975 (extrapolated from a count covering Umfolozi only), numbered some 30% more animals than I had projected; that is, a population reduction of only 24% had been achieved. This suggests that the true population size is some 10% greater than is estimated even by helicopter census. The age structure recorded by Brooks in 1974 comprised 55% "adults" over about six years of age, which is the same as that which I had projected for the 1975 population. Subsequently with the return of high rainfall conditions the annual removal quota was reduced to 89 animals in 1975, 53 in 1976, 20 in 1977, and 38 in 1978 (P. H. Brooks, personal communication).

One of the problems which arose was a marked preponderance of males in the population; the sex ratio recorded in 1974 by Brooks, based on a sample of 563 animals, was 157 males: 100 females. This could not be explained by the small excess of females among the animals removed between 1967 and 1974 (93 males:100 females, N = 1,475). It must have been due to the male-biased sex ratio of calves born into the population that I had noted during my study period. Recent removals were aimed mainly at correcting this sex imbalance. Despite the disruptions caused by the removal operations and the male-biased sex ratio, the white rhino population has continued to expand at a rate of 6.5%/annum over the period 1970 to 1978 (including animals removed).

Notwithstanding these efforts the problem of ultimately stabilizing the white rhino numbers in the Complex still remains. Caughley (1977) discusses possible strategies for the control of populations based on a simple logistic model. If the basic assumption of a linear relationship between the rate of population increase and population density is valid, then culling of a proportion equal to half the intrinsic rate of increase would stabilize the population at half the density that it would have attained in the absence of such removals (assuming no suppression of food production). Population oscillations would also be dampened, the effect being similar to that of adding a predator trophic level. In the case of the white rhino population this would require the removal of about 5% of the total population, or about 75 animals per annum from a total population of 1,500, if culling were unselective with regard to age and sex. This figure could be reduced by up to one third if removals were concentrated on subadults. With this culling strategy, the aim would be to reduce the rate of increase, and hence the degree of overshoot of their carrying capacity, thus suppressing or at least dampening oscillations in population size. The population would be allowed to move towards an equilibrium density similar to that which it would have attained in the presence of dispersal opportunities with whatever additional changes in natality and mortality rates that might be necessary.

The final problem to be considered is how best to carry out this culling. The ideal solution would perhaps be simply to remove the boundary fence, and to allow surplus animals to disperse into surrounding country where they could be harvested by the local people. Such a step seems to be politically impractical. The best alternative is to create dispersal sinks within the game reserve capable of absorbing the excess animals. This would entail creating vacuum zones, ideally situated near boundaries, from which all white rhinos that had settled there would be captured and removed at suitable intervals. The aim would be to create opportunities for dispersal processes to operate. Success would be dependent upon the adequacy of dispersal rates and also upon the size, situation, and habitat conditions offered by the vacuum zones. Culling methods would have to be carefully designed so as not to deter movements into vacuum zones. A possible strategy would be the complete removal of all resident rhinos from these zones at infrequent intervals, with several years elapsing between successive operations in the same area. It may still prove necessary to carry out some culling within the core zone to halt excessive habitat deterioration, but at least the extent of the interference with the population within this area would be minimized.
A vacuum zone culling policy is appropriate, especially for those species for which dispersal appears to be the main process by which population size is regulated. This includes elephants as well as white rhinos and perhaps a range of other species as well. Its effectiveness remains to be tested in practice.

TABLE III. *Suggested Removal Program to Reduce Biomass by One Third and Stabilize Population*[a]

Year	Removals: Adult ♂	Adult ♀	Immature	Juvenile	Total	Age structure of population: 0-1	1-2	2-3	3-4	4-5	5-6	6-7	Adult+Subadult ♂	Adult+Subadult ♀	Total
Early 1969															
Population						185	164	145	135	121	109	98	477	492	1926
Removed	13	36	109		158			32	29	26	22		13	36	
Remaining						185	164	113	106	95	87	98	464	456	
August 1970															
Population						207	179	159	110	103	92	84	498	500	1932
Removed	4	13	141		158			48	34	31	28		4	13	
Remaining						207	179	111	76	72	64	84	494	487	
August 1971															
Population						211	200	174	107	74	70	62	520	523	1941
Removed	26	49	183		258			74	46	32	30		26	49	
Remaining						211	200	100	61	42	40	62	494	474	
August 1972															
Population						196	204	193	97	59	41	39	509	499	1837
Removed	80	80	160	80	400	40	40	79	40	24	17		80	80	
Remaining						156	164	114	57	35	24	39	429	419	
August 1973															
Population						166	151	159	111	55	34	23	435	434	1568
Removed	80	80	160	80	400	41	39	71	49	24	16		80	80	
Remaining						125	112	88	62	31	18	23	355	354	

August 1974															
Population						133	121	119	85	60	30	17	353	359	1277
Removed	40	40	80	40	200	21	19	32	23	16	8		40	40	
Remaining						112	102	87	62	44	22	17	313	319	
August 1975															
Population						110	108	99	84	60	43	21	322	324	1171
Removed		5	80		85			28	24	17	12			5	
Remaining						110	108	71	60	43	31	21	322	319	
August 1976															
Population						119	106	104	69	58	42	30	322	325	1175

[a] *Population structure in early 1969; removals up to August 1972 are those actually accomplished except that immature sex ratio has been assumed even; natality of postparous female cohort = 36%, plus parturition by all females on reaching age 7; mortality rates of adult males = 3%, adult females = 1.2%, subadults (2.5 to 7 years) = 3%, juveniles (0 to 2.5 years) = 3.5%; immatures (2 to 6 years) assumed removed proportionately to existing ratios.*

VI. DISCUSSION

The problem of the white rhino highlights many of the issues raised with regard to overabundant populations of large herbivores within national park ecosystems. Caughley (1976b) outlined various hypotheses that have been advanced to account for overpopulations of elephants and advanced a "stable limit cycle" hypothesis as a general explanation. However, this explanation is not applicable to the white rhino situation, where the periodicity of response in the vegetation is very much shorter than that of the animal population. The "intrinsic eruption" hypothesis as outlined originally by Riney (1964) is quite adequate to explain the development of an overexploitation phase. Time delays in population response are inevitable because of age structure effects, and the degree of overshoot will be accentuated by the extent to which the population was initially depressed below carrying capacity. Time lags will be especially severe for megaherbivores because of their several-year delay between birth and attainment of breeding age. Dispersal is the only population process sufficiently sensitive in its operation to dampen oscillations of herbivore numbers and vegetation biomass. Thus any hindrances to dispersal, in the form of barrier fences or merely fences surrounding settlements, can exaggerate and sustain the period of over-exploitation of the vegetation with potentially severe consequences for persistence of the ecosystem. I consider that this explanation is also quite adequate to account for elephant overpopulations; the common factor is initial population depression, due in the case of elephants to the ivory hunting which took place all over Africa around the turn of the century (see Spinage, 1973).

However, while dispersal may regulate local population densities, it cannot control overall population expansion, unless compensatory mortality occurs elsewhere. It seems that regions of favorable habitat generate a population surplus, which moves out to colonize surrounding areas of more marginal but thinly settled habitat. Presumably individuals may either remain in saturated habitats, where as subordinates they are unlikely to replace themselves genetically; or they may move out in the expectation that more favorable conditions for breeding might occur elsewhere. If it were not for the dynamic nature of environmental conditions, such movements would be genetically suicidal. Marginal habitats may thus serve as population sinks (Lidicker, 1975), absorbing the surplus animals produced in more productive habitats. Local environmental catastrophes such as drought could open up a sufficient supply of favorable habitats for colonization so as to maintain the dispersal propensity, particularly for species which are capable of depressing the production of their own food resources in saturated habitats. If this account is correct, it implies that where free dispersal is inhibited, densities could rise to higher levels than those that would have prevailed under pristine conditions.

A policy of vacuum zone culling would seem to offer a number of advantages over alternative forms of management, whether by indiscriminate culling or by nonintervention:

(1) There would be no direct interference with the population in the core area where natural ecological processes could continue to operate.

(2) The animals dispersing into vacuum zones would presumably be the surplus individuals as determined by habitat conditions and social relationships with the core area, so that processes of natural selection could still operate.

(3) At the same time the ecosystem would be buffered against extreme fluctuations in herbivore density and vegetation resources. Such fluctuations would be likely to be exaggerated by the existence of a boundary, in contrast to the unrestricted conditions of the past, with the concurrent risk that critical thresholds for ecosystem persistence could be crossed.

(4) Within the core zone, wilderness qualities could remain untarnished by the roads or vehicle tracks that would otherwise be created by culling operations.

(5) If either live animals or their products -- such as meat, horns, or hides -- were of economic value, the vacuum zones could be conveniently situated near boundaries to facilitate the removal of these products for processing.

(6) A gradient in animal densities between the core area and the vacuum zones would be created, allowing greater opportunity to gain scientific understanding of the interactions between herbivores and vegetation -- a subject about which we are sorely lacking in knowledge.

(7) The creation of vacuum zones would enhance spatial heterogenity in the vegetation, while at the same time population levels would still be allowed to fluctuate in response to changing climatic patterns. In this way reservoir areas would be created for those species of plants and animals that are vulnerable under overgrazed conditions on the one hand, and for those that depend on periodic overgrazing for their persistence on the other. In consequence both species diversity and ecosystem resilience would be increased (see Holling, 1973; Walker *et al.*, in press).

The crucial unanswered questions concern the stability of the plant/soil system under the impact of intense grazing by white rhinos. It is easy to model the transformation of a white rhino population to a stable condition; but this is dependent on an assumption that a progressively worsening trend in the rate of replacement of its food resource is not initiated. While grazing may stimulate the production of leaf tissue, what is not observed is the compensatory reduction in root biomass that may be occurring. Smaller roots may leave plants vulnerable to mortality during periods of drought, and create openings in the soil that allow unpalatable plants to invade. Such trends could easily be self-reinforcing because of further consequences such as reduced fire intensity or increased soil erosion. Walker *et al.* (in press) describe cases where disturbing pressures can move a savanna community to an alternative stable state; for example the transformation by cattle overgrazing of an open savanna into a dense thicket with little grass cover. Exactly this type of change seems to have occurred in the Hluhluwe Game Reserve. Extensive areas, which in photographs taken during the 1930's appear as open grassland thronged by buffalo and zebra, now exist as thickets of woody shrubs such as *Acacia karoo* and *Euclea divinorum* occupied mainly by nyala and impala -- browsers which did not originally occur in the reserve. Apparently, this change was made possible by the absence of predators and led to the development of severe overgrazing before game culling was introduced as a policy during the late 1950's. Active management of both animal populations and fire has yet been unable to reverse this change.

So far the Umfolozi Reserve has undergone much less habitat change. Is this due to its lower rainfall, the less selective grazing impact of the white rhino, the occurrence there of browsers able to supress woody plant invasion in its early stages, different soil conditions less favorable to woody plant encroachment, or is it just a matter of time?

Simple population models are hopelessly inadequate if theoretical predictions are to be relied upon by the managers who have to implement the policy. What is needed are total ecosystem models, comprising a set of detailed process models which are capable of being interlinked to form a functioning whole. These should incorporate not only plant/herbivore interactions but also competitive relationships between different plant types, their interactions with external conditions such as fire regimes, nutrient recycling, and soil movement, and the influences of climatic fluctuations and spatial variability. This is the challenge.

ACKNOWLEDGMENTS

I am particularly grateful to P. M. Brooks and P. M. Hitchins for allowing me to quote their unpublished data.

REFERENCES

Botkin, D. B., Melillo, J. M., and Wu, L. S. (1981). How ecosystem processes are linked to large mammal population dynamics. *In* "Population Dynamics of Large Mammals" (C. W. Fowler, ed.) Wiley, New York.

Brooks, P. M. (1974). Population control report PMB 3. Natal Parks Board.

Caughley, G. (1976a). Plant-herbivore systems. *In* "Theoretical Ecology" (R. M. May, ed.), pp. 94-113. Blackwell, Oxford.

Caughley, G. (1976b). The elephant problem -- an alternative hypothesis. *East Afr. Wildl. J. 14*, 265-284.

Caughley, G. (1977). "Analysis of Vertebrate Populations." Wiley, New York.

Coe, M. J., Cumming, D. H., and Phillipson, J. (1976). Biomass and production of large African herbivores in relation to rainfall and primary production. *Oecologia (Berlin) 22*, 341-354.

Condy, P. R. (1973). The population status, social behaviour and daily activity pattern of the white rhinoceros in Kyle National Park, Rhodesia. M.Sc. Thesis, University of Rhodesia.

Field, C. R., and Laws, R. M. (1970). The distribution of the larger herbivores in the Queen Elizabeth National Park, Uganda. *East Afr. Wildl. J. 7*, 273-294.

Holling, C. S. (1973). Resilience and stability of ecological systems. *Ann. Rev. Ecol. Syst. 4*, 1-23.

Lidicker, W. Z. (1975). The role of dispersal in the demography of small mammals. *In* "Small Mammals: Productivity and Dynamics of Populations" (K. Petrusewicz, F. B. Golley, and L. Ryskowski, eds.), Ch. 5. Cambridge Univ. Press, London and New York.

McNaughton, S. J. (1979). Grazing as an optimization process: grass-ungulate relationships in the Serengeti. *Am. Nat. 113*, 691-703.

May, R. M. (1976). Models for single populations. *In* "Theoretical Ecology" (R. M. May, ed.), pp. 4-25. Blackwell, Oxford.

Owen-Smith, N. (1981). "The White Rhinoceros. The Ecology and Behaviour of a Megaherbivore." Garland STPM Press, New York.

Player, I. C., and Feely, J. M. (1960). A preliminary report on the square-lipped rhinoceros. *Lammergeyer 1*, 3-23.

Riney, T. (1964). The impact of introductions of large herbivores on the tropical environment. *Int. Union Conserv. Nat. Nat. Resour. Publ. New Series 4*, 261-273.

Spinage, C. A. (1973). A review of ivory exploitation and elephant population trends in Africa. *East Afr. Wildl. J. 11*, 281-289.

Walker, B. H., Ludwig, D., Holling, C. S., and Peterman, R. M. (In press). Stability of semi-arid grazing systems. *J. Ecol.*

LOCALLY ABUNDANT MARINE MAMMALS -- PROBLEMS AND ATTEMPTED SOLUTIONS

John Harwood

Natural Environment Research Council
Sea Mammal Research Unit
Cambridge
England

David M. Lavigne[1]

Department of Zoology
University of Guelph
Guelph, Ontario
Canada

I. INTRODUCTION

There are many instances where a particular species of marine mammal is sufficiently abundant locally to be perceived as a problem. Most of these cases involve interactions with fishery interests (such as the perceived competition between marine mammals and man for common prey species) which are beyond the scope of the present volume. There appear to be only five species of marine mammals which, although protected, are perceived as pests because of local overabundance. These include the polar bear (*Ursus maritimus)*, the sea otter *(Enhydra lutra)* -- see Estes (this volume), the Northwest Atlantic humpback whale (*Megaptera novaeanglia)*, the Pacific walrus (*Odobenus rosmarus divergens)*, and the British grey seal *(Halichoerus grypus)*. The latter three cases are the subject of this chapter. In the case of the humpback whale and, to an even greater

[1]*Present address: Natural Environment Research Council, Sea Mammal Research Unit, Cambridge, England*

ISBN 0-12-385280-3

extent, the walrus the exact nature of the problem is not entirely clear, but we have attempted to summarize the available information. In the case of the British grey seal the problems are more obvious and attempts to ameliorate the situation have met with some success. However, these management actions have resulted in secondary effects on the grey seal population which were not anticipated.

II. NORTHWEST ATLANTIC HUMPBACK WHALES

Humpback whales are an example of Caughley's (this volume) Class I overpopulation, animals whose activities conflict with those of man. Although classified as endangered and accorded protected status (Lien and Merdsoy, 1979) humpbacks as well as other baleen whales tend to congregate and feed in the inshore waters off Newfoundland where they collide with fishing gear causing a variety of economic losses to local fishermen. Man is not the only loser in this conflict, however. In 1979 alone at least 30 cetaceans including 13 humpbacks died as a result of these collisions (Lien and McLeod, 1980, Table I).

The humpback whale problem involves questions which are essentially nonbiological in nature. It is, however, a management problem which may be correlated with ecosystem changes, either natural or man-made, and which conceivably may be ameliorated by management decisions which take into account trophic interactions between humpback whales and their prey in the Northwest Atlantic ecosystem.

A. *The Problem*

The problem of local overabundance of baleen whales in the inshore waters of Newfoundland is a recent development. Over the last four to ten years there has been a marked increase in whale sighting reports (Lien and Merdsoy, 1979; Lien and Gray, 1980; Lien and McLeod, 1980). However, as there is little information on sighting effort, such reports are of limited value to assess changes in abundance. Also, during the same time period there has been a dramatic increase in the inshore fishing effort; the amount of various types of gear has tripled and man/months of fishing has quadrupled (Lien and Gray, 1980).

With the apparent increase in the number of whales and the amount of fishing effort, there has been a concomitant increase in collisions of baleen whales with fishing gear. Seventy percent or more of these collisions involve humpback whales (Lien and McLeod, 1980). Damage to fishing gear in 1979 alone has been estimated at $500,000 with resultant losses of catches of between $2,500,000 (Lien and Gray, 1980) and $3,000,000 (Lien and McLeod, 1980). Such losses apparently represent "1% of the value of the entire Newfoundland fishery" (Lien and Gray, 1980) or "about 3% of the total value of the Newfoundland inshore fishery" (Lien and McLeod, 1980).

Just as the distribution of whales is highly contagious, economic losses are also concentrated in certain areas. Economic hardships resulting from damage to fishing gear and subsequent down-time are thus borne by a relatively small number of fishermen (Lien and Gray, 1980; Lien and McLeod, 1980). For example, in St. Mary's Bay on the south shore of the Avalon Peninsula, fishermen have estimated that humpbacks damage 90% of their cod traps; in Trinity Bay fishermen estimated direct costs of repairing gear in the summer of 1978 to be $500,000, whereas costs accrued through down-time and reduced catches "would be many times that figure" (Lien and Merdsoy, 1979). That damage in Trinity Bay in 1978 appears equivalent to damage in all Newfoundland in 1979 is not discussed in the available literature.

Most collisions occur between June and August at a time when Mitchell (1973, 1974) found that baleen whales in the area feed principally on capelin (*Mallotus villosus*), and when fishermen are most active (Lien and McLeod, 1980). Whales which actually become entrapped in fishing gear tend to be small, immature animals although some large adults are also involved (Lien and McLeod, 1980).

B. Causative Factors

There are few quantitative data on the problem but several factors are correlated with the increased incidence of whale collisions, and any or all may be implicated to some extent.

1. Increased Fishing Effort. The increased fishing effort noted above involves increases in the amount of gear and in the number of man/months spent fishing. In the last three years alone, since Canada declared its 200-mile limit, the landed value of Newfoundland's inshore catch has reportedly increased by 80% while the number of fishermen has increased by about 60%. About one third, or 6,000, of these are apparently involved in the fishery on a part-time basis only.

2. Increased Whale Abundance. Humpback whales in the North Atlantic were heavily exploited from the eighteenth century; they were finally protected in 1955. Mitchell (1974) used cumulative catch data to estimate the initial size of the exploitable stock in the Northwest Atlantic as 1,500 animals. When Canada ceased whaling in 1972 there were an estimated 1,000 to 1,500 humpback whales in the Northwest Atlantic (Lien and Merdsoy, 1979). These authors noted that if the population had exhibited a steady growth rate of 5% to 7% since 1972, then the 1979 population should be between 1,600 and 2,400 animals. Current stock size is estimated by others to be between 800 and 1,500 whales (IWC, 1980) although figures as high as 3,000 have appeared (Whitehead, 1979).

In reality, available data are insufficient to provide an accurate estimate of stock size and to assess whether the current population is increasing, stable, or decreasing. Suffice it to recognize that humpbacks occur at low levels in the Northwest Atlantic, that absolute abundance is irrelevant to the current problem, and that the increase in reported sightings of whales off Newfoundland in recent years is greater than any reasonable estimate of possible increase in stock size (Lien and Merdsoy, 1980). Humpback whale sightings between 1974 and 1978 have increased by 300% (Whitehead, 1978, cited by Lien and McLeod, 1980); yet, in the absence of information on sighting effort these data cannot be used to indicate, unequivocally, an increase in whale abundance in inshore waters.

3. Changes in Distribution and Abundance of Prey Species. Assuming that there has indeed been an increase in the number of whales in inshore waters, this may have occurred as a result of recent changes in the distribution and relative abundance of their prey.

Humpback whales feed on herring, squid, and other nekton, including krill. In the last few years local mackerel cycles have reportedly peaked, and squid have also been unusually abundant off Newfoundland (Lien and Merdsoy, 1979). Although these apparent changes in prey density may be involved in attracting whales inshore, the most common explanation relates to changes in the distribution and abundance of capelin offshore. Capelin are thought to be the most important prey for cod *(Gadus morhua)* and harp seals *(Pagophilus groenlandicus)*, predators which, like whales, are currently being managed with a view to encouraging stock recovery after a period of overexploitation. There certainly appears to be an obvious correlation between the collapse of the capelin fishery and the increased incidence of whale sightings and collisions reported in inshore waters; nonetheless, the data are circumstantial.

During the 1970's there was a rapid development and subsequent crash in the capelin fishery in the Northwest Atlantic. Capelin catches increased from about 6,000 tons in 1971 to in excess of 360,000 tons by 1973. Catches were sustained at this level, although the total allowable catch was 500,000 tons, until 1977 when they declined to less than 250,000 tons (ICNAF, 1978). In the summer of 1978 capelin were absent off the south coast of Newfoundland and Canadian government scientists were unable to locate the previously well-defined capelin stock on the Southeast Shoal (Lien and Merdsoy, 1979); later that year the Canadian Minister of Fisheries closed the capelin fishery because of small catches and the predominance of immature fish in the catch (Fisheries and Environment Canada, 1978).

The apparent decline in capelin abundance observed in the nominal catch statistics has been corroborated by catch/effort analyses, accoustic surveys, and numerical population analyses (for a review, see Beddington and Williams, 1979). It has been suggested that, with the decline in capelin abundance offshore, baleen whales moved inshore where they found capelin in relative abundance, and this brought them into direct conflict with fishermen operating close to shore.

C. *Suggestions to Ameliorate the Problem*

Several methods have been suggested to reduce the interactions of humpbacks and other baleen whales with fishing gear, and the subsequent economic problems which ensue.

1. Renewed Whaling (Culling). The increase in whale sightings has led some fishermen to conclude that the whale population has increased substantially and that whaling should be renewed (Lien and Merdsoy, 1979). The perception of overabundance results, however, from the contagious distribution of whales around Newfoundland. Overpopulation is relative, i.e., it is a local problem. If the protected status of humpbacks were changed, the quota which would be allowed, using International Whaling Commission (IWC) criteria, would be so small as to have a negligible effect on the incidence of collisions. It has also been suggested that such quotas might even be less than the number of whales currently killed annually as a result of collisions with fishing gear (Anon., 1980a). As such accidental kills have not affected the number of collisions, neither would controlled whaling.

2. Current Measures.

a. Dealing with collisions. The Whale Research Group at Memorial University of Newfoundland maintains a hotline for fishermen to call when they have a whale entrapped in their gear. Trained divers and other experienced personnel are dispatched to the scene at no cost to the fishermen to free the whale. This service apparently minimizes damage to gear and whales alike (Anon., 1980a).

b. *Warning devices.* In 1979 a number of alarms were tested in an attempt to reduce the incidence of whale collisions. Thirty per cent of cod traps equipped with an underwater bell were damaged whereas the figure for control traps was 89% (Anon., 1980a). There was no evidence that the bells influenced the behavior of fish. Further research on the development of appropriate warning devices is reportedly in progress.

c. *Compensation to fishermen.* The Fishermen's Union has advocated an insurance program to assist fishermen in dealing with the economic losses associated with whale collisions with gear. Provincial and federal governments have discussed such a program but it seems that only gear damage or loss would be covered, not down-time losses which account for the bulk of the economic hardships (Anon, 1980a). Compensation programs alone will not, therefore, solve the current conflict.

3. *Ecosystem Considerations.* Circumstantial evidence suggests that the spatial distribution of baleen whales has been influenced to a large degree by the distribution and abundance of capelin. Consequently, it has been suggested that management decisions directed towards the recovery of offshore capelin stocks may also result in a redistribution of baleen whales, and hence, a reduction in the current conflict (Anon., 1980a; Lien and McLeod, 1980).

D. *Conclusions*

The problem of local overabundance of humpback and other baleen whales off Newfoundland is the result of biological, economic, and social factors (Lien and McLeod, 1980). Those working directly on the problem suggest that the present conflict between fishermen and whales may be resolved through adequate compensation programs for fishermen, continued development of devices which reduce or prevent collisions with fishing gear, and public education programs aimed at changing the attitudes of fishermen and reducing their present frustrations.

III. PACIFIC WALRUS

A. *The Problem*

The walrus "problem" has not been identified clearly in the literature; however, it appears that concern has been caused by the current size of the Pacific walrus population, the status of its food supply in certain areas, and the possibility that man may exploit this food supply. The situation seems to be perceived as a subtle blend of Caughley's (this volume) Class 3 and Class 1 problems -- there are too many walrus for their own good if man exploits their food supply -- with perhaps a hint of a Class 2 problem because, as Lowry *et al.* (1980) found, the diet of the walrus overlaps with that of the bearded seal *(Erignathus barbatus)*.

B. Causative Factors

The principal fear seems to be that the Pacific walrus population is now close to the carrying capacity of the environment. Evidence for this comes from three sources: estimates of the current size of the population with reference to its initial unexploited level, estimates of the impact of this population on its food supply, and the condition of individuals within the population.

1. The Size of the Population. The majority of the Pacific walrus population spends the summer in the Chukchi Sea and Bering Strait area and the winter in the Bering Sea. The distribution of these animals is intimately linked with that of the edge of the polar ice cap, although in late summer some animals haul out in large numbers at traditional sites along the Soviet coast (Estes and Gilbert, 1978). This restricted distribution makes it possible to survey a large proportion of the population from the air in a relatively short time. However, because the distribution extends into both Soviet and U.S. national waters, only rarely has the entire area been surveyed in one short period.

Although the walrus has been hunted for its meat, oil, hide, and ivory for several thousand years, this exploitation took place at a relatively low level until the nineteenth century when American and perhaps some European whalers began killing large numbers of animals. Up to 20,000 animals each year were removed in the period 1860 to 1880 (Fay, 1957). Not only did the population decrease dramatically under this onslaught, but its range became restricted so that by the early twentieth century walrus were rarely seen south of 60°N (Fay, 1957). Fay (1957) concluded that the population continued to decline until the 1950's although the United States forbade the commercial harvesting of walrus in 1941, and the USSR follows suit in the late 1950's (Estes and Gol'tsev, in press). Fay (1957) provided an estimate of 40,000 to 50,000 animals as the walrus population in 1954; this is based on a series of north/south transects made from an icebreaker in the United States sector of the Bering and Chukchi Seas. The counts from these transects were scaled up to provide an estimate of total population size, making allowance for the higher density of animals in Soviet waters. He also estimated that there must have been at least 200,000 animals in the eighteenth century and believed that this was the minimum population capable of withstanding the catches of the nineteenth century without being driven to extinction.

A number of other partial surveys have been carried out since 1954 and there is certainly evidence, summarized by Krogman *et al.* (1979), that the range of the population has expanded recently. However, the only complete survey of the entire distribution was made in September 1975 (Estes and Gol'tsev, in press). Transects were flown across the ice edge and aerial photographic surveys were made of the Soviet coastal hauling grounds. The total population was estimated at 140,000 to 200,000. The similarity of the upper value to Fay's (1957) estimate of the initial size of the population has not gone unnoticed and Krogman *et al.* (1978) concluded, on the basis of Estes and Gol'tsev's data and their own surveys, that the population is at or near the carrying capacity of the environment. However, all estimates of the size of the Pacific walrus population must be interpreted with caution. Although the spatial distribution of the animals is largely determined by the extent and location of the ice edge, they are not randomly distributed along the ice edge but are found in aggregations which vary greatly in size (Estes and Gilbert, 1978). Thus it is only possible to obtain a precise estimate of population size if a large proportion of the distribution is surveyed. In fact, Estes and Gilbert (1978) calculated that at least 40% of the total area under consideration must be surveyed in order to obtain an estimate with a confidence interval of $\pm$ 10%. They concluded that "estimates of abundance based on limited survey efforts will provide information of little reliability." In addition any estimate of population size based on surveys will be biased downwards to an unknown extent because an unknown proportion of the population will be in the water where they are less visible.

In summary, although there does seem to be evidence that the Pacific walrus has increased in numbers and range since the 1950's, the available data are not adequate to assess the magnitude of the increase. Even the data on range expansion must be treated with some caution because the southerly limit of the walrus distribution is largely determined by the southerly limit of the ice edge, and this may vary greatly from year to year. It is very difficult to determine what the equilibrium size for the population might be. Certainly Fay's (1957) calculation of the minimum initial stock size is not an estimate of equilibrium population size and was presumably never intended as such.

2. *The Status of the Food Supply.* After analyzing the contents of 107 walrus stomachs and reviewing the available literature, Fay *et al.* (1977) concluded that walrus feed primarily on mollusks, particularly bivalves. Evidence that the walrus population may now be having a considerable impact on its food supply comes from three sources.

Fay *et al.* (1977) took benthos samples from more than 200 sites in the Bering Sea, Bering Strait, and Chukchi Sea over the period 1970 through 1974. They found that standing stock biomass (expressed as grams of carbon per square meter) was the lowest in the most intensively used parts of the main summer and winter range of the walrus, i.e., at the northern and southern limits of the distribution and suggested that this might be the result of walrus predation pressure.

Fay *et al.* (1977) also used data from captive walrus and from stomach contents analysis to calculate the quantity of mollusks likely to be removed annually by walrus in different parts of their range. In some areas, the calculated removal rate was close to the estimated net annual production of the mollusk populations.

Lowry *et al.* (1980) analyzed the stomach contents of bearded seals shot at Nome and Diomede in the Bering Strait and at Wainwright on the east shore of the Bering Sea over the period 1964 through 1979. At Diomede, in particular, the number of stomachs containing bivalve mollusks and the amount of bivalve remains in the stomach contents appear to have declined. They reported that the number of walrus spending the summer in the Diomede area has increased of late, and they attributed the change in bearded seal stomach contents to the effects of walrus predation on the bivalve population.

Unfortunately all this evidence is circumstantial. Estimates of the standing stock of bivalves in the Bering and Chukchi Seas in the 1950's, which would allow an adequate assessment of the effects of increasing walrus predation, are not available. As it is, the observed variations in standing stock biomass over the range of the Pacific walrus cannot be attributed unequivocably to the effects of walrus predation. The data of Lowry *et al.* (1980) are tantalizing; however, the Bering Strait area had the highest standing stock biomass in the survey by Fay *et al.* (1977). In addition, the change from 1967 to the 1970's in the number of stomachs containing bivalve remains is not statistically significant. From the published data, it is not possible to test whether the change in the composition of the stomach contents is significant although there does appear to be considerable individual variation.

3. Changes in Individual Condition. Fay and Kelly (1980) reported that up to 900 walrus died on St. Lawrence Island in the Bering Sea in October and November 1978. They measured the thickness of the blubber on 38 animals and found it was significantly less than that of animals sampled in late winter and summer in previous years. The level of mortality in 1978 appeared to have been considerably greater than for any previous year, although only anecdotal information is available for many other years. There is no way of ascertaining whether the mortality level and individual condition observed at St. Lawrence Island in 1978 are a reflection of a more general problem in the population at large.

C. Suggestions to Ameliorate the Problem

Fay *et al.* (1977) concluded that the development of a commercial clam fishery in the southeastern sector of the range of the walrus would have a disastrous effect on the walrus population. Even if the available evidence is treated with due skepticism there can be no doubt that such a fishery can only be detrimental to the interests of the walrus population. No explicit suggestions appear to have been made, but obviously conflict between walrus and fisheries interests can be avoided by preventing the development of a fishery!

D. Conclusions

At present there appears to be no Pacific walrus "problem" which can be attributed to a locally abundant population. There are sufficient logistic and technical problems in estimating the current and historical size of the walrus population and its impact on the food supply to make it impossible to decide whether or not the population is approaching the carrying capacity of its environment. However, such arguments are largely irrelevant for any problems involving a conflict with fisheries interests. At present the only evidence that there might be too many walrus for their own good is the reported mortality and poor condition of animals at St. Lawrence Island in 1978; it is not entirely frivolous to point out that this affected less than 0.5% of the population.

IV. BRITISH GREY SEALS

British grey seals have been protected by law since 1914, when it was claimed, probably erroneously, that there were less than 500 animals (Bonner, 1971). Partly because of this protection, but also because fewer people now live on the offshore islands which are the principal breeding sites for grey seals, there are currently about 70,000 grey seals in British waters. This is about two thirds of the total world population (Summers, 1979). Most of the problems created by this rapid increase in population size have involved perceived effects on commercial fisheries and are, therefore, beyond the scope of this volume. However, at the Farne Islands off the northeast coast of England, problems that fall into Caughley's (this volume) Classes 2 and 3 have arisen.

A. The Problem

Over the period 1956 to 1971 the number of pups born each year at the Farne Islands increased from 750 to 2,040 (Summers, 1979). During the same period the measured levels of pup mortality doubled from 10% to over 20% (Bonner and Hickling, 1971a). Many of the dead and dying pups were emaciated, with purulent wounds and septic eyes (Bonner and Hickling, 1971b).

Two of the most important seal breeding islands in the Farnes group, Brownsman and Staple, are also important breeding sites for puffins (*Fratercula arctica*) which nest in burrows. By the late 1960's it appeared that the digging activities of the puffins, combined with vegetation removal by herring gulls (*Larus fuscus*) and lesser black-backed gulls (*L. argentatus*) and the wallowing behavior of the grey seals, were causing increased rates of soil erosion which could eventually make the islands unsuitable as breeding areas for puffins (Bonner and Hickling, 1971b).

B. Causative Factors

1. Increasing Numbers of Grey Seals. Levels of pup mortality at the Farne Islands are closely correlated with the density of breeding females on each island (Coulson and Hickling, 1964). As density increases, the probability increases that a pup will become separated from its mother and starve or that it will be trampled by other seals. Thus any attempt to reduce the number of seals on individual islands could reasonably be expected to reduce pup mortality.

∴ seal cull v. short term. alternative mechanisms

2. *Destruction of Vegetation by Sea Birds.* Piles of excavated soil accumulate at the mouths of puffins' burrows, which suppresses vegetation growth. Breeding gulls trample and tear up much of the surface vegetation around their colonies. As a result of these activities large bare areas of soil were recorded at the end of the breeding season in 1965 on both Brownsman and Staple (Bonner and Hickling, 1971a). Breeding grey seals like to wallow in the damp depressions which form in these denuded areas. This churns up the underlying soil into thick mud which is easily eroded by heavy winter rains. In addition these wallowing activities can cave in the roofs of existing puffin burrows, thus increasing puffin burrowing in subsequent years. Grey seals begin to come ashore to pup in October, almost as soon as the breeding birds have left, giving the vegetation virtually no time to recover. Although the seals were by no means entirely responsible for the observed acceleration of the erosion process, their contribution was not a minor one.

C. *Attempts to Ameliorate the Problem*

In 1971 Bonner and Hickling (1971b) drew up a management plan for the National Trust, which owns the Farne Islands. They suggested that the population should be reduced to 1,000 breeding female seals in an attempt to alleviate the erosion problem and to reduce the apparently high levels of pup mortality on the more crowded islands.

1. *Shooting of Females.* Therefore, 603 adult females were shot in 1972, as well as a number of bulls and pups, and a further 486 adult females were shot in 1975 (Bonner and Hickling, 1974; Hickling *et al.*, 1976). However, these culls did not have the desired effect. In fact levels of pup mortality were actually higher in 1973 than in 1971 (Bonner and Hickling, 1974) and a large number of seals still insisted on breeding on Staple in 1973, 1974, and 1976 (Bonner and Hickling, 1974; Hickling and Hawkey, 1975; Hickling *et al.*, 1977) with consequent damage to the soil cap.

2. *Use of Electric Fences.* In 1976 about 2,000 m^2 on the top of Staple were protected by a three-strand electric cattle fence (Hickling *et al.*, 1977). However, the fence was repeatedly flattened by a combination of seals and gales, and it provided little or no protection for the soil cap.

3. *Disturbance Measures.* In 1977 new tactics were adopted with the aim of deterring seals from breeding on Brownsman and Staple (Hickling *et al.*, 1978). Wardens lived on Brownsman throughout the breeding season, and very few seals bred there. In addition any seals attempting to breed on Staple's soil cap were shot. These measures were applied again in 1978 (Hickling and Hawkey, 1979), and they seem to be having the desired effect of reducing soil damage.

4. *Effects on the Seal Population.* The disturbances at the Farne Islands over the last eight years have not been without consequence for the seal population. Summers and Harwood (1978) describe the known secondary effects of breeding beach culls on grey seals. The disturbance associated with these culls can cause a redistribution of breeding seals in subsequent breeding seasons. Culling activities also deter some females from coming ashore to pup and may cause animals who have pupped to desert their young. These animals may not be impregnated in the season of the cull and may, therefore, fail to breed in the following season. It is not known whether these effects last for only one season or if they are permanent.

All these secondary effects have been observed at the Farne Islands. Pups are now born on many islands in the Farnes group which had not been used in the past (Hickling and Hawkey, 1979). In addition the decline in the number of pups born each year since the major culls in 1972 and 1975 is greater than would be expected simply on the basis of the numbers of females which were shot. Prime (1980) gave details of the changes in apparent fecundity (number of pups born divided by the number of sexually mature females calculated to be in the population) which has declined from about 90% before 1972 to less than 50% in 1979. It is not clear whether this is a real drop in pregnancy rate or is the result of increased emigration or juvenile mortality. A similar drop in pup production was observed at the grey seal colony on Gasker in the Outer Hebrides. Pup production fell by 40% in 1978 following a cull in 1977 and did not increase again in 1979 (Anon., 1980b).

D. *Conclusions*

The control measures enacted at the Farnes Islands have achieved some, but not all, of their avowed aims.

The current disturbance program, combined with other activities intended to reduce damage by birds, does appear to have reduced the rate of soil loss from the most vulnerable islands. However, the disturbance has had a profound, and not wholly understood, effect on the seal population.

On the other hand, none of the management methods has had much effect on pup mortality. This is primarily because grey seals insist on breeding at high densities on the more favored islands. Leaving aside the moral problems of when attempts should be made to interfere with "natural" mortality levels, research at other seal colonies since 1971 has indicated that the current levels of pup mortality at the Farne Islands are not unusually high; higher mortality rates have been recorded at North Rona in Scotland (Summers *et al.*, 1975) and at Ramsey in Wales (Anderson *et al.*, 1979).

REFERENCES

Anderson, S. S., Baker, J. R., Prime, J. H., and Baird, A. (1979). Mortality in grey seal pups: incidence and causes. *J. Zool., London 189*, 407-417.

Anon. (1980a). Questions and answers about the whale problem in the inshore fishery. Whale Research Group, Memorial University of Newfoundland. 8 pp.

Anon. (1980b). Annual assessment of the stocks of grey seals and common seals in Great Britain 1980. *NERC Newsjournal 2(9)*, 8-9.

Beddington, J. R. and Williams, H. A. (1979). The status and management of the harp seal in the North West Atlantic. Draft report. Marine Mammal Commission, Washington, D. C. 141 pp.

Bonner, W. N. (1971). Legislation on the seals in the British Isles. *The Salmon Net 7*, 30-33.

Bonner, W. N. and Hickling, G. (1971a). The grey seals of the Farne Islands: report for the period October 1969 to July 1971. *Trans. Nat. Hist. Soc. Northumberl. Durham Newcastle Upon Tyne 17(4)*, 141-162.

Bonner, W. N. and Hickling ,G. (1971b). Grey seals at the Farne Islands, a management plan. Natural Environment Research Council report. 21 pp.

Bonner, W. N. and Hickling, G. (1974). Seals of the Farne Islands: 1971-73. *Trans. Nat. Hist. Soc. Northumberl. Durham Newcastle Upon Tyne 42*, 65-84.

Coulson, J. C. and Hickling, G. (1964). The breeding biology of the grey seal, *Halichoerus grypus* (Fab), at the Farne Islands, Northumberland. *J. Anim. Ecol. 33*, 485-512.

Estes, J. A. and Gilbert, J. R. (1978). Evaluation of an aerial survey of Pacific walruses, (*Odobenus rosmarus divergens*). *J. Fish. Res. Board Can. 35,* 1130-1140.

Estes, J. A. and Gol'tsev, V. N. (In press). Abundance and distribution of the Pacific walrus (*Odobenus rosmarus divergens*): results of the first Soviet-American joint aerial survey, autumn 1975. *Fishery Bulletin Circulars.*

Fay, F. H. (1957). History and present status of the Pacific walrus population. *Trans. North Am. Wildl. Conf. 22,* 431-443.

Fay, F. H., Feder, H. W., and Stoker, S. (1977). An estimation of the impact of the Pacific walrus population on its food resources in the Bering Sea. Report MMC-75/06 and MMC-74/03. Marine Mammal Commission, Washington D. C.

Fay, F. H. and Kelly, B. P. (1980). Mass natural mortality of walruses (*Odobenus rosmarus*) at St. Lawrence Island, Bering Sea, autumn 1978. *Arctic 33,* 226-245.

Fisheries and Environment Canada. (1978). Conservation concern prompts closure of North Atlantic capelin fishery. News release FMS-HQ-NR-No 35, 23 November 1978. Fisheries and Oceans, Information Branch, Ottawa. 2 pp.

Hickling, G. and Hawkey, P. (1975). The grey seals of the Farne Islands: the 1974 breeding season. *Trans. Nat. Hist. Soc. Northumberl. Durham Newcastle Upon Tyne 42,* 93-97.

Hickling, G. and Hawkey, P. (1979). The grey seals of the Farne Islands: the 1978 breeding season. *Trans. Nat. Hist. Soc. Northumberl. Durham Newcastle Upon Tyne 43,* 35-44.

Hickling, G., Hawkey, P., and Harwood, L. H. (1976). The grey seals of the Farne Islands: the 1975 breeding season. *Trans. Nat. Hist. Soc. Northumberl. Durham Newcastle Upon Tyne 42,* 107-114.

Hickling, G., Hawkey, P., and Harwood, L. H. (1977). The grey seals of the Farne Islands: the 1976 breeding season. *Trans. Nat. Hist. Soc. Northumberl. Durham Newcastle Upon Tyne 42,* 119-126.

Hickling, G., Hawkey, P., and Harwood, L. H. (1978). The grey seals of the Farne Islands: the 1977 breeding season. *Trans. Nat. Hist. Soc. Northumberl. Durham Newcastle Upon Tyne 43,* 25-33.

International Commission for the Northwest Atlantic Fisheries (ICNAF). (1978). Provisional nominal catches in the Northwest Atlantic, 1977. *Int. Comm. Northwest Atl. Fish. Summ. Doc. 78/VI/28.* (Revised 15 June 1978) 51 pp + addendum.

International Whaling Commission (IWC). (1980). Report of the subcommittee on protected species and aboriginal whaling. *Int. Comm. Whaling Rep. 30,* 103-109.

Krogman, B. D., Braham, H. W., Sonntag, R. M., and Punsley, R. G. (1979). Early spring distribution, density and abundance of the Pacific walrus (*Odobenus rosmarus)* in 1976. Outer Continental Shelf Environmental Assessment Programs, final report RU14. Boulder, Colorado, 47 pp.

Lien, J. and Gray, S. (1980). Net loss. *Nat. Can. 9,* 4-9.

Lien, J. and McLeod, P. (1980). Humpback collisions with fishing gear in Newfoundland: arbitration and education in a whale fishermen dispute. Paper submitted to the Northeast Endangered Species Workshop, May 7-11, 1980. Provincetown, Massachusetts. 33 pp.

Lien, J. and Merdsoy, B. (1979). The humpback is not over the hump. *Nat. Hist. 88,* 46-49.

Lowry, L. F., Frost, K. J., and Burns, J. J. (1980). Feeding of bearded seals in the Bering and Chukchi Seas and trophic interactions with Pacific walrus. *Arctic 33,* 330-342.

Mitchell, E. D. (1973). Draft report on humpback whales taken under special scientific permit by eastern Canadian land stations, 1969-1971. *Int. Comm. Whaling Rep. 23,* 138-154.

Mitchell, E. D. (1974). Trophic relationships and competition for food in Northwest Atlantic whales. Proceedings of the Canadian Society of Zoologists annual meeting, June 2-5, 1974. Fredericton, New Brunswick. pp. 123-133.

Prime, J. H. (1980). The current status of the Farne Islands grey seal population. *Int. Counc. Explor. Sea Coop. Res. Rep. CM 1980/N:7.*

Summers, C. F. (1979). The scientific background to seal stock management in Great Britain. Natural Environment Research Council Publications Series C. 14 pp.

Summers, C. F., Burton, R. W., and Anderson, S. S. (1975). Grey seal (*Halichoerus grypus)* pup production at North Rona: a study of birth and survival statistics collected in 1972. *J. Zool., London 175,* 439-451.

Summers, C. F. and Harwood, J. (1978). Indirect effects of grey seal culls. *Int. Counc. Explor. Sea Coop. Res. Rep. 1978/N:4.*

Whitehead, H. (1978). Whales of the Bay de Verde Peninsula, Newfoundland. Unpublished manuscript, cited by Lien and McLeod, 1980.

Whitehead, H. (1979). Populations of humpback whales on Silver, Navidad and Mouchoir Banks during the winter 1977-78. *Int. Comm. Whaling SC/31/DOC. 6.*

THE CASE OF THE SEA OTTER

James A. Estes

National Fish and Wildlife Laboratory
Center for Coastal Marine Studies
University of California
Santa Cruz, California

I. INTRODUCTION

Most of what has been written in this volume concerns large mammalian herbivores. This focus probably was justified since most, if not all, examples of natural ecosystems in which collapse or destabilization can be convincingly argued to be the consequence of overabundance involve herbivores in terrestrial communities. What then can one say of the carnivores? The answer to that question seems to me to be little in most cases; mainly because nearly all populations of large carnivorous mammals have been disturbed or depleted by human activities. The problem is compounded by the fact that we have no real sense of how consistently biological communities respond to the removal of high trophic level consumers. One would hope that the behavior of every natural community is not unique in this regard. It is with this hope that my chapter is contributed to this volume.

Large carnivorous mammals occur in both terrestrial and aquatic communities throughout the world. The following examples are drawn only from my experience with sea otters. A critic might argue that sea otters occur only in a narrow band of inshore marine habitat, and that area is representative of typical situations neither on land nor in the sea. I cannot refute that argument. However, most other natural communities in which large mammalian carnivores feed are unmanageable as research areas, either because their associated biota are difficult to work with or they have been so disturbed by human activities that the natural interactions within them are extinct. In communities inhabited by sea otters many of the key species are accessible, observable, and relatively stationary. In this

ISBN 0-12-385280-3

sense the system is manageable at least. Furthermore, throughout much of the northern part of the sea otters' range, the inshore community probably has been largely unaffected by human activity beyond overexploitation of otters. This situation has provided a unique opportunity to observe ecosystem behavior as populations of a carnivorous mammal grow from low levels to eventual equilibria.

II. BACKGROUND

Historically sea otters occurred in the Pacific Ocean from the northern Japanese archipelago eastward through the Aleutian Islands to the central Pacific coast of Baja California (Kenyon, 1969). The Bering Expedition in 1741 marked the beginning of a period of intense overexploitation of the animals for their valuable furs. Sea otters were protected by international treaty in 1911, but by that time only a few remnant populations remained. Several of these populations declined to extinction. Others recovered under protection afforded by the treaty. Sea otters now occupy most of their historical range from the region of Prince William Sound southwestward through the Aleutian and Kurile Islands. They are largely absent southeast of Prince William Sound along the coast of North America, except for a small population in central California and translocated populations in Oregon, Washington, British Columbia, and southeast Alaska (Estes, 1980).

Sea otters forage nearshore in rocky and soft sediment habitats. Their prey consists of benthic invertebrates and, in some areas, fish. Before the fur hunting era, predation by sea otters probably limited populations of such forms as abalones, sea urchins, and several species of crabs and clams. After overexploitation of the otters by humans, populations of those invertebrates increased greatly. This was the state of affairs in the latter part of the nineteenth century when people began to exploit coastal marine resources of western North America. Commercial and sport fisheries developed for abalones, clams, crabs, and sea urchins. In central California the recovery and subsequent expansion of the range of sea otters has since damaged some of those fisheries (Wild and Ames, 1974; Miller *et al.*, 1975).

In this chapter I describe the current status of sea otters with particular emphasis on central California, and I discuss some direct and indirect effects of sea otter predation on nearshore community organization. Then follow some thoughts concerning the concept of carrying capacity in sea otters, and a possible mechanism for the maintenance of alternate points of stability in the plant/herbivore interaction in Alaska. Finally, I briefly discuss conflicting issues in management of sea otters in California.

III. POPULATION BIOLOGY OF THE SEA OTTER: WHAT IS WRONG IN CALIFORNIA?

At least seven remnant sea otter populations survived through the early twentieth century in the Aleutian archipelago and along the Alaska peninsula. As these populations recovered and repopulated old habitats, investigators described their growth rate characteristics at a number of locations. Open-ended populations (i.e., with access to adjacent, unoccupied habitat) increased at rates of 10% to 16%/year (Kenyon, 1969). The number of sea otters in Alaska is presently estimated at roughly 150,000 (Estes, 1980).

At the turn of the century, three remnant populations are known to have existed southeast of Prince William Sound along the coast of North America in the following areas:

(1) The Queen Charlotte Islands, British Columbia.
(2) Central California.
(3) The Islas San Benitos, Baja California.

Populations at the Queen Charlotte Islands and Islas San Benitos subsequently became extinct. The population in central California probably contained 50 to 100 animals at the beginning of this century, but by the late 1970's it contained about 2,000 individuals (Miller, 1978; California Department of Fish and Game, unpublished data) and had expanded its range to include about 300 km of coastline between Monterey Bay and Pismo Beach. Thus, the California population apparently increased at an average rate of about 4%/year. This is two to four times less than the estimated rates of increase of open-ended populations in the Aleutian Islands. Although the population in California is continuing to expand its range, recent surveys have failed to demonstrate that the number of animals in the population has changed during the last five to ten years. The uncertainty concerning the present status of the California population is, in part, a result of imprecise estimates resulting from variable weather and sea otter behavior, and observer error. Yet in my view there is no doubt that the population in California is growing at a lower rate than did populations in the Aleutian Islands. For these reasons, and because of the increasing threat of an oil spill in central California that might kill large numbers of otters, the Director of the U. S. Fish and Wildlife Service recently determined the California sea otter to be a "threatened population."

I am skeptical of Miller's (1980) view that the California population of otters is near carrying capacity. Miller's principal arguments in support of his position are that the density of otters is not increasing within the established range of the species and based on necropsies of beach cast animals, starvation is the principal source of mortality in the California population. On the other hand, suitable unoccupied habitat occurs at both ends of the present range of the population in California, and I am troubled by the fact that this has not accommodated a more rapid population growth. In the Aleutian Islands populations of otters have expanded from the point of recolonization by progressively advancing into unoccupied habitat around the islands' perimeters. I see no difference between that situation and the one in California, except that eventually on islands the two ends must meet. The distance around some of the larger islands in the Aleutian archipelago is similar to the length of coastline presently occupied by sea otters in California, yet apparently growth rates of populations on those islands did not decline until all available habitat was occupied (Kenyon, 1969). Therefore, if one is to accept the view that growth of the otter population in California is limited because it is near carrying capacity, one must also reconcile the fact that available habitat remains into which the population might be expected to grow further.

The comparatively low rate of increase of the otter population in California could be caused by:

(1) Lower birth rates.
(2) Higher mortality rates.
(3) Loss from the population by emigration.

Available evidence (not yet as full as desired) indicates no difference in overall birth rates between Alaskan and Californian populations. Counts made near the end of the pupping season (July in Alaska and March in California) show a ratio of 20 to 30 pups per 100 adults at both locations. Substantial loss from emigration is unlikely. Although extralimital sightings of otters have been made from Baja California to Oregon, these sightings are rare. If large numbers of animals were emigrating from the existing established range of the species, one should eventually see them accumulate at some other location. Since this has not happened, and in view of similar overall birth rates in Alaska and California, it seems likely that there is a difference in mortality rates between open-ended populations in Alaska and the population in California. Increased mortality might be ascribed to the following sources:

(1) Direct mortality caused by humans; there is considerable human activity in coastal areas inhabited by sea otters in central California. Shooting, collision with boats, and entanglement with fishing nets all are known sources of mortality to the animals.

(2) Indirect mortality from human activities; the habitat of the otter in California has been altered by human activities in numerous ways. For example, people and otters exploit the same food resources. Environmental contaminants such as heavy metals and pesticides have been found to occur at high levels in this system. It remains to be demonstrated, however, that these factors have a substantial detrimental effect on the animals.

(3) Competition with other natural predators; the diet of the otter in the Aleutian Islands does not overlap with any other species of consumer to any important extent. In California a number of vertebrate and invertebrate consumers have diets that overlap with that of the otter. Conceivably, these other natural predators increase mortality in the population of sea otters through competition for food.

(4) Direct mortality from natural predators; there have been a number of otter carcasses found on California's beaches with fragments of white shark teeth embedded in them (Ames and Morejohn, 1980). Since all individuals actually consumed by sharks are undetected, shark predation could be an important source of mortality that does not occur in Alaska.

(5) Marginal population effect; the known southern limit of the sea otter was near Punta Abreojos, Baja California. Perhaps sea otters in California have a higher mortality rate than do Alaska otters because California is a suboptimum habitat, near the natural limit of the species' range.

(6) Reduced fitness due to small initial population size. Because of its small residual population size in the early 1900's, the California population may have lost natural genetic variation. Associated with this loss might be reduced fitness in the form of decreased survival. A recent analysis of this possibility, however, suggests that it is unlikely (K. Ralls, personal communication).

These factors might act in concert, or there may be interactions among them that further affect rate of mortality. Unfortunately, establishing cause of mortality in natural populations of all marine mammals is a difficult task and at the present there seem to be no reasonable means of assessing the aforementioned potential sources.

IV. SEA OTTER PREDATION AND COMMUNITY ORGANIZATION

Effects of otters as predators to nearshore community organization is reasonably well understood in the Aleutian Islands (Estes and Palmisano, 1974; Estes *et al.*, 1978). Since the animals have not yet repopulated a small number of isolated islands, it has been possible to infer their role as a predator by comparing islands with and without otters. These studies have revealed that otters are important in limiting populations of herbivorous sea urchins. This interaction, in turn, maintains a productive assemblage of epibenthic and canopy-forming kelp in rocky bottom habitats. Kelp provides food and protection to many species of animals. At islands without otters, sea urchins have largely destroyed the kelp beds and those islands are comparatively depauperate in higher trophic forms, most notably nearshore fishes (C. A. Simenstad and J. A. Estes, unpublished data).

Sea otter predation also limits populations of benthic invertebrates throughout much of central California. Indirect effects of this interaction on kelp beds and their associated communities, however, are less clear. There is accumulating evidence that sea otters influence the composition of species in the kelp canopy. The two most abundant canopy-forming species are *Macrocystis pyrifera*, a perennial, and *Nereocystis leutkeana*, an annual. *Macrocystis* is thought to be a competitive dominant over *Nereocystis*, whereas *Nereocystis*, because it can carry out its life cycle in one year, is thought to be more successful than *Macrocystis* in heavily disturbed areas (Dayton *et al.*, 1980). There has been a gradual increase in the relative amount of *Macrocystis* subsequent to the range expansion of the otter, which may have been caused by the animals reducing the intensity of disturbance by herbivores on kelps (VanBlaricom and Jameson, 1979). These ideas presently are being investigated in more detail.

In southern California, sea urchin population explosions have greatly reduced the distribution of kelp beds at several locations (Leighton *et al.*, 1966). Since kelp is a commercially valuable resource, this has had a detrimental effect on the kelp industry. Although these population explosions have occurred too recently to be directly attributable to the disappearance of otters (Tegner, 1980), there is little doubt that the reestablishment of sea otters would reverse the process. Presently, quicklime (calcium oxide) is being used to control sea urchin populations.

V. SOME SPECULATIVE RELATIONSHIPS

Study of the otter and its habitat in the Aleutian Islands has led me to a number of hypotheses and speculations concerning the concept of carrying capacity, the existence of multiple stable equilibria in community structure, and the process of transition from one stable equilibrium to another. I include a brief discussion of these ideas here mainly because the subjects are of central interest to this volume. Furthermore, since it has been my experience that wildlife managers make overly simplistic assumptions concerning the nature of these subjects, I include these examples to point out some of the complexities which should be considered prior to making decisions with regard to the manipulation of otter populations.

A. *Maintenance of Stability in Systems Dominated by Sea Otters*

Rocky sublittoral communities with and without otters are distinctly different. Both appear to be reasonably stable in the sense that they have maintained their integrity over short periods of time (ten years of observation) and that they are resilient to natural perturbations (Simenstad *et al.*, 1978). The intent of an ongoing study at Attu Island, westernmost of the Aleutians, is to document changes in benthic community structure associated with recovery of the sea otter population that recently was reestablished there. One far-ranging objective is to determine if the structure and organization of the community at Attu will converge with that which we presently see at nearby islands where sea otters are at equilibrium density, and if they do converge, how will the change occur over time. I will limit this discussion to otters, sea urchins, and the kelp assemblage -- species which appear to be the system's most important components at the primary, secondary, and tertiary trophic levels. Observations at Attu indicate that as otters reoccupy new habitat they quickly eliminate large sea urchins so that the structure of the sea urchin population becomes similar to that in areas where otters are near equilibrium density (Simenstad *et al.*, 1978). Apparently this has resulted in increased survival of new sea urchin recruits, and consequently in increased sea urchin density. Following these changes, the community has persisted with no observable change over at least the past six years (the time during which I have been working at Attu), despite the fact that the otter population has increased from about 250 to over 800 animals in the

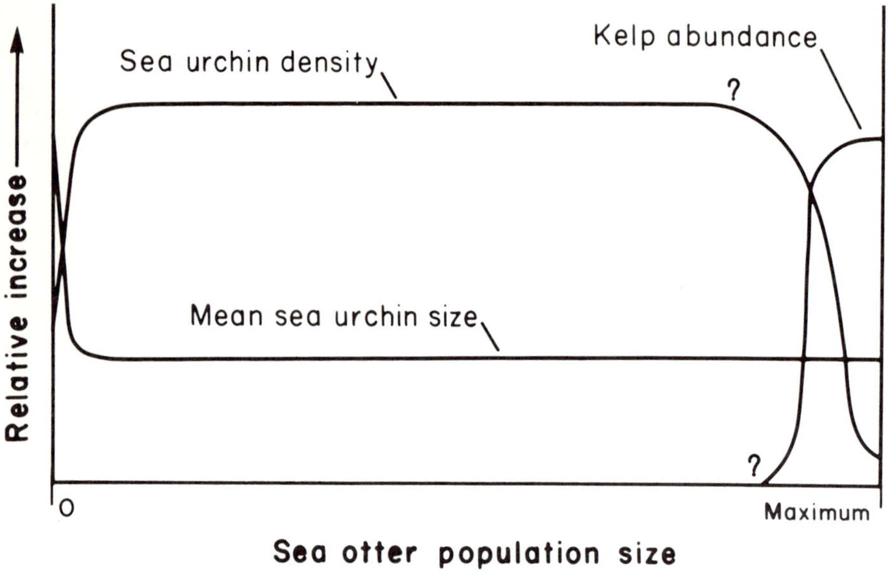

Fig. 1. Hypothesized changes in several aspects of community structure that will occur with the natural growth of the sea otter population at Attu Island. Values near zero have been determined from observations at Attu Island. Relative values at maximum population size are based on observations at Amchitka Island. Points of change indicated by question marks are less certain, but I hypothesize that they will occur shortly after the population becomes limited by competition for food.

same time period. If the system is to eventually attain a structure similar to that which we see in areas where otters are near equilibrium density, the changes will have to occur something like those shown in Fig. 1.

I expect that sea urchin density will remain high until the otter population has completely encircled Attu, at which time the intensity of competition for food should increase. As sea urchin densities decline, the intensity of herbivory also should decline until there are two points of stable equilibrium between the consumption of plants by herbivores and net primary productivity (Fig. 2). This mechanism for the maintenance of multiple points of stability in plant/herbivore systems was proposed in earlier papers (Noy-Meir, 1975; May, 1977) and has been discussed by Walker and Noy-Meir in this volume. The model requires that the net rate of plant production is a convex function of plant biomass (V), beginning at zero when V equals 0(V_0) and ending at zero when V equals its maximum value (V_{max}), with a maximum somewhere in between; and the consumption rate of plant

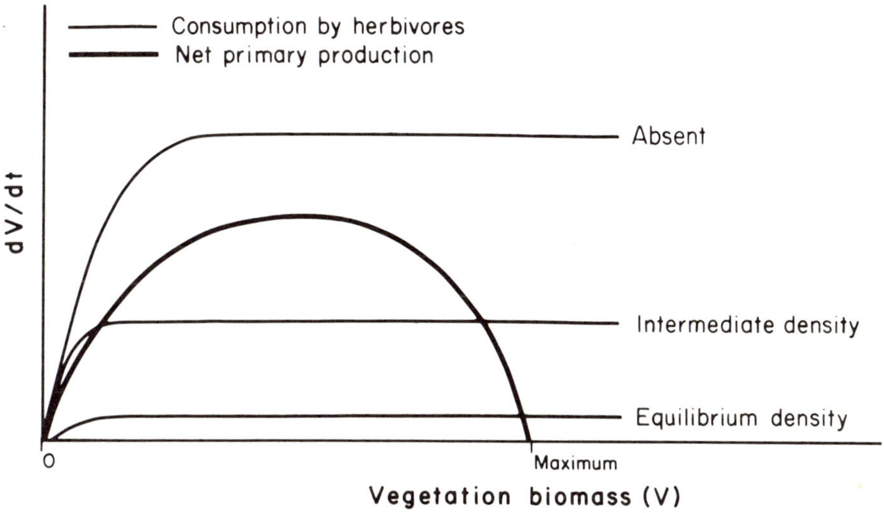

Fig. 2. Graphical model for the maintenance of alternate points of stability in the abundance of vegetation biomass. Hypothesized rates of consumption by herbivores are shown where sea otters are absent (highest values), at equilibrium density (lowest values), and at an intermediate density (intermediate values).

biomass by herbivores levels off at a constant value as a function of available plant biomass for each possible abundance of herbivores in the system. In communities with no otters, the rate of consumption by herbivores is always higher than the net rate of plant production and consequently the system is stable only at $V = 0$. Conversely, for communities in which otters are at equilibrium density, the rate of consumption by herbivores is less than the net rate of plant production at all values of V below V_{max}. As the sea otter population increases from low densities (or declines through culling from high densities), however, there will be a range of intermediate herbivore densities at which the rate of consumption and the rate of net production are in stable equilibrium near both V_0 and V_{max}. Under this circumstance, a perturbation to V (or to the intensity of herbivory) would be required to drive the system from one point of stability to the other. Although I do not know the exact forms of the production and consumption functions, it would seem that rapid changes in V (or dV/dt by herbivores) of the magnitude required to drive the system in one direction or the other beyond the point of unstable equilibrium are unlikely events. Therefore, the more important factor in determining community structure where multiple stable equilibria exist may be the direction of change in otter population size. That is, increasing populations might be expected to persist with a community at stable equilibrium near V_0 until the otter population nears maximum size. Communities in which otter populations were artificially reduced (declining populations), on the other hand, might be expected to remain in a stable equilibrium near V_{max} until population density was very low. The upshot of this is that the functional relation between sea otter population density and community structure might be quite different for increasing and declining populations.

B. *Carrying Capacity, Prey Switching, and Sea Otter Population Regulation in the Aleutian Islands*

One broadly held view by wildlife managers is that animal populations grow exponentially until food or other limiting resources are overexploited at which point they level off and decline. Kenyon (1969) believed this to be the typical pattern in island populations of otters that he studied. More recent studies at Amchitka Island (Estes, 1977) indicated that the population density of otters was:

(1) Larger than Kenyon's highest estimates.
(2) Near equilibrium with limiting resources (probably food).
(3) Relatively stable, at least in the sense that there is no evidence it has declined catastrophically in recent years.

These observations, together with those I have made on the habitat of the otter (discussed previously in this chapter), lead me to propose the following explanations for growth patterns of otter populations in the Aleutian Islands (mainly based on data from Amchitka Island), as they recover from small size to an eventual equilibrium. In defense against those who might accuse me of idle speculation, I point out that the predictions may be tested through continuing observations of the otter population and the nearshore community at Attu Island.

In early phases of recovery of a sea otter population, food was not a limiting resource and the population increased at a rate of about 15%/year. During this time, the benthic habitat was characterized by abundant sea urchins and little or no kelp. This situation persisted until available habitat was occupied by otters, at which time food and perhaps other resources limited further population growth. Intense competition for food was followed by a decline in the otter population, as well as a reduction in the size and density of sea urchins, in turn favoring (and eventually resulting in) establishment of dense kelp beds. (Expansion of kelp beds is limited largely by susceptibility of newly settled sporophytes to herbivores, and dispersal of spores or gametes from adult plants.) Once the algal assemblage had become established, a food web based largely on macroalgal production became an important trophic component of the system. Inshore fish populations were enhanced. Prey switching by otters to fish occurred (Estes *et al.*, 1978) although otters continued to eat large numbers of sea urchins. The new resource permitted a resurgence in growth of the otter population, which eventually came to a new equilibrium at an elevated size beyond that which was possible on herbivores alone as a prey resource (Fig. 3). Fish are not so susceptible to

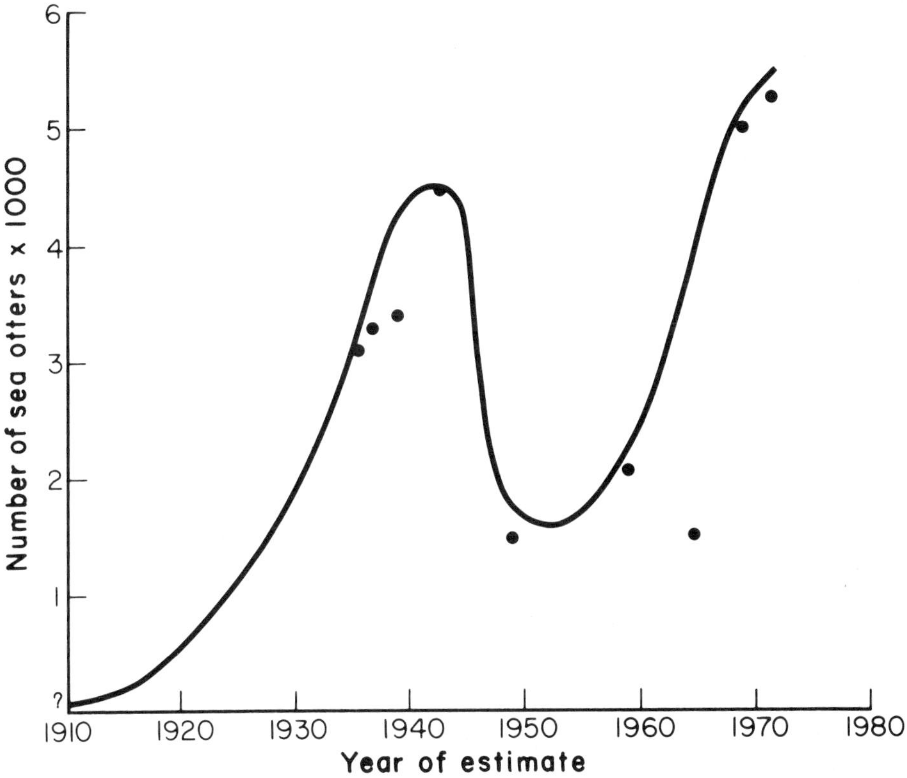

Fig. 3. Changes in size of the sea otter population at Amchitka Island. (From Kenyon, 1969; Estes, 1977). All estimates, particularly those made prior to 1945, are very rough. I believe only that the overall trend is accurate.

overexploitation by otters as are sea urchins because they are more difficult to capture and because juvenile fish leave the nearshore ecosystem as larvae and do not return until they reach sexual maturity. Otters again reached a point of limitation by food although in this instance it did not result in overexploitation of food. The mortality rate of otters increased, but this was limited mainly to juveniles and senescent adults. This occurred without an ensuing population crash and without degradation of the habitat.

VI. CONFLICTS AND ALTERNATIVES CONCERNING THE MANAGEMENT OF OTTERS

Conflicts over how to manage otters involve both biological and economic issues. Strictly on biological grounds, there is no reason to cull or artificially limit otters in my view. Although otters effectively reduce populations of herbivorous invertebrates, there is no evidence that this is detrimental either to the otter population or to other components of the ecosystem. This would seem to exclude otters from either Caughley's Class 3 or Class 4 problems (see Caughley, this volume). Our present understanding of these systems suggest that sea otter predation enhances the system's productivity. From the standpoint of economics we do not yet know enough to make quantitative determinations as to the effects of otters. On one hand, otters can eliminate some sport and commercial shellfisheries. I think it is unrealistic to imagine that otters and large-scale shellfisheries could coexist in the same areas. However, under some circumstances, sea otter predation probably enhances kelp production and quite possibly finfish populations associated with kelp beds. It is evident that "the sea otter problem" is a clear and simple example of Caughley's Class 2 problems of local overabundance. That is, there is no reasonable expectation that unculled otter populations will inevitably lead to catastrophic or irreversible change in the system. The real relevance biology holds to the problem is in making qualitative predictions concerning how the presence or absence of otters will influence the availability of resources deemed valuable in one way or another by humans.

Three options have been suggested to resolve problems with sea otters in California:

(1) Translocations have been suggested as a means of reducing the threat of a large oil spill to the population. There have been five attempted translocations of the species in recent years. Only one of those was clearly successful and at

least three were almost certainly unsuccessful (Jameson *et al.*, in preparation). Translocations are costly, and wherever suitable habitat might be found for a translocation, the introduced animals will compete with humans for shellfisheries.

(2) Range limitation, a second option, has been supported by sport and commercial shellfishermen as a means of stopping the further encroachment of the otters into productive fishery areas. The only practical present means of implementing this measure is by shooting. Considering the rather precarious present status of the otter in California, this alternative is unacceptable to most people.

(3) The third option is to take no action and to allow the population to expand or decline as it will. This option, although the easiest to implement, is not acceptable to many people because whereas the population has continued to increase its range, it does not appear to have increased in number. In addition, the threat of a large oil spill will increase over time as exploration, drilling, and tanker traffic increase in central California waters.

REFERENCES

Ames. J. A. and Morejohn, G. V. (1980). Evidence of white shark, *Carcharodon carcharias,* attacks on sea otters, *Enhydra lutris*. *Calif. Fish Game 66,* 196-209.

Dayton, P. K., Keller, B. D., and Ven Tresca, D. A. (1980). Studies of a nearshore community inhabited by sea otters. Final report MMC-78/14. Marine Mammal Commission, Washington, D. C. 91 pp.

Estes, J. A. (1977). Population estimates and feeding behavior of sea otters. *In* "The Environment of Amchitka Island, Alaska" (M. L. Merritt and R. G. Fuller, eds.), pp. 511-526. U. S. Energy Res. Devel. Adm. TID-26712, Springfield, Virginia.

Estes, J. A. (1980). *Enhydra lutris*. *Mamm. Species 133,* 8 pp.

Estes, J. A., and Palmisano, J. F. (1974). Sea otters: their role in structuring nearshore communities. *Science 185,* 1058-1060.

Estes, J. A., Smith, N. S., and Palmisano, J. F. (1978). Sea otter predation and community organization in the western Aleutian Islands, Alaska. *Ecology 59,* 822-833.

Jameson, R. J., Kenyon, K. W., Johnson, A. M., and Wight, H. W. (In preparation). The history and status of translocated sea otter populations. Submitted for publication to the *J. Wildl. Manage*.

Kenyon, K. W. (1969). The sea otter in the eastern Pacific Ocean. *North Amer. Fauna 68*, 352 pp.

Leighton, D. L., Jones, L. G., and North, W. J. (1966). Ecological relationships between giant kelp and sea urchins in southern California. *Proc. Int. Seaweed Symp. 5*, 141-153.

May, R. M. (1977). Thresholds and breakpoints in ecosystems with a multiplicity of stable states. *Nature (London) 269*, 471-477.

Miller, D. J. (1978). A descriptive autecological model of the sea otter population in California. Unpublished manuscript. Available from the California Department of Fish and Game, Monterey, California.

Miller, D. J. (1980). The sea otter in California. *Calif. Coop. Oceanic Fish. Invest. Rep. 21*, 79-81.

Miller, D. J., Hardwick, J. E., and Dahlstrom, W. A. (1975). Pismo clams and sea otters. *Calif. Dep. Fish Game Mar. Res. Tech. Rep. 31*, 1-49.

Noy-Meir, I. (1975). Stability in grazing systems: an application of predator-prey graphs. *J. Ecol. 63*, 459-481.

Simenstad, C. A., Estes, J. A., and Kenyon, K. W. (1978). Aleuts, sea otters, and alternate stable-state communities. *Science 200*, 403-411.

Tegner, M. (1980). Multispecies considerations of resource management in southern California kelp beds. *Canadian Technical Report of Fisheries and Aquatic Sciences 954*, 125-143.

VanBlaricom, G. R. and Jameson, R. J. (1979). Sea otter-canopy kelp relationships in central California: a historical review and a model. Paper presented at a workshop on the sea otter, Santa Barbara, California, August 23, 1979. Proceedings unpublished.

Wild, P. W. and Ames, J. A. (1974). A report on the sea otter, *Enhydra lutris* L., in California. *Calif. Dep. Fish Game Mar. Res. Tech. Rep. 20*, 1-93.

THE LEOPARD -- PROBLEMS OF AN OVERABUNDANT, THREATENED, TERRESTRIAL CARNIVORE

Stephen Cobb

Department of Zoology
University of Nairobi
Nairobi
Kenya

I. INTRODUCTION

Much attention in this volume is devoted to discussing the means of detecting apparent overabundance in large herbivore populations. This frequently takes the form of assessing the apparent carrying capacity of herbivores by an ecosystem or a vegetation community and then monitoring trespasses by the herbivores beyond that capacity. That herbivore overabundance is often illusory is succinctly explored by Caughley (1979). The next trophic level, by contrast, has been curiously ignored in this context. With the exceptions of the Isle Royale moose population and its wolf predators (Mech, 1966; Allen, 1974) and the Ngorongoro wildebeest population and its hyenas (Kruuk, 1972), evidence of large mammalian terrestrial carnivores controlling their large mammalian prey species, even on a temporary basis, is hard to find. Nevertheless, large carnivores are frequently considered overabundant. They are also frequently culled. This last statement begs a definition of the verb "to cull," for which Onions (1975) offers: "1) to choose, to select; 2) to gather, pick, pluck (flowers, etc.); 3) to subject to the process of selection." In this chapter I shall use the word in the limited sense of: "to kill animals according to selective criteria."

For the purposes of this chapter it is assumed the circumstances under which any culling (not necessarily of locally overabundant species) takes place, may be described by four categories:

ISBN 0-12-385280-3

(1) Ecological: culling in the apparent best interests of the ecosystem or species. Such culling practices are often implicit in the management policies of many national parks and other protected areas. In theory only biological criteria would be considered before a decision to cull was made.

(2) Economic: culling for the financial or nutritional gain of people. It is into this category that subsistence hunting, whaling, and fur exploitation fall, as well as culling to protect other economic interests.

(3) Recreational: culling in the guise of sport hunting. In well organized countries, the target animals and the level of annual cull are determined by research.

(4) Social: culling to reduce the local numbers of species that are thought to be a nuisance. This apparent nuisance value is frequently immeasurable. The types of nuisance include inferred disease transmission, crop damage, and the killing of domestic animals.

It is on the last category of culling that this chapter will concentrate. Mammals of many orders occasionally find themselves in this category, but the most persistent occupants of it are the Carnivora. It is no coincidence that those genera most frequently accused of being a nuisance are also those that are classified as being threatened. The species to which most attention is given in this chapter is the leopard (*Panthera pardus*) -- a widespread, threatened carnivore that finds itself culled according to all four types of criterion. The fact that the species is globally threatened serves only to heighten the problems associated with its local overabundance.

II. THE STATUS OF THREATENED CARNIVORES

Of the world's 4,200 species of mammals, 279 are classified by the International Union for the Conservation of Nature and Natural Resources (IUCN) Red Data Book (1978) as being *threatened*, a general description comprising the categories *endangered, vulnerable, rare,* and a fourth, *indeterminate*. If 6.6% of the world's mammalian species are threatened, it is worth considering first whether or not culling practices are in any way responsible for their current status, and secondly, whether or not culling practices should change when they deal with threatened species.

Table I shows that the most threatened mammalian order is the Sirenia (all four species threatened), followed by Proboscidea (both species threatened), and then the Perissodactyla (13 out of 17 species threatened). A relatively long way down the list come the Carnivora, with 46 out of 252 (18.3%) of its species threatened.

The role of the Carnivora as secondary consumers more or less guarantees that, in whatever environment they are found, they will be rarer than other mammals upon which they may prey. Thus, on purely ecological grounds, one would expect the Carnivora to contain a higher proportion of threatened species than other mammalian orders.

Table II, also drawn from the IUCN Red Data Book (1978) shows all those species of threatened terrestrial mammals that are found in ten or more countries. Clearly, this division has the disadvantage that not all countries are the same size. Nevertheless, this table has one straightforward message: Most of the threatened mammal species with a widespread distribution are Carnivora. The implication is that the Carnivora, being predominantly rather unselective feeders, have been able to establish wide distributions, but thereby are currently threatened. If the threats were of global environmental change, Table I would have shown that the Carnivora are one of the most highly threatened orders. That this is not so implies that nonecological processes are responsible for the widespread rarity of certain Carnivora. All the 14 carnivore species in Table II are either of direct economic interest to man (in that they carry valuable skins) or they conflict with his economic interests (they interfere with his fishing, his hunting, his livestock). The unifying factor that leads to their widespread rarity is the universal intolerance of them by man as close neighbors.

This analysis permits us to look at the threatened Carnivora in terms of the four categories of culling proposed in the introduction. The information in Table I and II shows that intrinsic ecological factors do not account for the reduced state of the populations of some carnivore species. That is not to imply that man does not cull carnivores for reasons of maintaining an apparent ecological balance. In East Africa, until the early 1960's, hyenas and leopards were considered vermin, not only in agricultural areas but also in national parks (Graham, 1973). They were "controlled" on account of the supposed deleterious effect they were having on the herbivore populations.

Culling of the Carnivora by means of recreation is almost invariably conducted under fairly tight control; where it is not, it merges almost imperceptibly into illegal hunting, which is better described as culling for economic reasons.

TABLE I. *Threatened Taxa within the Order Mammalia*

	Families		Genera		Species	
	Extant	Threatened	Extant	Threatened	Extant	Threatened
Monotremata	2	1 (50.0%)	3	1 (33.3%)	6	1 (16.7%)
Marsupialia	13	7 (53.8%)	63	18 (28.6%)	194	21 (10.8%)
Insectivora	8	5 (62.5%)	61	8 (13.1%)	373	8 (2.1%)
Dermoptera	1	0	1	0	2	0
Chiroptera	17	4 (23.5%)	183	7 (3.8%)	979	11 (1.1%)
Primates	13	10 (76.9%)	58	36 (62.1%)	213	60 (28.2%)
Edentata	3	3 (100.0%)	14	6 (42.9%)	32	6 (18.8%)
Pholidota	1	0	1	0	7	0
Lagomorpha	2	1 (50.0%)	9	4 (44.4%)	67	4 (6.0%)
Rodentia	32	7 (21.9%)	363	12 (3.3%)	1727	20 (1.2%)
Cetacea	10	5 (50.0%)	38	8 (21.1%)	91	9 (9.9%)
Carnivora	7	7 (100.0%)	96	27 (28.1%)	252	46 (18.3%)
Pinnipedia	3	3 (100.0%)	21	5 (23.8%)	33	9 (27.3%)
Tubulidentata	1	0	1	0	1	0
Proboscidea	1	1 (100.0%)	2	2 (100.0%)	2	2 (100.0%)
Hyracoidea	1	0	2	0	6	0
Sirenia	2	2 (100.0%)	2	2 (100.0%)	4	4 (100.0%)
Perissodactyla	3	3 (100.0%)	6	6 (100.0%)	17	13 (76.5%)
Artiodactyla	9	7 (77.8%	88	33 (37.5%)	196	65 (33.2%)

TABLE II. Threatened Terrestrial Mammals Found in Ten or More Countries

Species	*Number of countries*
Leopard	*70*[a]
Wolf	*35*[a]
African elephant	*35*
European otter	*33*[a]
African wild dog	*30*[a]
Cheetah	*25*[a]
Dugong	*24*
Chimpanzee	*22*
Ocelot	*21*[a]
Jaguar	*21*[a]
Black rhinoceros	*19*
Giant anteater	*17*
Asiatic wild dog	*16*[a]
Tiger	*14*[a]
Asiatic elephant	*14*
Clouded leopard	*13*[a]
Giant armadillo	*11*
Temminck's cat	*10*[a]
Marbled cat	*10*[a]
Bush dog	*10*[a]
Giant otter	*10*[a]

[a]*Order Carnivora.*

In all events, it is most unlikely that culling of this type is directly responsible for the present rarity of any of the threatened Carnivora, except perhaps the tiger *(Panthera tigris)* on the Indian subcontinent.

Culling of the threatened Carnivora for economic reasons is rather widespread. By and large, it is because those carnivores carry attractive and thus very valuable skins. Of the species of large carnivores in Table II, none except the wolf *(Canis lupus)* is culled under a legally constituted program. Most culling is thus haphazard, illegal, and unquantified. Such trade in the skins as subsequently takes place is now largely illegal (Inskipp and Wells, 1979), owing to the effects of the Convention on International Trade in Endangered Species of Wild Fauna and Flora (CITES). Opportunities for rationalizing the culling of leopards in Africa on an international basis are explored by Myers (1976).

Culling of the Carnivora for social reasons has a long history. The Carnivora conflict with almost all of man's agricultural, pastoral, and domestic interests. Hence, for example, the virtual disappearance of the large carnivores from Europe over the last 2,500 years. Even in parts of the world with a less lengthy tradition of agricultural development, sustained persecution of the Carnivora is commonplace and has been practiced for a long time. The large carnivores are thought to be a nuisance; they seem a nuisance because they kill domestic animals, because they frighten and occasionally kill people, and because they exploit wild animal resources that people would prefer to exploit themselves. The extent of this nuisance has rarely been quantified, nor has the threshold nuisance level below which people will tolerate the presence of large carnivores in their vicinity. Meanwhile the Carnivora are shot, poisoned, and trapped, both by private individuals and by governments, even in countries where they are fully protected. There is always room for exceptions. It is to this category of culling and the quest for alternatives to it, that the next two sections of this chapter are primarily addressed.

III. NUISANCE CARNIVORES: WHAT ALTERNATIVES TO CULLING?

A. *The Leopard in East Africa*

The leopard as a species is classified by IUCN (1978) as vulnerable, while certain of its Eurasian subspecies are endangered. It is so classified because of the threats posed to its long-term survival and also because of its apparent rate of decline. The threats are threefold:

(1) The leopard is killed for its valuable skin.

(2) It is killed because it conflicts with man's domestic interests.

(3) It is threatened by ever-diminishing available wild lands.

This third threat clearly increases the number of times that the second can come into play. An important additional point about leopard conservation in Africa is that a large proportion of wild leopards, perhaps the majority, occur outside protected areas. Thus the areas in which many leopards are to be found do not have any stated wildlife management objectives.

Being both the bearer of an extremely desirable skin and an efficient killer of domestic animals, the leopard is subjected to culling for both economic and social reasons. If CITES works in the long term and the economic pressure on leopards is relieved, it is possible that this might serve only to increase their potential nuisance value, thereby increasing the pressure to cull. The alternatives to culling are relatively few. Those that exist and have been tried, and their relative chances of success, are described below.

B. Removal to Protected Areas

Removal trapping of leopards has been widely practiced in East Africa. An interesting example of the problems associated with this as a conservation option come from Meru National Park, Kenya. Over a period of 11 years until 1979, a total of 108 leopards had been released into the national park. They had been trapped elsewhere, up to 200 km away, on account of their nuisance value to farmers. To discover whether or not this policy of release within the safety of the park was a sensible one, Hamilton (1979) attached radio collars to 12 leopards (number 97 to 108) prior to their release. Of the total, four radios quickly ceased to work. Contact was maintained with the remaining eight for between one week and eighteen months. All eight left the park. Subsequently only one, after ten months outside, returned and stayed within the park. By African standards the park is relatively small, so that a walk of 25 km in almost any direction would have taken a leopard from the release point out of the park. Perhaps a larger park should have been chosen. A further imperfection of the operation was that all the leopards were released in the same place -- an area that P. H. Hamilton (personal communication) demonstrated to have its own, albeit small, resident leopard population. The residents may, of course, have established themselves from among the earlier releases. Even if that were true, it would appear that the translocation of nuisance leopards is a conservation option of limited value. This could have been discovered earlier had the management authorities taken the trouble to find out the fate of the leopards, rather than top out a stream of 96 of them before starting research on the matter.

C. Removal to Captivity

Excepting the fact that CITES makes the export of leopards relatively difficult, the next option open to the leopard conservationist as an alternative to a physical cull of nuisance animals is live export, particularly to zoos in other countries. The horizons of this are relatively limited and raise two problems.

The first is that a disproportionate number of nuisance leopards are adult males, the category of animal least desirable to zoos. The second problem is that, as with a number of other species of large carnivores, many zoos find it all too easy to breed leopards. Despite all the advantages of increased genetic heterogeneity, most zoos would prefer to buy, or to receive, captive-bred rather than wild-caught leopards. They do so because captive-born leopards exist and are on offer from other zoos, facilitating many different exchange arrangements where a leopard might be swapped for other animals. Thus, the market is glutted. Imports from the wild, despite the fact that the species is both popular with the public and also endangered, are not highly desired.

An indication of how overabundant leopards are in captivity is that in several British zoos vasectomy is practiced on adult males. In addition cubs are killed at or shortly after birth or up to one year later. Responsible zoos will now seek a market for potential cubs before allowing their adults the opportunity to breed.

Meanwhile, despite the fact that the species is threatened, there is no international studbook for it in captivity, there is no coordinated captive breeding program, and in Britain, for example, few zoo directors are aware of the provenance of their leopards (A. Harris, in preparation). From the point of view of introducing new genetic variability from the wild to the captive populations, there is a great deal to be said for offloading nuisance leopards on the captive population. But cost alone inhibits this. In 1979, the retail price of a female leopard in Britain was £280; that of a male much less. This is only twice the market price of a laboratory cat. Yet the cost of transport and quarantining a wild-caught leopard could easily be ten times this price.

D. Other Leopard Management Practices

Another form of leopard management exists, consisting of attempts to reintroduce previously captive animals into the wild. Although of considerable popular appeal, this option appears to benefit the leopards but little and, therefore, has no intrinsic merit. The late Joy Adamson tried in her latter years to reintroduce Penny, a female leopard, to Shaba National Reserve in northern Kenya. Although Penny did breed, she is now reported to be starving (Mawson, 1980) because nobody is feeding her. As a form of leopard conservation, this has no evident advantages over straightforward captive breeding in zoos.

IV. FURTHER OPTIONS IN LEOPARD CONSERVATION

People find leopards an intolerable nuisance. They have done so for thousands of years and they are not liable to stop doing so in areas of conflict, despite a growing global interest in conservation. It follows, therefore, that leopards are going to continue to be killed or removed from conflict areas. The former is the simpler option and the more attractive to the private individual, but it is inevitable that leopards will continue to be killed on a wide scale. It is, therefore, urgent to find sensible alternatives to make use of trapped nuisance leopards.

The sorry state of the organization of the world's captive leopard population suggests at least the following options:

(1) The provision of new animals for captive breeding programs. The high cost of this would need some kind of subsidy.

(2) The provision of blood for genetic studies.

(3) The provision of semen and, possibly in the future, ova, to be used for attempts at artificial insemination and embryonic implants. The former technique is already established in the Carnivora, for example the wolf (*Canis lupus*) (Seager *et al.*, 1975). This is a potentially cheaper way of exporting the genetic heterogeneity of wild leopards to the captive population than live transport.

Meanwhile, more general action along two lines would also benefit leopard numbers and the health of captive populations:

(1) The Convention on International Trade in Endangered Species has not yet been signed or ratified by all the 70 countries in which the leopard occurs. If that ever occurs, trade in leopards and their skins would, in theory, be more controllable and might follow the lines suggested by Myers (1976).

(2) Starting an international studbook for the nominate race is a step that needs to be taken. Currently, the vast majority of captive leopards in Britain are captive born, their

precise genealogies unknown (A. Harris, personal communication). The situation is probably no better in most other countries, although studbooks for the Eurasian subspecies do now exist (Shoemaker, 1979). Detailed genetic studies on the differences between subspecies have yet to be conducted.

The existence of leopards is not known to conflict with other aims of wildlife management in any conservation area, for example by threatening the numbers of other endangered species. Except in areas where this might occur, such as with the Arabian tahr *(Hemitragus jayakari)* in Oman, it is not currently likely that leopards will have to be culled for conservation purposes to maintain an intended equilibrium between wildlife species. Small fenced reserves are a possible exception. The same probably cannot be said for other large threatened carnivores, such as the tiger in India.

I cannot but end on a realistic note. The three options of uses for captured leopards that are suggested above are biologically sensible and full of optimism, but they do not account for very many leopards. Indeed in most cases the leopard is used to provide a service, but the problem remains of what to do with it afterwards. It might be possible to continue to release leopards into expectant national parks, but the evidence is that this action is futile. Within protected areas in Africa at least, the leopard, on account of its opportunism and broad spectrum of habitat tolerance, has probably as secure a future as any large carnivore. Outside those areas man will not coexist with leopards, and we must accept that he will continue to kill them and that the rational alternatives are of limited scope.

ACKNOWLEDGMENTS

I am grateful to Alison Harris for deservedly harsh criticism of an earlier draft of this chapter.

REFERENCES

Allen, D. L. (1974). Of fire, moose and wolves. *Audubon 76*, 38-49.

Caughley, G. (1979). What is this thing called carrying capacity? *In* "North American Elk: Ecology, Behavior and Management" (M. S. Boyce and L. D. Hayden-Wing, eds.) Univ. Wyoming Press, Laramie.

Graham, A. (1973). "The Gardeners of Eden." Allen and Unwin, London.

Hamilton, P. H. (1979). Translocated leopards: where do they go? *Wildlife News 14(3)*, 2-6. African Wildlife Leadership Foundation, Nairobi and Washington, D. C.

Harris, A. E. (In preparation). Management methods for the breeding of endangered species of mammals in captivity in the British Isles. Fellowship Thesis, Institute of Animal Technicians, London.

Inskipp, T. and Wells, S. (1979). "International Trade in Wildlife." Earthscan, London.

International Union for the Conservation of Nature and Natural Resources (1978). "Red Data Book, Vol. 1: Mammalia" (J. Thornback, ed.) IUCN, Gland, Switzerland.

Kruuk, H. (1972). "The Spotted Hyena." Univ. of Chicago Press, Chicago, Illinois.

Mawson, P. (1980). Letter to the editor. *Africana 7(2)*, 5. Nairobi.

Mech, L. D. (1966). "The Wolves of Isle Royale." U. S. National Park Service Fauna Series 7, Washington, D.C.

Myers, N. (1976). "The Leopard *Panthera pardus* in Africa." Monograph 5. IUCN, Gland, Switzerland.

Onions, C. T., ed. (1975). "Shorter Oxford English Dictionary." Oxford Univ. Press (Clarendon), London and New York.

Seager, S. W. J., Platz, C. C., and Hodge, W. (1975). Successful pregnancy using frozen semen in the wolf. *Int. Zoo Yearb. 15*, 140-143.

Shoemaker, A. H. (1979). 1977 report on the rare leopard studbooks. *Int. Zoo Yearb. 19*, 286-293.

THE NORTHERN FUR SEAL -- AN EXAMPLE OF COMPLEXITY

Douglas G. Chapman

College of Fisheries
University of Washington
Seattle, Washington

The exploitation of marine mammals by man dates back many thousand of years and perhaps even primitive exploitation led to local extinction of some marine mammal stocks. In recent centuries (particularly the eighteenth and nineteenth) western societies acquired the means to hunt and travel more efficiently. The pursuit of marine mammals became worldwide and was carried to extremes that led to the destruction of many populations. The results have been the same over several oceans, though the available records of exploitation may vary.

Modern biologists are fortunate in respect to the exploitation of the northern fur seal (*Callorhinus ursinus*). Its exploitation, especially the population of the Pribilof Islands in the Bering Sea, has been reasonably well documented. More recently, it has been as carefully studied as any other marine mammal (Kenyon *et al.*, 1954).

The northern fur seal breeds and gives birth on relatively few islands. In addition to the Pribilofs, breeding herds are found on the Commander Islands, the Asian extension of the Aleutian Chain, Robben Island, some islands of the Kurile Chain, and since the 1960's on San Miguel Island off California. There is ample evidence that some degree of intermixing takes place between all the breeding groups so that they are genetically one population, though for management purposes they can be treated as several.

The largest subpopulation is found on the two main Pribilof Islands, St. Paul and St. George, and shows substantial intermixture, particularly at subadult ages (Griben and Gallucci, 1980), though also a large degree of homing (Kenyon, 1960) of females to the rookery of their birth. In this chapter the main

ISBN 0-12-385280-3

discussion is with the St. Paul subpopulation which is about 80% of the Pribilof total. Reference will be made to the others as these cast some light on St. Paul population dynamics.

The history of exploitation can be divided into a small number of discrete periods as follows:

1787-1834	Unmanaged and apparently excessive exploitation by Russian companies on land, though records of actual numbers are rather uncertain.
1835-1867	A very restricted kill followed by a modest level of kill of males which should have allowed the herd to rebuild to close to pre-exploitation population levels. (This assumes that carrying capacity did not change in the 80 years between 1787 and 1867.)
1868-1870	Uncontrolled, excessive exploitation in the interim regime prior to effective U.S. control. The 1868-1869 kill has been estimated at 329,000 (Osgood *et al.*, 1915).
1871-1891	Alaska Commercial Company given a monopoly of land take but restricted to taking males older than one year. During this period commercial pelagic sealing began and by the second decade began to have a significant effect. Total kill these two decades on both islands was 1,981,623 on land and 310,410 at sea, not including many unrecorded losses.
1892-1911	Reduced kills on land and at sea because the herd had been substantially reduced. (Pelagic sealing remained uncontrolled during this period.) The 1911 herd was subsequently estimated at 200,000 to 400,000 seals.
1912-1917	Total protection from commercial exploitation after the signing of a convention by the United States, Great Britain (on behalf of Canada), Japan, and Russia.

1918-1955	Exploitation of surplus males only, with steady rebuilding of the population and steady increase of the kill of young males until about 1943 with stable kill (on average) thereafter.
1956-1963	Reduction of female herd and increased exploitation of males. The latter was accomplished by increasing the size limits and lengthening the season.
1964-1968	Female harvest set at the estimated sustainable yield level with continued fairly heavy male harvest.
1969-1973	Discontinuation of female harvest. Extended size limits still used in kill of males, but season length about the same as in period prior to 1956.
1973-present	St. George Island set aside as a sanctuary; continued male harvest on St. Paul at slightly reduced levels.

The precise season lengths and size limits for the taking of subadult males are given in Engel *et al*. (1980). It is seen that the lengths have varied slightly but frequently. However, what is not known is the exact effect of the length of the season and the effect of variations in size on the escapement of males from the kill, i.e., the numbers that are spared to become potential recruits to the breeding herd. There is on average a five-year period from escape at age five to full recruitment to the breeding herd at age ten. Certainly it is true that following the increase in season length and size limits beginning about 1956, the number of harem and idle bulls peaked in 1961 and thereafter dropped sharply (Table I). The 1961 St. Paul count of both classes was 22,954. In 1971 and 1972 total number of adult males on St. Paul Island was barely over 6,000 but has increased since.

The highly polygamous nature of the fur seal and hence the apparent surplus of males, at least for immediate breeding purposes, is obvious so that most of the management regimes focused on the procedure of harvesting excess young males up to a level that would maintain sufficient breeders. That immature males provided the most desirable skins for luxury furs turned out to be a happy coincidence from the point of view of maximizing revenue to be obtained from the exploitation. What proportion of males was actually surplus for immediate breeding purposes was arrived at by trial and error procedures, if at all.

TABLE I. Counts and Estimates of Harem and Idle Bulls, St. Paul Island, 1921 to 1979

Year	Harem	Idle	Total	Year	Harem	Idle	Total
1921	3237	651	3888	1951	9434	3581	13015
1922	3005	466	3471	1952	9318	4717	14035
1923	2885	287	3172	1953	9848	5912	15760
1924	2950	352	3302	1954	9906	6847	16753
1925	2908	256	3164	1955	9034	8650	17684
1926	3196	334	3530	1956	9384	9016	18400
1927	3596	786	4382	1957	9582	10060	19642
1928	4707	1133	5840	1958	9970	9510	19480
1929	5621	1259	6880	1959	10003	11485	21488
1930	6431	1469	7900	1960	10247	10407	20654
1931	7158	1439	8597	1961	11163	11791	22954
1932	7863	1854	9717	1962	10332	9109	19441
1933	7929	1847	9776	1963	9212	7650	16862
1934	8438	1774	10212	1964	9085	7095	16180
1935	9015	1995	11010	1965	8553	5616	14169
1936	9619	2168	11787	1966	7925	5931	13856
1937	10237	2431	12668	1967[a]	7230	4439	11669
1938	10243	1727	11970	1968[a]	6176	3100	9276
1939	8721	2551	11272	1969[a]	5928	2535	8463
1940	9261	3803	13064	1970	4945	1666	6611
1941	9689	4894	14583	1971[a]	4200	1900	6100
1942	10115	4175	14290	1972[a]	3738	2384	6122
1943	10548	3358	13906	1973[a]	4906	2550	7456
1944	10680	2374	13054	1974[a]	4563	1782	6345
1945	10350	3890	14240	1975	5018	3535	8553
1946	10166	3440	13606	1976	5324	4041	9365
1947	14035	3166	17201	1977	6457	3845	10302
1948	9986	3235	13221	1978	6496	3908	10404
1949	9154	2811	11965	1979	6242	4457	10699
1950	9292	3102	12394				

[a]*Estimated from partial counts.*

No other management goal was given any consideration until the discussions that were held in 1954 and 1955 relative to the renegotiation of the 1911 agreement. This convention had been abrogated by Japan in 1941, which cited among other things damage to Japanese fisheries by fur seals, some of which came from the Pribilofs (though the larger number adjacent to Japan came from Asian breeding grounds). This concern was expressed indirectly in the new convention which the original four countries finally ratified in 1957. The aim of management as formulated in this treaty was to maximize the total yield "taking into account other living marine resources." The precise nature of this condition has never been clearly enunciated.

During the negotiations over the treaty renewal, the idea was expressed that the maximum harvest could be obtained with a kill of some females as well as of the surplus of males. This consideration led in part to the decision to reduce the female herd and, as indicated above, substantial reduction was carried out from 1956 to 1963. The numbers killed in this reduction program were influenced by the estimates of the population from tagging that appear now to have been seriously biased (Chapman, 1964).

It was the expectation of the North Pacific Fur Seal Commission that the reduction in the number of females would be compensated by increases in pregnancy rates or juvenile survival rates or both and that there would result a smaller but more productive herd. This greater productivity has not been realized to date, and the population has shown no measurable response. Pregnancy rates have been monitored through pelagic samples that are complicated because of segregation of the animals by sex and by age class, which vary between areas and months. Comparatively large collections of animals have been taken over the period 1958 to 1974 (and some in 1952), but the situation is not resolved. On the basis of preliminary analysis, Bigg (1979a) states (p. 129), "The trend [of pregnancy rates for females $\geq$ 7 years] within regions is toward a slightly decreasing rate. This was not an expected result from the herd reduction program to produce maximum sustainable yield (MSY)." Another factor bearing on net reproduction is the age of first pregnancy. On this point Bigg (p. 130) notes, "The average age at first pupping varied little between year classes. There appears to be a slight increase in the average from 5.7 years for the 1954 class to 6.0 to 6.2 years for 1956 through 1962 year classes."

The survival rate of immature males is not completely known since there is no satifactory way of estimating the proportion of animals that escape the kill to become potential recruits for the breeding herd. Lander (1975) has shown that because this is small, relative to the total kill, it is possible to obtain an approximate estimate and has given these from 1950 through 1970 year classes in a recent paper (1979). These undicate no significant trend for this time period. Another way of looking at the combined results is through the numbers in the commercial harvest, taking note to recall that from 1956 to the early 1960's the length of the harvest season was extended and the acceptable size limits were also extended, which resulted in a larger proportion of the year class being harvested. The kill by year is given in Table II for 1930 to 1949 and by year class from 1947 to 1975. The vast majority of the kill occurs at years three and older so these are essentially nonoverlapping periods. The maximum average kill occurred during the decade 1940 to 1949 and has been decreasing since that time. While the decline may have now ceased, there is no evidence that there has been any turnaround. The same pattern is displayed by the estimates of the number of pups born (Table III). Interpretation of this table is complicated by the varying methods that have been used to obtain the estimates and by the gap in the records, but again it is clear that there is no evidence for a population response in this variable on either St. Paul Island or St. George Island where all killing terminated at the end of 1972.

TABLE IIA. Kill of Male Seals by Year Class, St. Paul Island, Alaska (in thousands)

Year Class	*Kill*	*Year Class*	*Kill*	*Year Class*	*Kill*
1947	54.7	1957	43.0	1967	39.5
1948	46.3	1958	66.2	1968	30.3
1949	42.3	1959	45.2	1969	27.8
1950	57.2	1960	27.7	1970	34.2
1951	54.9	1961	36.9	1971	26.7
1952	64.6	1962	35.0	1972	25.0
1953	48.1	1963	40.1	1973	25.1
1954	32.5	1964	44.9	1974	26.8
1955	39.5	1965	33.2	1975	24.2
1956	14.9	1966	32.4	1976	27.5[a]
Average	45.5		40.5		28.5?

[a]1981 component estimated.

TABLE IIB. Kill of Male Seals by Year, St. Paul Island, Alaska (in thousands)

Year	Kill	Year	Kill
1930	34.4	1940	51.6
1931	40.0	1941	79.3
1932	39.5	1942[a]	.1
1933	44.4	1943	95.3
1934	43.0	1944	39.8
1935	45.8	1945	58.7
1936	43.5	1946	52.4
1937	44.1	1947	49.8
1938	46.1	1948	58.3
1939	47.6	1949	57.5
Average	42.8		54.3

[a]Harvest essentially eliminated due to military action, though many of these seals that would normally have been taken in 1942 were killed in 1943.

Several possible causes have been advanced for this failure of the population to respond to herd reduction in the first place and its failure to exhibit a growth in either the number of pups born or in the harvest in the twelve years since female killing was terminated altogether. The most obvious possible cause is the large fishery that developed in the Bering Sea and the Gulf of Alaska beginning in the late 1950's. The catch in the eastern Bering Sea alone reached almost 2,000,000 metric tons in 1970 and 1971. To put this in perspective, we note that it has been estimated that annual consumption of fish in the Bering Sea by fur seals is 388,500 tons (Anon., 1980). Estimates of total food consumption (fish plus all other items) in all areas by fur seals is given as 1,500,000 metric tons (McAlister and Perez, 1977). What is not clear from these simple estimates of prey consumption is the effect of the fishery on the energetics of the fur seal community, i.e., the intake of food per unit of energy expended.

The results of this might be reflected in the growth of the seals, but here the picture is also fairly confused. Two preliminary studies have been carried out on the extensive pelagic samples of fur seals that were taken from 1958 to 1974

TABLE III. Estimated Number of Pups Born, St. Paul Island, Alaska, 1950-1979

Year	St. Paul Island	St. George Island	Year	St. Paul Island	St. George Island
1950	451.0[a]	-	1965	253.8	-
1951	447.0[a]	-	1966	298.9	69.4
1952	438.0[a]	-	1967	300.2[b]	-
1953	445.0[a]	-	1968	238.1[b]	-
1954	450.0[a]	-	1969	232.9	-
1955	461.0[a]	-	1970	230.5	54.4
1956	453.0[a]	-	1971	-	-
1957	420.0[a]	-	1972	269.0[b]	67.1[b]
1958	387.0[a]	-	1973	294.0[b]	60.4
1959	335.0[a]	-	1974	266.0[b]	-
1960	320.0[a]	-	1975	278.3	-
1961	284.5	-	1976	294.0	-
1962	227.1	-	1977	-	43.4
1963	262.5	-	1978	-	47.2
1964	283.9	-	1979	245.9	-

[a]From Chapman (1964; estimated by accumulating number of mature females and assuming pregnancy rate equals 0.6).
[b]Estimated from partial samples using ratio estimate with adjacent year counts as base.

by Lander (1979) and Bigg (1979b). Lander notes that many comparisons were made which gave inconsistent results, and concludes, "In these circumstances it is difficult to reach general conclusions concerning relative growth conditions in the different years." Bigg on the other hand concludes from his Fig. 4, showing lengths of pregnant females five and over by age and years of collections (in three periods -- 1958 to 1962, 1963 to 1968, and 1969 to 1974), "the growth rate for five to eleven year olds appears to have gradually increased in recent years." Even if there has been a gradual increase in growth rate, what is not clear is what the increase would have been had there been no human removals of prey fish.

One other piece of confusing evidence has been noted by Gentry (1977). He analyzed the length of time that St. George Island females spend at sea on feeding trips. The average duration of a feeding trip was compared with that of St. Paul Island females observed by Peterson (1968). The average trip duration has not increased and may even have decreased slightly. As Gentry noted, there are difficulties in interpreting this. We note first that two different islands are involved 40 miles apart. Secondly, the observations that Peterson made were after the major part of the female reduction and do not represent a comparison between the female herd at its peak level and its recent low level. Thirdly, the duration of the feeding trip might be genetically determined with the variable factor being the amount of milk produced for which no information is available.

A second factor that has been noted is the possible impact of active salmon gillnets and of floating debris from various types of fishing nets on associated activities. Engel *et al.* (1980) provide a table (82) showing the rate of entanglement of seals taken in the commercial kill since 1967. The rate has averaged 0.0046 in the decade 1970 to 1979. These animals are the survivors of entanglement and provide no clue as to whether other animals become entangled and do not survive. More certain is that some fur seals are captured in and killed by active salmon gillnets operated in the North Pacific by Japanese motherships. Estimates of the catch were made by Fukuhara (1974) as "3,150 to 3,750 animals per year." Jones (1980) has given estimates for both Japanese motherships and land-based high seas salmon fishing for the period 1975 to 1978. These estimates vary from a high of 7,342 in the first of these years to a low of 546 in 1978. While this level of kill clearly has some effect, it cannot account for the recent population dynamics of the Pribilof fur seal herd.

A third factor that is very difficult to quantify but may play a role is human activity. Prior to 1947 the main human disturbance on land was the killing drive on the hauling ground every fifth day together with the once-a-year bull count. In recent years human activity has increased substantially -- the numerous scientific programs involve some disturbance. Also there are many tourists on St. Paul Island and while tourists are restricted, there is undoubtedly some impact on some rookeries. Further, restrictions on native activities were much more severe in the period prior to 1960 than in more recent years. No one has attempted to specifically quantify the level or frequency of disturbance of the seals, and it would be even more difficult to measure what effect this might have on reproductive potential.

Of course it is to be noted that any willful or even inadvertent harassment is now illegal under the United States Marine Mammal Protection Act, unless it is part of activity under scientific permit or as part of the harvest activities which are themselves exempt under the Act. While the Marine Mammal Protection Act established a moratorium on the taking of marine mammals within the United States, the fur seal was specifically exempted since it was covered by the North Pacific Fur Seal Convention. Because of this the fur seal has been less affected by the Marine Mammal Protection Act than any United States controlled marine mammal.

Up to now the fur seal herd has also been generally free of controversy. The latter situation may change. In the treaty negotiations that led to the current North Pacific Fur Seal Convention, the Japanese government expressed concern over the competition of fur seals with its fishery. At that time the United States fishery did not express any particular concern over fur seal competition, but since then United States fishing activities in the Bering Sea and Gulf of Alaska have increased. Further, there is increasing interest in pollock and in the bottom fish resources of those areas where fur seals might be regarded as a serious competitor. Fishery concerns have been expressed about damage to catch and gear and through competition of sea lions and harbor seals. On the other hand, many animal welfare and environmental groups believe that the fur seal harvest is, for a variety of reasons, undesirable. Legislation has been introduced into the United States Congress that would bring about its termination. These different viewpoints are on a collision course; the final outcome is not yet discernible.

In the meantime we can summarize what the extensive harvests and research on fur seals have shown us about marine mammal management. First, it is apparent that we do not yet know the mechanics which control this population. Smith and Polacheck (1981) reach the same conclusion. They state in a conclusion of a paper that examines fur seal life tables, "These analyses strongly suggest that this population must be regulated by changes in at least two and possibly all three of these major components of its life table. This fact may be responsible for the failure of the available data to provide convincing evidence for density-dependent changes and implies that an understanding of the mechanism regulating this population may be difficult to obtain."

As noted, St. George Island was set aside as a sanctuary beginning in 1973 with the aim in part that basic behavioral research could be carried out there in order to arrive at a better understanding of the population. Whether this will lead to determination of the population mechanisms is not certain. In any case it will not be a short-term process. In the meantime the Pribilof Island fur seal studies and harvest results seem to show that population management, at least of marine mammals, is more complex than was thought a generation ago.

REFERENCES

Anon. (1980). Final environmental impact statement on the Interim Convention on Conservation of North Pacific Fur Seals. U. S. Department of Commerce, Washington, D. C.

Bigg, M. A. (1979a). Preliminary comments on pregnancy rate and age at first pupping. *In* "Preliminary Analysis of Pelagic Fur Seal Data Collected by the United States and Canada during 1958-74" Ch. 4. Submitted to the 22nd Annual Meeting of the Standing Scientific Committee, North Pacific Fur Seal Commission, Washington, D. C.

Bigg, M. A. (1979b). Preliminary comments on female growth by length in northern fur seals. *In* "Preliminary Analysis of Pelagic Fur Seal Data Collected by the United States and Canada during 1958-74" Ch. 5. Submitted to the 22nd Annual Meeting of the Standing Scientific Committee, North Pacific Fur Seal Commission, Washington, D. C.

Chapman, D. G. (1964). A critical study of Pribilof fur seal population estimates. *Fish Wildl. Serv. Fish Bull. 63*, 657-669.

Engel, R. M., Lander, R. H., Roppel, A. Y., Kozloff, P., Hartley, J. R., and Keyes, M. C. (1980). Population data, collection procedures, and management of the northern fur seal, *Callorhinus ursinus* of the Pribilof Islands, Alaska. Northwest and Alaska Fisheries Center processed report 80-11. National Marine Fisheries Service, Seattle, Washington.

Fukuhara, F. (1974). "Estimated mortality of seabirds, fur seals and porpoise in Japanese salmon drift net fisheries and sea lions in the Eastern Bering Sea trawl fishery." Unpublished report. Northwest and Alaska Fisheries Center, Seattle, Washington.

Gentry, R. L. (1977). Female feeding cycles, a 12 year comparison. *In* "Fur Seal Investigations, 1976" pp. 34-37. Northwest and Alaska Fisheries Center processed report, Seattle, Washington.

Griben, M. and Gallucci, V. F. (1980). The intermixture of subadult male fur seals (*Callorhinus ursinus*) between the Pribilof Islands of St. George and St. Paul, Alaska and its possible relation to declining harvests on St. Paul. (To be published).

Jones, L. L. (1980). Estimates of the incidental take of northern fur seals in Japanese salmon gillnets in the North Pacific, 1975-79. Submitted to North Pacific Fur Seal Commission, Washington, D. C.

Kenyon, K. W. (1960). Territorial behavior and homing in the Alaska fur seal. *Mammalia 24,* 431-444.

Kenyon, K. W., Scheffer, V. B., and Chapman, D. G. (1954). A population study of the Alaska fur seal herd. *Fish Wildl. Serv. Spec. Sci. Rep. Wildl. 12.*

Lander, R. H. (1975). Method of determining natural mortality in the northern fur seal (*Callorhinus ursinus*) from known pups and kill by age and sex. *J. Fish Res. Board Can. 32,* 2447-2452.

Lander, R. H. (1979). Role of land and ocean mortality in yield of male Alaska fur seal, *Callorhinus ursinus. Fish. Bull. 77,* 311-314.

McAlister, W. B. and Perez, M. A. (1977). A marine mammal ecosystem model for the Bering Sea. Northwest and Alaska Fisheries Center, Marine Mammal Division manuscript report, Seattle, Washington.

Osgood, W. H., Preble, E. A., and Parker, G. H. (1915). The fur seals and other life of the Pribilof Islands, Alaska, in 1914. *Bull. U. S. Bur. Fish. 34,* 172 pp. (Doc. 820).

Peterson, R. S. (1969). Social behavior in pinnipeds with particular reference to the northern fur seal. *In* "The Behavior and Physiology of Pinnipeds" (R. J. Harrison, R. C. Hubbard, R. S. Peterson, C. E. Rice and R. J. Schusterman, eds.), pp. 3-53. Appleton, New York.

Smith, T. and Polacheck, T. (1981). Re-examination of the life table for northern fur seals with implications about population regularity mechanisms. *In* "Population Dynamics of Large Mammals" (C. W. Fowler, ed.), Ch. 5. Wiley, New York.

NOTES ON SOME TOPICS IN THEORETICAL ECOLOGY, IN RELATION TO THE MANAGEMENT OF LOCALLY ABUNDANT POPULATIONS OF MAMMALS

Robert M. May[1]

Biology Department
Princeton University
Princeton, New Jersey

John R. Beddington[2]

International Institute for
Environment and Development
London
England

I. INTRODUCTION

This volume contains many careful presentations of information about specific instances of natural populations of mammals that are locally abundant, along with detailed accounts of various theoretical ideas. Our task at the meeting from which this book sprang was to attempt some kind of review of the pertinent theoretical literature, and, where approriate, to show its bearing on empirical studies.

What follows is, however, more in the form of notes than a comprehensive review. In these notes, most aspects are touched on very sketchily, because they have already been presented in full, for a nonmathematical audience, in *Theoretical Ecology: Principles and Applications* (May, 1981). Some new ideas that might be relevant to understanding the dynamical behavior of confined populations of terrestrial mammals are developed a bit more fully.

[1]*Work supported in part by NSF under grant DEB 81-02783.*
[2]*Work supported by IUCN and World Wildlife Fund.*

ISBN 0-12-385280-3

The chapter begins with an outline of the rich array of dynamical behavior that single populations can exhibit, and the role of thresholds and breakpoints in ecosystems with a multiplicity of stable states. Ways in which the genetic structure of a population needs to be considered in management decisions are noted. The likely importance of spatial patchiness in the distribution of many populations of terrestrial mammals and their food supplies is noted. After a brief survey of approaches to the problem of defining "overabundance," and of deciding what to do about it, the theory of patch dynamics is taken up again, and its possible implications for the management of confined populations are pursued. The chapter concludes with some notes on the need to take explicit cognizance of the way political and economic considerations inevitably interweave with biological ones, clouding many management decisions.

II. INTRINSIC DYNAMICS OF POPULATIONS: TIME LAGS AND OSCILLATIONS

Theoretical studies over the past decade have shown that natural and laboratory populations of animals are capable of manifesting an exceedingly rich spectrum of dynamical behavior. Depending on the values of the biological parameters that characterize their life history strategies, single populations can exhibit a stable equilibrium point, or stable cyclic oscillations, or apparently chaotic fluctuations. Such a range of dynamical behavior is possible for populations with discrete, nonoverlapping generations (corresponding, for example, to univoltine temperate zone insects, and described by first-order difference equations), for the opposite extreme of populations with continuously overlapping generations (corresponding, approximately, to some long-lived mammalian populations, and described by first-order differential-delay equations), and for the intermediate case of populations with many discrete but overlapping generations. These theoretical studies have been successful in explaining either detailed data, or general trends, for specific field and (more commonly) laboratory populations. A general account of this theoretical and empirical work is in May (1981 Ch. 2); for a discussion specific to insects and to whaling, see Hassell *et al*. (1976) and May (1980), respectively.

Of special relevance to the themes of this volume, however, is the observation that many populations exhibit patterns of damped oscillation, or overshoot and overcompensation, as they return to equilibrium after disturbance. Owen-Smith gives an analysis of his white rhinoceros data on this basis. Caughley has given several discussions of the phenomenon in connection with populations of terrestrial mammals -- particularly newly introduced populations of ungulates (Caughley, 1970, 1976, 1977). The overshoot phenomenon can be invoked as a reason for not culling an overabundant population, on the grounds that it will settle back to its equilibrium value if left to itself. Models of single species interactions with a time delay essentially subsume an interaction between a species and its food supply. Caughley's model explicitly considers the interaction between vegetation and a grazing herbivore and produces the phenomenon of the overshoot of the ultimate herbivore equilibrium. A single species version of such a model would be a first order differential-delay equation. More on this follows.

III. THE POSSIBILITY OF MULTIPLE STABLE STATES

There are various mechanisms whereby ecosystems can exhibit more than one alternative stable state. Noy-Meir (1975, and this volume) has proposed specific such mechanisms for populations of herbivores interacting with the vegetation in pastures or on rangelands. For general reviews, including many different examples and mechanisms, see May (1977, 1981 Ch. 5).

In addition to the work presented by Noy-Meir, this volume contains suggestions that two or more alternative stable states arise from the interaction between wildebeest and their woodland/grassland habitats (Walker). On a grand scale, Botkin explains major differences between the herbivore/plant associations in North America and Africa on this basis.[1] Elsewhere, Hilborn and Sinclair (1979) have suggested that multiple stable states in the association between East African ungulates and rinderpest explain major biogeographical features in this region.

Once we admit the ease with which multiple stable states can arise, we can no longer be cheerful about allowing systems autonomously to return to their "pristine equilibrium." There is always the possibility that disturbance has carried a system across the watershed or "breakpoint" dividing one stable configuration from an alternative, so that the "recovery" will be to a wholly new state. Of course, this need not necessarily be the case; but it is a possibility to be kept in mind.

[1]*See Workshop Report.*

IV. GENETICS OF SMALL OR GEOGRAPHICALLY DISTINCT POPULATIONS

The above considerations have to do with the dynamics of populations. Their genetical structure is also important, in at least two ways.

First, local subpopulations of species that are, or once were, widely distributed are likely to have evolved gene complexes that are adapted to their local environment (Williams, 1975; Endler, 1977). Well-intentioned programs that move "excess" animals from a locally abundant population, to replenish a small (and often dwindling) population elsewhere, run the risk of diluting out local adaptations in the latter population. Again, this is not necessarily to decry reallocation programs, but simply to suggest they warrant prior thought (preferably backed by some isoenzyme or other analysis of the possible differences in genetic structure between the two populations in question).

Second, it has recently been shown that the effects of inbreeding depression pose problems for the preservation of small populations of endangered species, in zoos or reserves (Ralls *et al.*, 1979). Although clearly not problems of local overabundance, these problems deserve consideration in this general context. It seems that there will be circumstances when worries about inbreeding in small populations must be weighed against worries about dilution of locally adapted gene complexes by introduction of animals from other places.

V. SPATIAL HETEROGENEITY AND "PATCH DYNAMICS"

A recent theme in theoretical population biology is that spatial heterogeneities in the distribution of prey and predator populations are often of central importance in stabilizing prey/predator interactions. These ideas have been developed independently in several of the different areas encompassed by the general rubric "prey/predator" -- for parasitic infections and their hosts (Crofton, 1971; Anderson and May, 1978; Price, 1980); for insect host/parasitoid systems (for a review, see Hassell, 1978, 1981); for some kinds of plant/herbivore interactions, particularly those involving herbivorous insects

(Lawton and McNeill, 1979; Caughley and Lawton, 1981); for plants and gall-forming insects (Price, 1980); and more abstractly for associations between vertebrate and invertebrate predators and their prey (see, for example, May, 1981 Ch. 5). Most of this work combines theoretical models with extensive collections of field or laboratory data. In particular, many associations between plants and insect pests have been argued to depend on the dynamical effects of spatial patchiness (see, for example, Krebs, 1972 Ch. 17; Huffaker, 1971).

The mechanism common to essentially all these theoretical studies (and to the field and laboratory observations) is as follows. Let the prey be patchily distributed, and let the predators tend to aggregate in regions of relatively high prey density. The regions of low prey density then constitute a kind of implicit refuge, whereby the prey population is maintained; conversely, the predator population flourishes in the regions of relatively high prey density. If there is too much implicit refuge, the prey population will tend to "run away," growing exponentially until controlled by factors other than predation. If there is too little refuge, the system will tend to exhibit the diverging population oscillations (ultimately extinguishing the predator population, or even both prey and predator populations), that characterize most simple and spatially homogeneous prey/predator models.

Several recent studies have elaborated these ideas, considering a hypothetical environment made up of many discrete patches, and showing how the dynamic interplay among empty patches, patches containing only prey, and patches containing both prey and predators could -- under certain circumstances -- lead to overall stability of the system (Maynard Smith, 1974; Hilborn, 1975; Hastings, 1977; Zeigler, 1977; Gurney and Nisbet, 1978; May, 1978).

The classic laboratory study, however, remains that of Huffaker (1958), in which the predatory mite *Typhlodromus occidentalis* sought its prey, the phytophagous mite *Eotetranychus sexmaculatus*, in an experimental universe wherein oranges, variously arranged, constituted the discrete patches. With few oranges, or with easy communication among all the oranges, the system rapidly collapsed; with relatively large numbers of oranges, Huffaker was able to keep both predator and prey populations going for some time. Notice that individual oranges never exemplified steadiness; they went from being empty, to being inhabited by prey, to occupancy by predator and prey, and then (after the predators had devoured their prey) back to empty. But the system as a whole was relatively steady, once the oranges (patches) were sufficiently numerous, and, in theory, may have been quite steady if the number of oranges had been much larger than was ever the case experimentally.

Notice that all the above work focuses on organisms where the spatial and temporal scales of the study are relatively small. We speculate, however, that the same dynamic processes may be at work in many of the associations between plants and large mammalian herbivores in Africa. That is, we suggest the essential mechanism stabilizing the interaction between vegetation and, for example, the elephants or rhinoceros eating it may be the patchy aggregation of herbivores, and the shifting, patchy patterns of subsequently recovering vegetation.

If there is truth in this speculation, it incidentally follows that the size of the region needed for this system stably to persist is not easily guessed. It could be vast. The number of Huffaker's oranges needed for his system of prey and predator mites to be maintained is clearly very large, and not easily deduced from the "trivial movements" (*sensu* Southwood, 1978) of the mites; by analogy, the size of the region needed for the vegetation/elephant association to maintain itself (on the basis of this conjecture) bears no intuitive relation whatsoever to the "trivial movements" of elephant groups, or to the size of local patches of destruction (analogous to individual oranges). We plan to develop these ideas more fully elsewhere.

The interplay between spatial heterogeneity and population dynamics is, of course, ineffably more obvious when the movement is in the form of large-scale annual migrations, rather than more-or-less random wanderings from "orange to orange." Thus the role of spatial and seasonal heterogeneity in the overall dynamics of the Serengeti ecosystem is plain to see.

VI. WHAT IS OVERABUNDANCE, AND WHAT TO DO ABOUT IT?

Elsewhere in this volume, various schemes for defining and classifying "overabundance" have been set out. Thus Cobb offers a classification according to biological, economic, and social factors; Sinclair a classification based on the types of dynamical behavior; Laws a scheme related to the metaphor of r- and K-selection; and Caughley a 4-class scheme that blends biology with social and economic factors. Botkin attempts a rapporteur's appraisal of these definitions of overabundance, and of the recommended courses of action.[1]

[1]*See Workshop Report.*

To our way of thinking, one major dichotomy supervenes above all these classificatory schemes, and overrode much of the discussion at the workshop. This dichotomy is between the view (which could be called "East African") that, to a first approximation, the best approach to management is to leave things alone, and the view (which could be called "South African") that purposive management, sometimes including culling, is usually required in parks and reserves.

In favor of the laissez faire view is the observation that many of our problems first arose from interfering with natural processes, and the fact that all sorts of economic and political pressures are more easily kept at bay by such a policy. But the possible existence of multiple stable states means that the disturbed system, if left alone to follow its natural trajectory, may end up as a disaster; given that for many parks and refuge areas the future circumstances are very different from those of the past, there is no reason to believe the "equilibrium" to which the system is tending bears any simple relation to its pristine or "natural" equilibrium state.

If a philosophy of active management is adopted, it should be deliberately designed to yield the maximum possible amount of information about the system, so that future management can be more scientifically based. This point is forcefully made by Sinclair, and is brought out in the discussions in this volume; Holt has been urging it on the International Whaling Commission for years. To this end, one particular strategy is to use different (but coherently integrated) management rules in different subareas of a park, or in different parks, to facilitate comparisons.

VII. PATCH DYNAMICS AND MANAGEMENT

Of the many kinds of "overabundance" alluded to above, many clearly hinge on economic and social considerations that have little or nothing to do with biological "overabundance." At the heart of this volume, however, are the poignant problems that arise when populations appear to become (by one or other definition) too abundant in areas set aside explicitly for their preservation and/or protection.

Why do these populations become locally overabundant? If we carve out an island refuge, and give its environs over to the uses of man, why should the population dynamics in this "part of the main" be different from what it previously was?

In some circumstances, the answer is, of course, trivial. It can be that animals flooding in from the surrounding countryside, or other kinds of gross or subtle disturbance, cause our park or refuge area to be quite unlike its pristine self. Although clearly debatable, this is arguably the case in several East African parks.

Less trivial is the case of many of the fenced South African reserves. Here it can be argued that they should, by design, be like small pieces of the original whole. Whence, then, the phenomenon of overabundance, as so clearly demonstrated for Owen-Smith's rhinoceros?

One possible answer lies in our speculations about patch dynamics, outlined above. It could well be that the reserves are too small for the dynamic interplay among patches to create overall stability; there are, as it were, not enough oranges. This could be so even for a very large reserve (remember, there is no easily-intuitable relation between the "trivial movements" of the animals and the size of the region needed for patch dynamics to stabilize the system).

If the explanation lies along these lines, several things follow.

The idea of culling those animals entering "vacuum zones" needs reappraisal. If the finite size of the reserve is the problem, substituting Dirichlet boundary conditions (populations equal to zero at the boundary) for Neumann boundary conditions (reflecting walls) will not usually solve it! Such a killing in "vacuum zones" could fail to halt the abundance problem, or could solve it, or could gradually extinguish the population in question; there simply is no direct relation between the policy and the conjectured "patch dynamics" mechanism producing the population expansion. This is not to condemn the "vacuum zone" idea, but rather to call for its reexamination in the light of this and other ideas about what is causing the population increase.

For mammals, these ideas about patch dynamics are pretty much peculiar to large herbivores. Marine mammals are unlikely to present analogous problems; restriction to a few breeding sites may produce great crowding there, but there is every indication that this is natural, and their food supply is not endangered thereby -- it can be, by fishing it out from under the marine mammals, but that is a different problem (May *et al.*, 1979). For many bird species, and for many mammalian carnivores, population density is regulated by territorial mechanisms (see Krebs, 1972); we would thus expect problems of local overabundance in reserves rarely to arise for these animals, and this is the case. (Indeed, for birds one essentially never hears of such problems, and their conservation literature focuses almost exclusively on species/area relations [Soulé and Wilcox, 1980; Diamond and May, 1981].)

VIII. POLITICAL AND ECONOMIC CONSIDERATIONS

As mentioned above, decisions about culling locally abundant mammals can in some instances be colored by economic motives masquerading as biological concerns. The case of the vicuna, discussed in this volume by Otte and Hofmann and by Eltringham and Jordan, is an example: The culling program has often been represented as necessitated by local overabundance; but Otte and Hofmann plainly state the basic reason for the program is to generate sustainable harvests that may act as an economic incentive for the indigenous peoples not to extinguish the vicuna. This, indeed, may well be the best available compromise in this and other situations, but it would be better to avoid misrepresenting economic and political realities as being biological necessities.

More generally, it seems desirable to give explicit recognition to the economic, political and social factors that almost invariably swirl together with population biology in the conservation or management of natural populations. Clark (1976, 1981) has given an incisive account of the essential dynamical features of some of these complications. In particular, Clark emphasizes that economic considerations tend to discount the value of future harvests, creating a tendency to overexploit harvested populations rather than to maximize the sustainable biological yield; this can happen even when there is a sole owner (so that "tragedy of the commons," open access effects are eliminated). These insights help explain why many agencies (such as the International Whaling Commission), in principle dedicated to preserving stocks, in fact tend to overexploit them.

In short, virtually all conservation problems -- including those of most locally abundant wild mammals -- involve political and economic complications along with purely biological ones. For this reason, it is usually helpful to spell out, very fully and explicitly, the main purpose that a given park or reserve is meant to serve.

REFERENCES

Anderson, R. M. and May, R. M. (1978). Regulation and stability of host-parasite population interactions: I, regulatory processes. *J. Anim. Ecol. 47*, 219-247.

Caughley, G. (1970). Eruption of ungulate populations, with emphasis on Himalayan tahr in New Zealand. *Ecology 51*, 53-72.

Caughley, G. (1976). Wildlife management and the dynamics of ungulate populations. *In* "Applied Biology" Vol. I. (T. H. Coaker, ed.), pp. 183-246. Academic Press, New York.

Caughley, G. (1977). "Analysis of Vertebrate Populations." Wiley, New York.

Caughley, G. and Lawton, J. H. (1981). Plant-herbivore systems. *In* "Theoretical Ecology: Principles and Applications, Second Edition" (R. M. May, ed.), Ch. 7. Blackwell, Oxford and Sinauer, Sunderland, Massachusetts.

Clark, C. W. (1976). "Mathematical Bioeconomics." Wiley, New York.

Clark, C. W. (1981). Bioeconomics. *In* "Theoretical Ecology: Principles and Applications, Second Edition" (R. M. May, ed.), Ch. 16. Blackwell, Oxford and Sinauer, Sunderland, Massachusetts.

Crofton, H. D. (1971). A model of host-parasite relationships. *Paras. 63,* 343-364.

Diamond, J. M. and May, R. M. (1981). Island biogeography and the design of natural reserves. *In* "Theoretical Ecology: Principles and Applications, Second Edition" (R. M. May, ed.), Ch. 10. Blackwell, Oxford and Sinauer, Sunderland, Massachusetts.

Endler, J. A. (1977). "Geographic Variation, Speciation and Clines." Princeton Univ. Press, Princeton, New Jersey.

Gurney, W. S. C. and Nisbet, R. M. (1978). Predator-prey fluctuations in patchy environments. *J. Anim. Ecol. 47,* 85-102.

Hassell, M. P. (1978). "The Dynamics of Arthropod Predator-Prey Associations." Princeton Univ. Press, Princeton, New Jersey.

Hassell, M. P. (1981). Arthropod predator-prey systems. *In* "Theoretical Ecology: Principles and Applications, Second Edition." (R. M. May, ed.), Ch. 6. Blackwell, Oxford and Sinauer, Sunderland, Massachusetts.

Hassell, M. P., Lawton, J. H., and May, R. M. (1976). Patterns of dynamical behaviour in single-species populations. *J. Anim. Ecol. 45,* 471-486.

Hastings, A. (1977). Spatial heterogeneity and the stability of predator-prey systems. *Theor. Popul. Biol. 12,* 37-48.

Hilborn, R. (1975). The effect of spatial heterogeneity on the persistence of predator-prey interactions. *Theor. Popul. Biol. 8,* 346-355.

Hilborn, R. and Sinclair, A. R. E. (1979). A simulation of the wildebeest population, other ungulates, and their predators. *In* "Serengeti: Dynamics of an Ecosystem" (A. R. E. Sinclair and M. Norton-Griffiths, eds.), pp. 287-309. Univ. of Chicago Press, Chicago, Illinois.

Huffaker, C. B. (1958). Experimental studies on predation: dispersion factors and predator-prey oscillations. *Hilgardia 27,* 343-383.
Huffaker, C. B., ed. (1971). "Biological Control." Plenum Press, New York.
Krebs, C. J. (1972). "Ecology: The Experimental Analysis of Distribution and Abundance." Harper and Row, New York.
Lawton, J. H., and McNeill, S. (1979). Between the devil and the deep blue sea: on the problem of being a herbivore. *In* "Population Dynamics" (R. M. Anderson, B. D. Turner, and L. R. Taylor, eds.), pp. 223-244. Blackwell, Oxford.
May, R. M. (1977). Thresholds and breakpoints: ecosystems with a multiplicity of stable states. *Nature (London) 269,* 471-477.
May, R. M. (1978). Host-parasitoid systems in patchy environments: a phenomenological model. *J. Anim. Ecol. 47,* 833-843
May, R. M. (1980). Mathematical models in whaling and fisheries management. *In* "Some Mathematical Questions in Biology" Volume 13 (G. F. Oster, ed.), pp. 1-64. American Mathematical Society, Providence, Rhode Island.
May, R. M., ed. (1981). "Theoretical Ecology: Principles and Applications, Second Edition." Blackwell, Oxford and Sinauer, Sunderland, Massachusetts.
May, R. M., Beddington, J. R., Clark, C. W., Holt, S. J., and Laws, R. M. (1979). Management of multispecies fisheries. *Science 205,* 267-277.
Maynard Smith, J. (1974). "Models in Ecology." Cambridge University Press, Cambridge.
Noy-Meir, I. (1975). Stability of grazing systems: an application of predator-prey graphs. *J. Ecol. 63,* 459-481
Price, P.W. (1980). "Evolutionary Biology of Parasites." Princeton University Press, Princeton, New Jersey.
Ralls, K., Brugger, K., and Ballou, J. (1979). Inbreeding and juvenile mortality in small populations of ungulates. *Science 206,* 1101-1103
Soulé, M. E. and Wilcox, B. A., eds. (1980). "Conservation Biology: An Evolutionary-Ecological Perspective." Blackwell, Oxford and Sinauer, Sunderland, Massachusetts.
Southwood, T. R. E. (1978). The components of diversity. *In* "Diversity of Insect Faunas" (L. A. Mound and N. Waloff, eds.), pp. 19-40. Blackwell, Oxford.
Williams, G. C. (1975). "Sex and Evolution." Princeton University Press, Princeton, New Jersey.

Zeigler, B. D. (1977). Persistence and patchiness of predator-prey systems induced by discrete event population exchange mechanisms. *J. Theor. Biol.* 67, 687-713.

LARGE MAMMAL FEEDING STRATEGIES AND RELATED OVERABUNDANCE PROBLEMS

Richard M. Laws

British Antarctic Survey
Natural Environment Research Council
Cambridge
England

Animals are adapted to the environment they inhabit and one type of adaption is the adjustment of their size, lifespan, annual cycle, and other traits to characteristic features of their food supply. These aspects of their ecology are intimately related, through time scales and recovery rates, to practical considerations of management by culling or other methods. Laws (1981) gave a few examples of relationships between large and very large mammals and their food bases. These ideas are developed in the present chapter in the context of overpopulation, a few selected case studies are given illustrating some contrasting relationships, and other relevant biological and management considerations are touched on. I have chosen to set these in the framework provided by the concept of *r*- and *K*-selection (Macarthur and Wilson, 1976; Pianka, 1970).

I. A POSSIBLE CLASSIFICATION

A first step in any approach to local overabundance problems is to make an appreciation of the animal's broad population dynamics strategy. This might take the form of a deliberately oversimplified classification in terms of *r*- and *K*-selection (May, 1978), since rational conservation management should be geared to the range of demographic properties broadly described by this concept. No animals are thought to be completely *r*- or *K*-selected but the terms are useful in a comparative sense.

PROBLEMS IN MANAGEMENT OF LOCALLY
ABUNDANT WILD MAMMALS

ISBN 0-12-385280-3

There is a continuum that encompasses small mammals such as rodents and shrews at one extreme, through intermediate forms to elephants and whales at the other extreme. Usually only the larger animals are identified with overpopulation problems and they all tend toward the *K*-strategy end of the spectrum.

No less important than the animal's general bionomic strategy is that of its principal food resource and it is useful to extend the classification in terms of *r*- and *K*-selection to plants. The range stretches on land from terrestrial algae, through annual grasses and herbs, perennials, bushes, woodland trees to forest giants such as the mahoganies, living longer than any mammal. In aquatic habitats it encompasses phytoplankton (supreme opportunists), benthic algae, seaweeds, sea grasses and the giant kelp.

There are difficulties in the application of the theory of *r*- and *K*-selection (Pianka, 1970) but nonetheless it is useful -- we all know what is broadly meant by an *r*-selected species as compared with a *K*-selected species (Wilson, 1975). These terms are used below within the range of examples to imply relative tendency to *r*- or *K*-selection.

In Table I are set out some examples of these relationships. The examples chosen range upwards from large mammals, such as deer and seals. Overpopulation of smaller mammals seem to be less of a problem. Examples of their possible local overabundance, in the sense of the present discussion, that immediately spring to mind include lemmings, rabbits, coypu, and beavers. No doubt other examples could be found. (Conservation problems involving rabbits in England, particularly the maintenance of closely-grazed grasslands, often tend to be due to underabundance, following myxomatosis.) I am not aware of good examples of problems posed by lemmings; rabbits in Australia, coypu in England, and beavers in Tierra del Fuego pose a different set of problems to those considered here.

Terrestrial predators are not incorporated either, nor are small prey species. These sets would range from insects/insect eaters; through worms/shrews; small mammals and birds/foxes, jackals; to moose/wolves; antelopes, buffalo/hyenas, large felids. It is for consideration whether smaller mammals and terrestrial predators need to be incorporated in any theoretical approach that might be developed.

Table I provides a classification within which to discuss case studies and I propose briefly to discuss three examples. Other examples are discussed by other contributors in this volume and there is an extensive literature for some of them.

I am conscious that this framework is only one of several possible ones. It omits consideration of (among other factors) soils, climate, and nutrient pools, but I suggest that these variables could be superimposed on the basic plant/animal or prey/predator relationships.

Nutrient cycling is a factor in arguments advanced by opponents of culling. While it cannot be ignored, discussion about it is likely to be sterile in the absence of quantitative data on rates of change due to natural processes as compared with man-made perturbations. Some data exist for specific areas (e.g., Weir, 1972). More research is needed to fill in gaps and to substantiate statements about the effect of translocations, or culling and removal of carcasses. It is likely that in many situations natural rainfall run-off and soil erosion would be more significant. Again, in culling to remove the effects of compression it has been argued that imported minerals and nutrients were being removed. Hippopotamus, which graze on land and defecate and die largely in the water, are an example of a natural one-way transport of minerals, nutrients, and energy from a terrestrial system; the seasonal migrations of whales (Laws, 1977), seals, and caribou provide others.

K-selected species of animals are adapted to stable environments and have evolved rather subtle and more delicate mechanisms to respond to ecosystem changes than have the more opportunistic *r*-selected species. When out of balance they can be expected to influence ecosystems in ways that *r*-selected species probably do not. Their role in structuring ecosystems may be great, while *r*-selected species, being opportunists, would be expected to have less influence on the structure of mature ecosystems (Ray, in press). Because of their more refined regulatory mechanisms they are less likely to overshoot carrying capacity than *r*-selected species. But when they do, the effect will be sustained and possibly catastrophic. Three examples illustrate this point -- the overpopulation of hippopotamus and elephant, and a somewhat different example of the crash of an *r*-selected species (caribou) overeating a slow growing, long-lived food resource (lichen) is also outlined.

II. SELECTED CASE STUDIES

A. Elephant

Much has been written about "the elephant problem" and several explanations have been advanced for elephant/habitat interactions. They may be briefly summarized as:

Natural systems:

1. Climatic cycles with 50-year period (Phillipson, 1975).
2. Stable limit cycles, density-dependent with c. 200-year period (Caughley, 1976).

TABLE I. *Some Examples of Suggested Relationships between Sizes and Longevity of Large Mammals and Their Food Resources* [a]

Environment	Terrestrial	Aquatic	Terrestrial	Aquatic
Controlling food resources	Small, short-lived units High productivity Low unit biomass (r-selected)		Large, long-lived units Low productivity High unit biomass (K-selected)	
	Grass Herbs Softwood	Zooplankton Sea grass	Trees Bushes Lichens[b]	Large inver-tebrates -Benthic -Pelagic Macroalgae
Life span (approximate)	years	days/years	decades/ centuries	years/ decades
Large mammals (r- and K-selected)	Antelopes -Wildebeest -Impala -Uganda Kob Deer -Wapiti	Crabeater seals Ringed seals Antarctic fur seals Minke whales	Antelopes -Kudu -Lesser kudu -Gerenuk -Nyala	Dolphins Seals -Weddell seals -Leopard seals -Grey seals -Elephant seals

	-Red deer *-Reindeer* *Vicuna* *Buffalo* *Bison* *Tahr* *Zebra* *Feral horses*		*Reindeer*	*-Northern fur seals* *Walrus* *Sea otters*
Life span (approximate range)	*15-30 years*	*20-40 years*	*15-20 years*	*20-40 years*
Very large mammals (K-selected)	*Hippopotamus* *White rhino* *Moose*	*Baleen whales* *Dugong* *Manatee*	*Elephant* *Giraffe* *Black rhino*	*Sperm whales* *Killer whales* *Steller's sea lions*
Life span (approximate range)	*30-50 years*	*40-100 years*	*40-70 years*	*30-80 years*

[a]*In practice many species have diets which embrace more than one type of food resource; their placings represent my assessment of the key interactions.*

[b]*Lichens are placed in this column because their most significant features in the present context (suggesting K-selection) are their longevity and low productivity.*

Systems influenced by man:

3. Compression or displacement aggravated in arid regions by climatic cycles of c. 10-year period (Laws, 1969, 1970).

4. Eruptive increase in recovery phase (Hall-Martin, 1980).

5. Combination of 3 and 4 above.

With the exception of 1 the predicted changes are expected to be density dependent and related to food resources. In all these situations (except 4 where knowledge is insufficient for us to say whether theoretically an equilibrium might be achieved) marked long-term habitat changes are involved. Their nature depends on the soils and rainfall regime as does the feedback effect on the elephant population dynamics and the populations of associated fauna (Laws, 1970; Laws *et al.*, 1975).

Explanations 2 and 4 may be applied in the pristine state but there seems to be little good evidence for 1. The current situations covered by 3, 4, and 5 have been caused by human interference in the past 50 to 100 years and presumably would not have been possible earlier, when human populations were much smaller.

Hypotheses 1 and 2, if they can be confirmed, imply that there can be no stable equilibrium between elephants and trees in a given locality. However, in a natural state before the explosion of human populations, over a large enough area, cycles of types 1 and 2 could have been out of phase in local sub-populations and so the total regional population of elephants and the total forest/bush cover might well have been relatively constant for long periods. Whichever of the five explanations advanced above (or possibly others) are correct, in the pristine state Africa could be regarded as a mosaic of habitat types, influenced by grazing and browsing (particularly by elephants).

Thus, in precolonial days before the rapid increase of the human population, there was probably a dynamic mosaic of habitats, each piece of the mosaic undergoing changes in human or elephant densities, with consequent changes in the habitat. Overall the population densities of man and elephants in a region, and the total vegetation resources on which they depended, were probably broadly stable for centuries at a time. But in any particular place, the environment went through a cycle of woody vegetation -- subsistence agriculture (i.e., clearings) -- bush or woodland -- elephants increasing (promoting more open habitats) -- grassland/domestic stock -- eleplant decrease -- bush or woodland -- human cultivation and so on, as land use by man and elephants changed.

As the human populations increased and modern settled farming replaced shifting cultivation and pastoralism, so individual components of the mosiac became trapped in one phase of the cycle. Elephants too were restricted and unable to move out of areas that had become unsuited to their needs. The complementary distribution of man and elephant populations changed from one characterized by human islands in a sea of elephants to increasingly smaller islands of elephants in a sea of people. These islands have increasingly centered on national parks and reserves. In some of them management is now necessary.

The extreme examples, and very interesting too, are the tiny relict elephant populations in South Africa. The Knysna Forest elephants have remained stable at only 10 to 14 animals since 1920; the Addo population, reduced to a nucleus of only 11 animals in 1931, fenced in and protected from 1954, has been increasing since then at 7% per year to 102 animals in 1979 (Hall-Martin, 1980), the greatest rate of increase known for any elephant population. These two small herds show that, given the right conditions, elephants can survive near extinction for decades and the Addo population shows they can increase again, but they are unique examples and their numbers are still perilously low. Most others have continued towards extinction.

The elephant can be considered to be an extreme *K*-selected species and while it exhibits a wide range of regulatory adjustments to declining habitat resources (lengthened calving interval, deferred maturity, increased calf mortality), these have not usually, if at all, been sufficient to compensate for the compression (increased population density), habitat change, and consequent changes in carrying capacity. The *K*-selected elephant exhibits an inertia of the adult age classes; even with complete elimination of births the biomass locked up in the population persists in the area until mortality or emigration removes it. Elephants continue to feed, and change the habitat, at a rate influenced by the short-term rainfall cycles.

There seems to be no evidence of overgrazing by elephant; the problem is usually one of overbrowsing of bushes and trees. Here the crucial factor is that these food resources are *K*-selected (relative to grass) and their replacement is slow, even in the absence of browsers. The previously mentioned inertia of the adult component of elephant biomass means that feeding pressure on bushes and trees is sustained for years. This effect is enhanced in drier areas, in dry seasons, and in drought years. The proportion of browse required in the diet is greater in dry situations (Field, 1971) -- the grass is not nutritious in these conditions. Elephants in semi-arid environments, or worse, may be totally dependent on browse for their food over several months of the year. The impact of a slowly diminishing elephant biomass (e.g., 2% annually) on a more rapidly decreasing standing stock of trees (e.g., 6% annually) would progressively increase. In wetter periods the rate of change may be temporarily slowed or reversed, but would continue at an accelerated pace in subsequent droughts.

Several case studies show that, in both wet and dry habitats, where there is elephant overpopulation destruction of trees proceeds as a zone of damage on a front radially outwards from the sources of surface water (lakes, rivers, waterholes -- natural and artificial) (e.g., Laws *et al.*, 1975).

Although the elephant shows a great plasticity in reproductive response, a typical *r*-strategist would adjust much more rapidly. For example, in an antelope first breeding annually on average from age two years, puberty can quickly be deferred to age three years in response to habitat or climate change, with massive effect on birth rates, and vice versa. In an animal first breeding at twelve years and with a calving interval of three years, a large effect on birthrate is not immediately apparent. As the *K* qualities have been selected for in evolution, it seems probable that the present habitat changes evident in some areas, and consequent effects on the elephant populations (high calf mortality, inadequate adjustments of age at puberty, and calving intervals) are unnatural. In such situations management aimed at reducing population density is indicated.

Of the alternatives available, increasing the range is not politically feasible, translocation is not practically feasible, and we seem to be left with culling as a management tool. Laws (1969) proposed that management should be directed to unit populations or subpopulations. Once identified these could be treated in different ways as the circumstances require. However, the complex social behavior of elephants means that lacunae created by cropping tend not to be rapidly filled by immigration (Laws *et al.*, 1975), in contrast to experience with hippopotamus. The social behavior does, however, make possible the elimination of complete family units, bull herds, or clans, with a minimum of disturbance to the remaining population.

B. Hippopotamus

Similar hypotheses can be advanced to explain hippopotamus population changes. Again natural predation is a negligible factor and population regulation is likely to be mediated by food supply. I studied hippo in the Queen Elizabeth National Park in Western Uganda, now the Rwenzori National Park. It had the highest known biomass density of large mammals in the world at 21,000 kg/km^2, compared with 19,000 kg/km^2 for the next highest, the very small Lake Manyara National Park in Tanzania (Leuthold and Leuthold, 1976). I chose to work on hippopotamus because it provided a very good point of entry into a complex tropical ecosystem. American Fulbright scholars, notably Petrides, Swank, and Longhurst, as a result of earlier work, had recommended a management cropping scheme to reduce overgrazing. The cropping program began in 1957 and provided opportunities to collect quantitative information on age structures, growth, reproduction, nutrition, and disease. The hippo is central in the food web, and it was thought to be overabundant. The chosen approach was to study its ecological relations and the dynamics of its populations and try to determine the optimum grazing density for this area. This had both academic and practical value.

The hippo spends the day kneeling or lying in the water with little expenditure of energy and at night follows trails, marked by fecal deposits, inland to close-grazed areas of short grasses -- the hippo lawns. It feeds by plucking grass with its massive lips, as much as 50 cm wide. The grazing area extended about 3.5 km from the shore on average except where there were wallows that enabled it to penetrate further inland and enlarge the range. The highest density populations in the world were found in the lakes and rivers of the western Rift Valley. Maximum grazing densities were formerly over 31/km^2, equivalent to a stocking rate of 62 steers/km^2, which is very high indeed. They created an energy and nutrient sink from land to water, removing organic matter by grazing and defecating in the water thus fertilizing the lakes (Laws, 1968a).

An experimental management scheme, aimed at maintaining a range of grazing densities by regular cropping and monitoring by counts, was set up. The maintained grazing densities ranged from near zero to 23/km^2 and changes in the habitats were followed by recording vegetation transects, quadrats, and exclosures to see how the grasslands were influenced by different hippo grazing pressures and by counts of other large mammals. The reduction of the hippo populations was accompanied by analysis of the crop for population parameters (age at maturity, pregnancy rate, growth rate, and mortality rate) as well as nutrition and disease. About 1,000 hippopotamus annually were shot in the water, then hauled out on land and examined, as local butchers, to whom the carcasses were sold, cut them up. A method of age determination was developed and full data of high quality obtained from the carcasses (Laws, 1968b). The population size was monitored by aerial and boat counts. Unfortunately, unlike elephants, the method of cropping was very biased because the larger animals are easier to shoot, but found jaws from natural deaths provided an independent indication of the population dynamics. The life span of the hippopotamus is 45 to 50 years.

An instructive example of one of the experimental areas was the Mweya Peninsula -- 4.4 km^2 -- where in 1957 there had been 90 hippopotamus (21.4/km^2). At that time, it was largely bare ground except for some bush thickets. By 1963, following the removal of hippo (to 1/km^2), the grass had come back and erosion was halted; at the same time the numbers of other grazing species had increased substantially (from 40 to 179) and the total standing stock animal biomass had also increased by 7.7%. Buffalo numbers in their population growth closely followed a logistic curve and increased from 4.6 to 27.8/km^2. The changes up to 1967 are discussed by Laws (1968a), Field and Laws (1970), and Eltringham (1974). With the cessation of management, hippopotamus numbers built up again to 25.3/km^2 by 1973; buffalo, having reached a density of 40.4/km^2 in 1968, declined to 22.8/km^2 in 1973. Elephant, waterbuck, and warthog numbers also changed and vegetation changes were monitored.

The initial experimental maintenance of hippo densities only lasted for about five years but established that in the Queen Elizabeth Park in the climate then current the optimal grazing density for hippopotamus was about 8/km^2 if vegetation and mammal diversities were to be maintained. As the cropping reduced the overall population size, the mean age of sexual maturity declined from 12.1 years to 9.7 years, the pregnancy rate increased slightly, and the proportion of calves increased from 5.9% to 14.0% (Laws, 1968a; Laws and Clough, 1965).

Studies in Murchison Falls National Park (now Kabalega Falls National Park) begun in 1966, showed that a hippo population of about 19/km^2 on average was associated with overgrazing and substantial erosion along the banks of the Nile River. A similar cropping scheme was instituted there which also produced a mass of data on the population. Some 4,000 hippopotamus were taken on land at night, because the strong currents ruled out shooting in the river.

Unfortunately, owing to the identification in 1964 of more pressing problems related to elephant overpopulation (described in the previous section) only part of the results of the hippopotamus studies have been published. These studies demonstrated that management by cropping could rapidly reverse the habitat change. Grasses, which are *r*-selected, respond much more quickly than *K*-selected bush or trees to reduction in grazing or browsing pressure.

The hippopotamus studies also showed that the creation of vacuum areas as recommended by Owen-Smith (this volume) for white rhinoceros encourage immigration, so that cropping could be confined to selected localities. Cropping was by shooting animals in the water; complete schools could not be taken and the social behavior was disrupted. The disturbance due to cropping was, therefore, more severe than in the case of the elephant, which lend themselves to removal of complete social groups without disturbance to the remaining population. In hippo areas where heavy sustained cropping was practiced this disturbance was manifested by the formation of abnormally large schools -- numbering 150 or more. This can be compared with the similar effect of selective shooting on elephant populations -- the formation of herds of over 1,000 animals due to breakdown of the normal social structures. Unfortunately, the progress of events in the Queen Elizabeth (Rwenzori) National Park also demonstrated another requirement, that management areas should be integrated units; the cropping program created partial vacuum areas that appeared to attract hippopotamus from the neighboring Albert (Kivu) National Park, where no management was practiced.

C. *Reindeer/Caribou*

A third example, reindeer or caribou, drawing on the work of Klein (1968) and Leader-Williams (1980) illustrates a species that is more *r*-selected and that, in some irruptive situations where it was dependent on a *K*-selected food resource (lichen) and not subject to predation, has become extinct. Introduced to islands, the numbers increase initially in response to the discrepancy between the carrying capacity of the environment and the present population. Where lichens were the principal food resource (e.g., St. Matthew Island) they were initially abundant but were rapidly eaten out, and because they recover slowly were not replaced. A rapid population crash of caribou ensued. In other islands, where vascular plant resources exist, these were less subject to overgrazing and the reindeer, or caribou, initially overshot the long-term carrying capacity, but then came into equilibrium with the habitat at a lower level. Introduced reindeer feeding on the subantarctic tussock grass (*Poa flabellata*) at South Georgia provides a very clear example (Leader-Williams, 1980).

III. OTHER BIOLOGICAL CONSIDERATIONS

The importance of man probably overrides all else in terms of direct (poaching, crop protection, etc.) and indirect (competition for space) conflicts. These conflicts tend to remove the opportunity for dispersion and migration and to bring populations in the different cells of a habitat mosaic, previously expected to be out of phase, into phase, when they may concurrently become trapped in the troughs of population fluctuations. The smaller the range occupied by the animal, the more severe the effect to be expected, and the greater chance of irreversible changes. Systems can reach an equilibrium in large enough areas but not in too small areas. It is, therefore, better to maintain a few large areas than many small ones of the same cumulative size. Island biogeographical theory is relevant and determines species survival (see Soulé *et al.*, 1979).

The social and other behavior of the animal in question is important -- whether it is organized as solitary, in small groups, harems or family units, or in large herds; whether it is resident, territorial, or migratory.

Limiting factors are not necessarily water or energy, but may be vegetation type, soils, minerals, nitrogen, or other chemicals.

IV. MANAGEMENT CONSIDERATIONS

Definition of policy is all important. The objectives for a conservation area need to be stated (for example, "to maintain populations of stated species," "to maintain existing habitats," or "to maintain ecological diversity" -- though expanded in greater detail) and sustained. A laissez faire policy or one which allows objectives to be changed too readily in response to events may be inadequate because there is a tendency to take the easy way rather than the course necessary to meet the original objective. Overpopulation may be defined indirectly with respect to permissible limits of change in habitat types, population number, or some other criteria.

The important thing is to stick to the objective unless there is a positive, well-substantiated cause to change, otherwise conservation and management aims can rapidly be lost sight of.

The time scale for management response is an important consideration. There is a need to develop pragmatic rules, otherwise acceptable research findings may come too late to resolve the problem, and trends may become irreversible in some cases.

It is necessary to balance theory with practice. For example, Fowler and Smith (1973) proposed selected culling of elephant family units with suitable age distributions that would permit a stable age distribution to be achieved. Laws *et al.* (1975) tested this on their Kabalega elephant data and found that, while this might be possible for bull herds, selective cropping of family units was not practicable as a means of altering the female age structure in the direction of a stable age distribution.

Where possible one should avoid disruption to other forms of land use (especially tourism) by the designation and use of buffer zones (Wheater, 1968) or vacuum areas (Owen-Smith, this volume) for culling. This also minimizes interference with the behavior and social organization of the managed species.

Subpopulations or unit populations or clans should be identified and managed in compartments (as proposed by Laws [1969] for the Tsavo ecological unit).

Some management practices in conservation areas have already had adverse effects. An example is the provision of artificial watering points in some regions, especially dry areas such as Tsavo East. Natural climatic cycles exercise some constraints on population numbers. Dry years lead to increased mortality and reduced natality; in wet years the population climbs back. If the constraints of the dry years are removed, the recovery in wet years is more likely to overshoot.

The hazards of overabundance of a species are not necessarily so severe for that species (especially an unspecialized species such as the elephant) as for others more dependent on the habitats that have changed. There is a tendency to become preoccupied with the more conspicuous species and to forget dependent species.

Habitat changes can make the system more susceptible to other factors -- such as fire, drought, and poaching. The impact of predators (including poachers) is influenced by the opening or closing of habitats. On opening up bushed habitats animal predators and primitive hunting man may be less effective owing to the lack of cover, but modern man armed with rifles is more effective in open habitats. (This has been an important factor in the recent explosion of poaching in East Africa, where the opening up of bush by elephants has sown the seeds of their own destruction at the hands of men with automatic rifles.)

Effects of overpopulation, even for a single species in different habitats, are very different in different climatic regimes (e.g., elephant in Kabalega Falls National Park and Tsavo National Park East).

V. CONCLUSIONS

The purpose of this chapter, at the request of the organizers of the workshop, was to outline a possible framework for discussion of perceived large mammal overpopulation problems which acknowledged that the outcome is critically dependent on the properties of both the animals and the plants. Only a brief and superficial discussion has been possible but the outline classification, based on *r*- and *K*-selection concepts, and the chosen case studies, indicate that both the short-term and long-term consequences of overpopulation (acute and chronic) are likely to differ according to the nature of the limiting food resource and the turnover rate of the mammalian population. Although *K*-selected elephants require substantial quantities of grass, it is the availability of browse in the dry season that imposes the greatest constraints on the population. Where elephants are less dependent on browse, as in higher rainfall regions, compression leads to a population crash that is more extended in time and a new equilibrium may be more easily attained.

Even a relatively *r*-selected species such as caribou appears to be critically dependent on the time scale of its interactions with different food types. In situations where compression or irruption with a limited range apply, such as an island, the outcome in the absence of predators seems to depend on whether the food is *K*-selected (lichens) or *r*-selected (vascular plants). Time lags are important and the worst effects are found where both animal and plant populations exhibit them.

Hippopotamus and white rhinoceros can reach very high densities when dispersion is prevented and even dramatic effects on the environment can be reversed on a relatively short time scale, because of the short time lag in grassland recovery in periods of wetter climate. In the case of elephant, dependent on trees, or caribou, dependent on lichens, when dispersion is prevented, the time lag introduced by the growth rate and longevity of the food resource means that recovery is likely to be very slow and both the animal population and the ecosystem are put at greater risk. In the absence of culling local extinctions may occur, not necessarily of the species primarily responsible for the habitat change, but also of other indigenous species.

REFERENCES

Caughley, G. (1976). The elephant problem--an alternative hypothesis. *E. Afr. Wildl. J. 14,* 265-283.

Eltringham, S. K. (1974). Changes in the large mammal community of Mweya Peninsula, Rwenzori National Park, Uganda, following removal of hippopotamus. *J. Appl. Ecol. 11,* 855-865.

Field, C. R. (1971). Elephant ecology in the Queen Elizabeth National Park, Uganda. *E. Afr. Wildl. J. 9,* 99-123.

Field, C. R. and Laws, R. M. (1970). The distribution of the larger herbivores in the Queen Elizabeth National Park, Uganda. *J. Appl. Ecol. 7,* 273-294.

Fowler, C. W. and Smith, T. (1973). Characterizing stable populations: an application to the African elephant population. *J. Wildl. Manage. 37,* 513-523.

Hall-Martin, A. (1980). Elephant survivors. *Oryx J. Fauna Preserv. Soc. 15(4),* 355-362.

Klein, D. R. (1968). The introduction, increase, and crash of reindeer on St. Matthew Island. *J. Wildl. Manage. 32,* 350-367.

Laws, R. M. (1968a). Interactions between elephant and hippopotamus populations and their environments. *East Afr. Agri. For. J. 33,* 140-47.

Laws, R. M. (1968b). Dentition and aging of the hippopotamus. *E. Afr. Wildl. J. 6,* 19-52.

Laws, R. M. (1969). The Tsavo Research Project. *J. Reprod. Fertil. Suppl. 6,* 495-531.

Laws, R. M. (1970). Elephants as agents of habitat and landscape change. *Oikos 21,* 1-15.

Laws, R. M. (1977). Seals and whales in the southern ocean. *In* "Scientific Research in Antarctica." (Discussion meeting organized by V. E. Fuchs and R. M. Laws.) *Philos. Trans. R. Soc. London Ser. B 279,* 81-96.

Laws, R. M. (1981). Experiences in the study of large mammals. *In* "Population Dynamics of Large Mammals" (C. W. Fowler, ed.) Wiley, New York.

Laws, R. M. and Clough, G. (1965). Observations on reproduction in the hippopotamus, *Hippopotamus amphibius. Symp. Zool. Soc. London 15,* 117-40.

Laws, R. M., Parker, I. S. C. and Johnstone, R. C. B. (1975). "Elephants and Their Habitats: The Ecology of Elephants in North Bunyoro, Uganda." Oxford Univ. Press (Clarendon), London and New York.
Leader-Williams, N. (1980). Population dynamics and mortality of reindeer introduced into South Georgia. *J. Wildl. Manage. 44(3)*, 640-657.
Leuthold, W. and Leuthold, B. M. (1976). Density and biomass of ungulates in Tsavo East National Park, Kenya. *E. Afr. Wildl. J. 14*, 49-58.
Macarthur, R. H. and Wilson, E. O. (1967). "The Theory of Island Biogeography. Monographs in Population Biology, 1." Princeton Univ. Press, Princeton, New Jersey.
May, R. M. (1978). Factors controlling the stability and breakdown of ecosystems. *In* "The Breakdown and Restoration of Ecosystems" (M. W. Holdgate and M. J. Woodman, eds.), pp. 11-25. Plenum, New York.
Phillipson, J. (1975). Rainfall, primary productivity and "carrying capacity" of Tsavo National Park (East) Kenya. *E. Afr. Wildl. J. 13*, 171-201.
Pianka, E. R. (1970). On *r*- and *K*-selection. *Am. Nat. 104*, 592-597.
Ray, G. C. (In press). The role of large organisms. *In* "Analysis of Marine Ecosystems" (A. R. Longhurst, ed.) Academic Press, London.
Soulé, M. E., Wilcox, B. A., and Holtby, C. (1979). Benign neglect: a model of faunal collapse in the game reserves of East Africa. *Biol. Conserv. 15*, 259-272.
Weir, J. S. (1972). Spatial distribution of elephants in an African National Park in relation to environmental sodium. *Oikos 23*, 1-13.
Wheater, R. J. (1968). Land use proposals for Murchison Falls National Park and surrounding area. *East Afr. Agri. For. J. 33*, 19-22.
Wilson, E. O. (1975). "Sociobiology: The New Synthesis." Harvard Univ. Press, Cambridge, Massachusetts.

RESPONSES OF VEGETATION TO THE ABUNDANCE OF MAMMALIAN HERBIVORES

Imanuel Noy-Meir

Botany Department
Hebrew University
Jerusalem
Israel

I. INTRODUCTION

Vegetation is essential for the preservation of the fauna in any nature reserve (excluding zoos), being both the primary food resource for all animals and a necessary microhabitat for many species of mammals, birds, and insects. This is perhaps obvious to ecologists but may sometimes be underestimated by wildlife managers and conservationists. To decide when populations of mammalian herbivores are overabundant and in need of culling we must observe, understand, and predict the responses of vegetation to variations in the abundance of those herbivores.

The following discussion is based mainly on experience with grazing systems with domestic livestock and the literature that appertains to them. There is much evidence from such systems concerning the response of vegetation to ungulate herbivores over a wide range of situations and abundances, including rather high abundances. There are also some elements of theory explaining and predicting these responses.

This evidence and theory are relevant to wild mammalian herbivores (or wild ungulates at least), since the basic processes of herbivore feeding behavior and vegetation growth responses are similar. In most wildlife/vegetation systems (except recent introductions) and in many livestock/vegetation systems (mainly in the "Old World") these processes have been shaped by a long history of mutual adaptation. Evidence about the effects of herbivores in Africa has recently been reviewed by Cumming (1979).

ISBN 0-12-385280-3

The main differences between livestock and wildlife grazing systems is the dynamics of herbivore populations in time and space, which interacts with the responses of the vegetation. Some of the consequences will be briefly discussed at the end of this chapter.

II. GREEN LEAF DYNAMICS

When a herbivore removes some amount of plant material, it obviously reduces the future food supply to the herbivore population(s). But, if this plant material is alive and green, the reduction of future supply is not simply equal or proportional to the amount taken. This is mainly because green foliage produces more green foliage at a rate which depends on the amount (leaf area), spatial arrangement, and quality (age, photosynthetic ability) of the existing foliage.

Moderate or intermittent defoliation in the growth season of vegetation which is too dense (leaf area above optimum, shading of lower by upper leaves) or senescent, will be largely compensated for by new leaf growth. This may be produced by the photosynthesis of new materials in the remaining foliage or from photosynthate reserves in other parts (stems, roots, rhizomes). Several different processes may be involved in this compensatory growth or stimulation (reviewed by McNaughton, 1979). Repeated or continuous moderate defoliation by herbivores will thus result in only a moderate reduction in the cover and standing biomass of the vegetation and usually in an increase in its photosynthetic efficiency and productivity. In such a situation the vegetation may be said to be "undergrazed." The damage done by the herbivore population to its food supply is less than what it eats. The herbivore may actually be improving its nutrition by causing an increase in the proportion of young leaves.

However, frequent or continuous heavy defoliation in the growth season may reduce green leaf area to a point where its regrowth rate can no longer keep up with the rate of defoliation; this will induce further reductions in leaf area and growth rate. This is particularly likely to happen in vegetation where green foliage is already sparse (leaf area below or near optimum). Once the rate of grazing and browsing by herbivores exceeds a certain critical level, the ability of plants to compensate by regrowth may break down and the deterioration of green leaf area becomes self-accelerating. The result may be a "pasture crash," a drastic reduction of plant cover and productivity for the rest of the season. In a critical situation any additional amount of green foliage taken by herbivores may cause a much greater than proportional damage to their future food supply.

A sparse vegetation, with suboptimal leaf area and low productivity, may maintain itself even after a crash and under continuous heavy herbivory. Its maintenance depends on photosynthesis by leaves which are virtually inaccessible to the grazers (too low, too high, protected by branches or spines); in addition, the plants keep producing some accessible green foliage which is taken almost as soon as it appears. Plants which have no residual foliage may be grazed to death. The vegetation may be said to be "overgrazed."

The transition from an undergrazed to an overgrazed vegetation under continuous grazing (and the corresponding loss of productivity) can be quite rapid. It may be triggered by a temporary reduction in leaf area (e.g., due to bad weather or disease) or a temporary increase in herbivore abundance, when both are already near the critical threshold. Once the crash has occurred, it may not be easily reversible. Vegetation which has been overgrazed from the beginning of the season, or following a crash, can "get away" ("erupt") out of herbivore control to attain an undergrazed and more productive state. But this is likely to happen only if grazing pressure is reduced for a period ("deferred grazing") or when conditions for plant growth improve considerably (e.g., from winter to spring).

"Crashes" and "get-aways" across critical thresholds do not occur in all vegetation/herbivore combinations. In some, plant cover and productivity may vary regularly and continuously in response to herbivore abundance in the growth season, and the transitions between undergrazing and overgrazing may not be sharp. However, theoretical reasons supported by field evidence, suggest that threshold responses of green foliage to herbivore pressure are quite common in herbaceous vegetation grazed by ungulates (Morley, 1966; Noy-Meir, 1975, and unpublished; Caughley, 1976a; May, 1977; Walker *et al.*, in press). These responses are more likely if the herbivore is efficient in feeding on the vegetation even when it is sparse and short (e.g., with sheep more than with cattle, with tall tussock grasses more than with small creeping grasses). As a rough rule, a threshold response can occur if the herbivore finds and eats its fill (or nearly so) at a density (leaf area, green biomass) of its food plants which is lower than that at which their maximal growth rate is attained. Otherwise, the response will be continuous. A threshold response will be more likely if the grazers select strongly for the more productive young leaves (as they often do).

III. PERENNIAL RESERVES

This mechanism of a herbivore-induced crash in plant cover and productivity has been postulated and observed for the dynamics of green foliage during a growth season. It may well carry over into the dynamics of plant photosynthate reserves over years. The details of the processes involved are not well understood but the general effect seems to be as follows: Under moderate defoliation a certain perennial balance is established between the mobilization of resources from rhizomes, stems, and seeds to green leaves which start the season's growth, and the replenishment of these reserves by translocation of photosynthates in the growth season. Heavy defoliation during the season reduces replenishment of perennial reserves (including seed production), which in turn reduces the initial green leaf area next season. These depletions may become amplified, over several years, to the point at which there is a breakdown of the ability of plants for perennial regeneration, leading to an overgrazed state of the herbage with low production that may involve heavy mortality or extinction of sensitve species. Sensitivity to this effect depends on life history, growth form, and energy allocation strategy of the species -- in general, erect leafy perennial grasses are probably most susceptible, while tall woody plants and (for different reasons) annuals tend to be quite resistant to attrition of their reserves by herbivores.

IV. PLANT POPULATIONS

Most mammalian grazers (both "grazers" and "browsers") characteristically feed by taking one or a few bites from each plant at each visit; this is rarely enough to kill an adult plant. Pushing trees over or ringbarking them, and pulling up grasses by the roots, are rather the exceptions.

As herbivore pressure increases, each individual plant is defoliated more frequently. Under frequent heavy defoliation a plant may die through the exhaustion of its energy reserves or of its growing points, particularly if this occurs in a critical stage of the annual cycle. If the adult individuals of a plant species are susceptible to being killed by attritive grazing, the population may become extinct in a fairly short time in response to an increase in herbivore abundance.

The dynamics of such a process have hardly ever been recorded in detail. But this process was probably responsible for the virtual extinction of species and whole communities of perennial grasses upon introduction of sheep and cattle into parts of North America, South Africa, and Australia.

However, the adult plants of many herbaceous and woody species appear to be able to regrow and survive indefinitely under very frequent and heavy herbivory. Not surprisingly, such are the dominant perennials in regions with a long history of ungulate grazing, e.g., evergreen oaks of the Mediterranean and grasses of East African savannas (McNaughton, 1979). This remarkable resistance is apparently due to a combination of attributes: Some inaccessible residual foliage, reserve organs, and meristems which are well protected and which are rather cautiously committed to new growth. An increase in the abundance of herbivores can cause reduced biomass and productivity but no plant mortality and no immediate danger of extinction in such populations.

Over a longer time period, persistent high herbivore pressure can endanger the populations of plants which are resistant as adult individuals by killing most seedlings (which are usually more susceptible) and preventing replacement of adult plants which die from other causes. This process is most dangerous for populations of shrubs and trees that mature slowly and for which environmental conditions for successful reproduction and recruitment occur infrequently, e.g., in semi-arid Australia (Crisp, 1978).

V. BOTANICAL COMPOSITION

The plant species in any community differ greatly in palatability and accessibility to herbivores and in their responses to defoliation, so that some species are reduced by herbivory much more than others. On the other hand, species which are excluded (or suppressed) by competition may invade (or increase in abundance) when grazing pressure is increased. Range managers and ecologists assess the condition of the vegetation by the amounts of decreaser, increaser, and invader species and adjust livestock density accordingly (Dyksterhuis, 1949; Ellison, 1960; Stoddart *et al.*, 1975). As shown by Westoby (1980), there are many observed exceptions to the simple theory on which this practice is based; vegetation composition may respond to grazing pressure in several different ways, depending on the kind of vegetation and herbivore and the nature of their interaction.

It is often taken for granted in range management that the species encouraged by grazing are less palatable and less nutritive to herbivores than the decreasers, and it is inferred that the higher the grazing pressure the lower the quality (as well as quantity) of feed produced from the range.

This is true in some cases, particularly at extreme pressures, when quite unpalatable plants may become dominant. Often, no difference in animal condition and production is found between pastures of very different botanical composition, judged from "good" to "poor" (e.g., Snaydon, 1978; Gutman and Seligman, 1979). Most large herbivores, although they select for certain preferred plants, can still do just as well on a wide range of less preferred ones. Sometimes the species encouraged by grazing actually are of greater nutritive value than the grazing-sensitive species they replace, e.g., annual grasses and forbs versus tropical perennial grasses (Kelly and Walker, 1976; Breman *et al.*, 1980).

VI. DEAD HERBAGE AND LITTER

Most herbivores also take dry plant material when not enough green fodder is available. Since dry material does not contribute to plant growth, any amount removed just means an equivalent reduction in the food available later. There can be secondary effects, because the dead herbage and litter accumulated influences the environment of living plants. In general it has beneficial effects: reduced evaporation, enhanced germination and establishment, increased infiltration and reduced runoff, protection from wind and water erosion. Therefore, it is often recommended that a range should be grazed so as always to leave a substantial amount of dry herbage and litter, and bare soil is considered a sign of overgrazing (reviewed by Ellison, 1960). But too much dead herbage (attributable to undergrazing) inhibits plant regrowth (e.g., Weaver and Rowland, 1952; Kelly and Walker, 1976), and germination can be optimal on a surface with quite sparse litter (e.g., Gutman, 1978).

The amount of dry herbage largely determines the frequency and intensity of fires, which may in turn affect vegetation cover and composition in ways which are specific to each ecosystem.

VII. SOIL EFFECTS: TRAMPLING

The effects of herbivores on soils that have the greatest impact on vegetation are the modification of soil structure by trampling and of soil nutrient processes by excretions. Both effects are usually strongest near tracks, water points, and rest areas (Lange, 1969). The nutrient effects are complex and are discussed by Botkin (this volume)[1]. The effects of herbivore trampling also are not quite simple.

In clayey and loamy soils, particularly when they are wet, trampling causes compaction and hardening of the upper layers of soil with damaging effects on infiltration, aeration, plant establishment, and growth (Marshall, 1972; Walker *et al.*, in press). In sandy soils the main effect is loosening of the surface, particularly where it is kept firm by an algal/lichen crust (Rogers and Lange, 1971). Also in silty and some clayey soils, dry season trampling breaks the physical or biological crust into a loose "bull-dust." Surface loosening can initiate or accelerate wind and water erosion in situations susceptible to those processes. In certain circumstances trampling may have a role in mixing seeds into the soil and promoting germination.

VIII. OTHER EFFECTS

For some plant species, seed dispersal by herbivores is important. Other more subtle ecological and genetical effects of herbivores on plants are still being revealed as detailed field studies develop.

IX. SOME GENERALITIES ABOUT THE OVERALL RESPONSE

The mechanisms by which herbivores affect plants, and by which plants respond, are many and complex. The details of the effect of an increased abundance of herbivores on vegetation, reacting in turn on the herbivores themselves, will be different in each situation. Uncritical transfer of management practices from one ecosystem to another can lead to complete failure, as has often been experienced in livestock systems. The management

[1] *See Workshop Report.*

of each system should be based on a thorough study and analysis of the critical plant/herbivore/soil interactions in the system and on the results of direct experiments for purposes of management wherever possible (Sinclair, this volume).

Sometimes, decisions must be taken when time or resources are not sufficient to allow manipulative experiments or research on ecological processes. Then -- rather than make decisions which are entirely arbitrary, habitual, or borrowed -- it may be useful to consider the relevance and applicability of any general principles emerging from the accumulation of evidence and theory on vegetation/herbivore systems. Such principles as are formulated here are not intended to substitute for problem-oriented research and experiment; they may perhaps help to guide it.

To summarize the preceding discussion in another way, the overall response of the vegetation to herbivore abundance can roughly be divided into three "phases," at low, medium, and high herbivore abundance:

(1) Permanently undergrazed: At low herbivore abundance, compensatory processes (stabilizing negative feedbacks) dominate at all levels (green foliage, reserves, populations, species composition) and in all seasonal conditions. Increasing herbivore abundance causes small and reversible changes; plant cover and animal nutrition may be slightly reduced or improved, productivity and species diversity (Harper, 1969; Whittaker, 1976) generally increase.

(2) Transitional: At higher herbivore abundance, any of several self-accelerating processes (destabilizing positive feedbacks) may be initiated in a season of poor growth (or temporarily increased herbivory) and overwhelm the compensatory processes (attrition of leaf area, reserves or meristems, invasions by unpalatable plants, accelerated runoff and erosion). This can be reversed in favorable periods but recovery may take some time. In these circumstances the vegetation changes irregularly between undergrazed and overgrazed conditions. As herbivore abundance is increased, overgrazed conditions are likely to prevail for longer periods.

(3) Permanently overgrazed: At very high herbivore abundance, once all possible accelerating deterioration cycles have run their course, vegetation and soil may again stabilize at low productivity. This is maintained by the more resistant plant species and landscape components that are insensitive to variations in weather and herbivory. In general, plant productivity and the condition of animals in the overgrazed state are poor, particularly so in vegetation that has been exposed suddenly to overgrazing by an introduced herbivore (North America, Australia). In vegetation long exposed to herbivory

it is likely that plant resilience to grazing has built up by ecological and genetical processes (Africa, Mediterranean); the overgrazed state may be of low but respectable productivity and high species diversity, and capable of faster recovery in suitable conditions.

X. MANAGEMENT OF ABUNDANCE IN LIVESTOCK SYSTEMS

The manner in which herbivores are managed, and density maintained, is largely determined by the objectives. In commercial livestock (or game) systems the objective is to maximize yield and income. In the short term this is usually attained at livestock densities where vegetation is in the unstable transitional phase. Thus operators are tempted to increase livestock numbers into that phase at the risk of long-term deterioration (or fluctuations). This has in fact happened in large parts of North America, South Africa, and Australia. The recommendations of range scientists and the practices of responsible managers (with the support of ecologists) are aimed at keeping livestock abundance well in the safe undergrazed phase, to prevent large and risky fluctuations. Accordingly, some gain in the short term (or even in the long term) is sacrificed for ecological and economic stability. The diagnostics used (increaser/decreaser plants, bare soil) are intended to define the safe range and give early warning of any possible deterioration. Given the management objectives this is probably a sensible policy. Two serious reservations must be made. First, a "play safe" policy which is not occasionally tested by taking some risks tends to become overly conservative in its estimates of allowable abundance; in particular it tends to underestimate the recovery ability of the vegetation and the nutritive value of increaser species (Sinclair, this volume). Secondly, in permanently undergrazed vegetation, plant species and genotypes which are resilient to overgrazing (but usually inferior competitors in undergrazed conditions) will gradually disappear, so that if eventually an overgrazing crisis does occur, the crash of the long undergrazed system will be the more drastic (Walker *et al.*, in press).

In economically strong livestock systems, management can be intensive enough to avoid this ever happening. In other situations this risk of permanent undergrazing must be weighed against the risks of temporary overgrazing.

In traditional pastoralist societies, the main objectives of the herder are not maximal yield and income but rather a subsistence supply and a large herd (capital, status). Grazing land is held in common and exploited competitively. Unless regulated by tribal law, livestock numbers will tend to increase until the vegetation is just about permanently overgrazed. Further increases are checked by poor nutrition and high mortality in critical seasons. In most of the regions concerned, the vegetation has had thousands of years to adapt and is moderately productive in the traditional overgrazed state. However, where supplementary feeding becomes economically feasible (as it does today in parts of the world) this can induce further increases in the herd and lead to the maintenance of a highly overabundant herbivore population on a severely overgrazed range.

XI. CONSEQUENCES FOR MANAGEMENT OF WILDLIFE SYSTEMS

Generalizations about wildlife conservation systems are even more difficult to make than about livestock production systems. Wildlife conservation deals with a wider variety of systems in terms of environments, vegetation types, herbivore species and communities, and spatial patterns. Also, the objectives of management for conservation are less clearly defined than for production. Several possibly conflicting objectives are usually involved, to which different people attach different values -- preservation of one or several animal species, of a plant/animal/soil ecosystem, biological and aesthetic diversity, ecological stability, minimal interference with nature, animal welfare. This divergence was made clear in Caughley's (this volume) four classes of perceived overabundance and in the discussion of the workshop. There is perhaps one point of general agreement: Any management policy (or lack of it) for a nature conservation area that leads to the extinction of species in the area (which could have been prevented by a different policy) has failed all the basic aims of conservation.

The generalization made above -- that the overall response of vegetation to herbivore abundance may be divided into three phases (undergrazed, transitional/unstable, overgrazed) -- is likely to be applicable and useful for wild as well as domestic herbivores. But in wildlife systems herbivore abundance itself (rather than being controlled by man near a constant level) varies in response to the condition of the vegetation. The regulating process is a dependence of herbivore natality and mortality on the amount or quality of vegetation or, most likely, on the amount of high quality (green?) plant material in a critical season (Sinclair, 1975). This process has a potentially stabilizing effect (a negative feedback) on the vegetation/herbivore system. However, as predicted from predator/prey theory, and as can be observed in nature, this response of the herbivore population in itself does not guarantee that a vegetation/herbivore system will always (or ever) regulate itself naturally towards a stable equilibrium. This is only one of several possible alternative dynamic behaviors (Rosenzweig and MacArthur, 1963; May, 1974, 1977; Caughley, 1976a, 1976b; Walker *et al.*, in press; Klein, this volume; Sinclair, 1979 and this volume).

If the herbivore population response becomes effective at vegetation conditions which are already unstable (efficient herbivores) or if it lags behind vegetation changes (large, long-lived herbivores), it may result in persistent (limit cycle) fluctuations in plant and animal abundances or in cycles of transitions between undergrazed and overgrazed conditions. In addition, such cycles may be the natural dynamics of plant/herbivore systems that have had a long time to coevolve. Theoretically, selection of herbivores for feeding efficiency tends to promote cycles, while the contrary selection of plants to reduce the ability to graze may not be sufficient to ensure stability (Rosenzweig, 1973). However, in coevolved systems plants would have been selected for resilience (ability to recover after overgrazing). In all probability plant resilience and herbivore dispersal have prevented extinctions of plant and animal species (except in rare events) in "natural" systems in the past, even where large fluctuations occurred. But it cannot be taken for granted that these mechanisms are equally effective in the wildlife and nature reserves of today, with confinement and other man-made changes.

Unfortunately, the theory of plant/herbivore systems provides no simple general answers to the practical problems -- when is a herbivore population overabundant and should it be regulated by culling or left alone? However, theory can help to sharpen the questions and the search for information needed to answer those questions in each specific case.

The ultimate answers depend, apart from value and priority decisions, on two critical questions to which objective answers are possible:

(1) If no culling is done, is the vegetation/herbivore community likely eventually to approach a stable situation (with only seasonal fluctuations), or to go into crash/eruption cycles?

(2) If no culling is done, is the community likely to change in such a way that the risk of species extinctions is increased?

The questions refer essentially to the stability and resilience (Holling, 1973) of the ecosystem, which qualities may depend on quite different characteristics. The appropriateness of these questions depends on management objectives.

The managers of a conservation area may have perfectly valid reasons (Caughley's Classes 1,2,3) for preferring that the state of the ecosystem which would be achieved by a culling plan (fewer herbivores, more vegetation) would be more in line with their objectives than any other state into which the ecosystem might develop in the absence of culling. In this case neither of the two questions is of great practical relevance. Or, managers of conservation areas may legitimately prefer any reasonably stable ecosystem over one which changes in drastic cycles (Caughley's Class 4); in this case the first question is relevant and should be approached by research and experiment. Often there may be no good reasons, related to the specific objectives of the conservation area, for either of these preferences. Then only the second question (what would be the effect of intervention on the probabilities of species extinctions in a particular ecosystem?) is really crucial to decisions about the management of herbivore populations. This is probably one of the most difficult questions to answer in ecology, but apparently one which will require more attention in the future.

REFERENCES

Breman, H., Cisse, A. M., Djiteye, M. A., and Elberse, W. T. (1980). Pasture dynamics and forage availability in the Sahel. *Isr. J. Bot. 28.*

Caughley, G. (1976a). Plant herbivore systems. *In* "Theoretical Ecology" (R. M. May, ed.), pp.94-113. Blackwell, Oxford.

Caughley, G. (1976b). The elephant problem -- an alternative hypothesis. *East Afr. Wildl. J. 14,* 265-283.

Crisp, M. D. (1978). Demography and survival under grazing of three Australian semi-desert shrubs. *Oikos 30,* 520-528.

Cumming, D. H. M. (1979). The influence of large herbivores on savanna structure in Africa. *In* "Symposium on Dynamic Changes in Savannah Ecosystems." S.C.O.P.E., Pretoria.

Dyksterhuis, E. J. (1949). Condition and management of rangeland based on quantitative ecology. *J. Range Manage. 2,* 104-115.

Ellison, L. (1960). Influence of grazing on plant succession of rangelands. *Bot. Res. 26,* 1-78.

Gutman, M. (1978). Primary production of transitional Mediterranean steppe. *Proc. Int. Rangeland Congr. 1*, 225-228, Denver, Colorado.
Gutman, M. and Seligman, N. (1979). Grazing management of herbaceous Mediterranean foothill range in the Upper Jordan River Valley. *J. Range Manage. 32*, 86-92.
Harper, J. L. (1969). The role of predation in vegetational diversity. *In* "Diversity and Stability in Ecological Systems" (G. M. Woodwell and H. H. Smith, eds.) *Brookhaven Symp. Biol. 22*, 48-61.
Holling, C. S. (1973). Resilience and stability of ecological systems. *Ann. Rev. Ecol. Syst. 4*, 1-23.
Kelly, R. D. and Walker, B. H. (1976). The effects of different forms of land use on the ecology of a semi-arid region in south-eastern Rhodesia. *J. Ecol. 64*, 553-576.
Lange, R. T. (1969). The piosphere: sheep track and dung patterns. *J. Range Manage. 22*, 396-400.
McNaughton, S. J. (1979). Grazing as an optimization process: grass-ungulate relationships in the Serengeti. *Am. Nat. 113*, 691-703.
Marshall, J. K. (1972). Effects of stock on microenvironments in Australian rangelands. U. S.-Australia Range Science Seminar. *Dep. Agric. Misc. Publ.*
May, R. M. (1974). "Stability and Complexity in Model Ecosystems." Princeton Univ. Press, Princeton, New Jersey.
May, R. M. (1977). Thresholds and breakpoints in ecosystems with a multiplicity of stable states. *Nature (London) 269*, 471-477.
Morley, F. H. W. (1966). Stability and productivity of pastures. *Proc. N. Z. Soc. Anim. Prod. 26*, 8-21.
Noy-Meir, I. (1975). Stability of grazing systems: an application of predator-prey graphs. *J. Ecol. 63*, 459-481.
Rogers, R. W. and Lange, R. T. (1971). Lichen populations on arid soil crusts around sheep watering places in Australia. *Oikos 22*, 93-100.
Rosenzweig, M. L. (1973). The evolution of the predator isocline. *Evolution 27*, 84-94.
Rosenzweig, M. L. and MacArthur, R. H. (1963). Graphical representation and stability conditions of predator-prey interactions. *Am. Nat. 97*, 209-223.
Sinclair, A. R. E. (1975). The resource limitation of trophic levels in tropical grassland ecosystems. *J. Anim. Ecol. 44*, 497-520.
Sinclair, A. R. E. (1979). Dynamics of the Serengeti ecosystem: process and pattern. *In* "Serengeti: Dynamics of an Ecosystem" (A. R. E. Sinclair and M. Norton-Griffiths, eds.) Univ. of Chicago Press, Chicago, Illinois.

Snaydon, R. W. (1978). Indigenous species in perspective. *Proc. Br. Weed Control. Conf. 14*, 906-913.

Stoddart, L. A., Smith, A. D., and Box, T. (1975). "Range Management, Third Edition." McGraw-Hill, New York.

Walker, B. H., Ludwig, D., Holling, C. S., and Peterman, R. M. (In press). Stability of semi-arid savanna grazing systems. *J. Ecol.*

Weaver, J. E. and Rowland, N. W. (1952). Effect of excessive natural mulch on development, yield and structure of native grassland. *Bot. Gaz. (Chicago) 114*, 1-12.

Westoby, M. (1980). Elements of a theory of vegetation dynamics in arid rangelands. *Isr. J. Bot. 28.*

Whittaker, R. H. (1976). Animal effects on plant species diversity. *Intern. Ver. Veget.-kund, Rinteln.*

ENVIRONMENTAL CARRYING CAPACITY AND THE EVIDENCE FOR OVERABUNDANCE

A. R. E. Sinclair

Institute of Animal Resource Ecology
University of British Columbia
Vancouver, British Columbia
Canada

I. INTRODUCTION

It is a recognized objective of conservation to set aside representative samples of biomes in the world so that they can continue to function naturally with a mimimum of human interference. In essence they are the environmental control areas necessary to monitor the changes that have been imposed by man on other areas through agriculture and urbanization. Such natural areas fall into Caughley's Class 4, and it is these I shall consider in this chapter.

Many parks, such as those in South Africa and Malawi, are set aside specifically to protect representative sets of species. Although this is a valid objective, it is a purely arbitrary decision which species are represented and what level their populations should be allowed to reach. They are, in principle, no different than zoos. Management to prevent overpopulation is decided according to value judgments as to what is a safe, or economical, or aesthetic population level. These situations fall into Caughley's Class 2 criteria. Countries with parks or reserves of this nature only, however, are ignoring their responsibility to set aside natural areas of minimum interference.

PROBLEMS IN MANAGEMENT OF LOCALLY
ABUNDANT WILD MAMMALS

ISBN 0-12-385280-3

II. EVIDENCE FOR OVERPOPULATION

To use the term "overabundance" correctly within natural areas, we must already know what population size a particular environment can support on a long-term basis -- a population often called the "carrying capacity" for that environment.

There are many attempts to provide evidence for overabundance and some examples are presented in this volume: Hanks *et al.* suggest that soil erosion outside the parks in South Africa, resulting from overgrazing by cattle, is likely to occur within the parks from indigenous grazers. Cumming considers the destruction of trees by elephants in Zimbabwe parks is evidence that grasslands will result. Klein points to changes caused by nondispersing deer in North America that, by implication, will end in destruction of habitats.

These examples are consistent in that the purported evidence is, in fact, pure speculation, or in other words hypothetical; it is not known that indigenous ungulates in South Africa will cause soil erosion and the collapse of ecosystems, nor that elephants will cause desertification, nor that deer will cause irreversible change in North American forests, for none of these things have yet been followed to their conclusion.

Hypotheses such as these are perfectly valid suggestions as to what might happen. But there are equally valid alternative hypotheses which lead to opposite predictions, and some of these are outlined in the following sections.

III. PERSISTENCE

Models of the interaction between predator and prey or herbivore and plant, such as those of Caughley (1976a), show an inverse relationship between the biomasses of the two populations. Following a perturbation in either population, the two approach some sort of equilibrium. Fig. 1 depicts what might happen when a herbivore population increases after a reduction -- perhaps through excessive harvest, poaching, or a bad winter. The biomass of plants is initially high but falls steeply because the herbivore population rises with the abundant good food. After a number of dampening oscillations, the two populations settle to a region of small fluctuations. Fig. 2 illustrates the same thing starting with a high herbivore population decreasing perhaps due to compression into a smaller area, or from an adverse change in environment lowering the carrying capacity of the area, or from the introduction of a predator to the system.

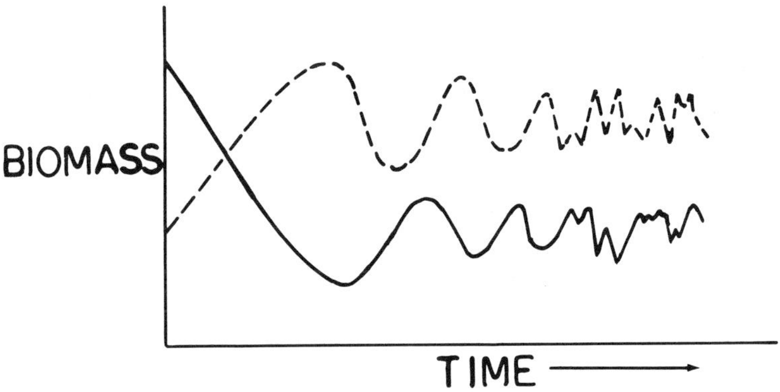

Fig. 1. Hypothetical relationships between plant community (solid line) and a herbivore population (broken line), starting with low herbivore numbers.

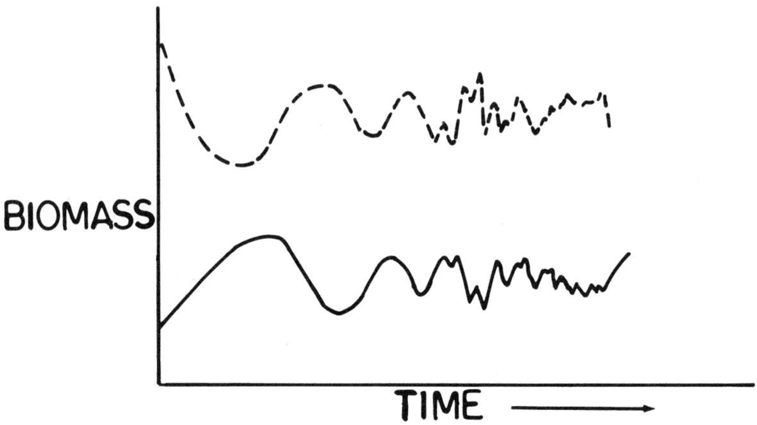

Fig. 2. Hypothetical relationships between plant community (solid line) and a herbivore population (broken line), starting with high herbivore numbers.

A population at carrying capacity should not be thought of as one with a stable or constant level. Rather, it is one that is fluctuating, often extensively, between certain boundaries. We should also beware of considering the carrying capacity as an average of these random fluctuations (though for mathematical simplicity it is sometimes necessary) because half of the time the population is above the average. In general, one can say that a population is at equilibrium if it can persist while fluctuating within certain limits in that environment over a long period of time -- the emphasis is on persistence rather than steady state.

Simple models can also be unstable; a predator can capture prey faster than the latter can reproduce, driving the prey so low that it either becomes extinct (and so also the predator) or the predators cannot find enough prey and become extinct, leaving the prey to increase to some new higher level (Fig. 3).

In the herbivore/plant scenario, the herbivore removes photosynthetic material faster than the remaining leaves can obtain energy to replace it. Consequently, the plant draws on its root reserves to grow new shoots, causing the roots to regress. Eventually, the roots cannot reach water or the plant is starved of energy and dies.

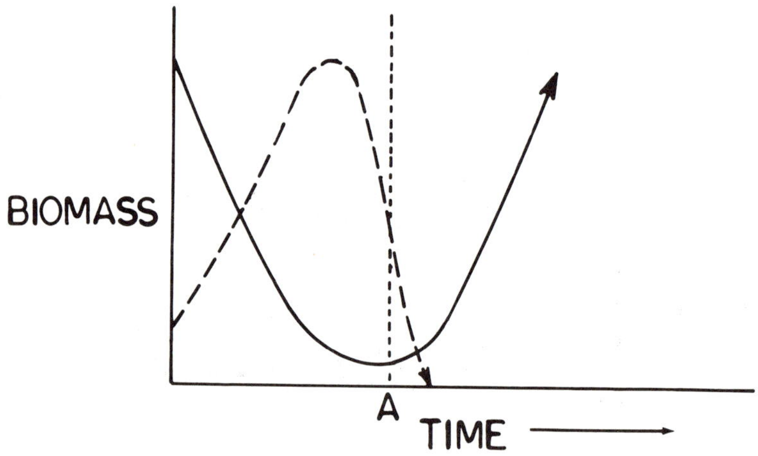

Fig. 3. Hypothetical unstable interaction between a plant community (solid line) and a herbivore population (broken line), which goes extinct. Before time A the behavior of the system is similar to the initial stages of Fig. 1.

An important point drawn from Fig. 3 is that the behavior of the system up to time *A* is apparently similar to that in Fig. 1. Before *A*, it is not possible to tell from the behavior what the outcome will be. If there is sufficient historical and experimental knowledge of the system that Fig. 1 is already known, then it would be possible to predict the outcome in Fig. 3 once the equilibrium level of Fig. 1 is passed. If no information about Fig. 1 is known, then the outcome of Fig. 3 is unpredictable.

A good example of Fig. 1 is that of the Yellowstone National Park elk population. Until the 1930's, elk *(Cervus canadensis)* were relatively well protected and numbers were around 12,000 to 15,000. As a result of a management decision that there were too many elk in the park, reduction shooting began and in the subsequent 30 years it became increasingly severe, progressively reducing the population. In the mid-1960's, the population was only 3,000 and new managers reversed the decision, providing complete protection. A decade later the elk had increased and stabilized at 12,000 again.

Scheffer (1951) describes the history of two reindeer *(Rangifer tarandus)* herds introduced to the Pribilof Islands in the Bering Sea in 1911. On St. Paul Island the population grew from 25 to 2,000 in 1938, then declined to 8 in 1950, resembling the model in Fig. 3. On St. George Island the population grew from 15 to 200 in 1922, then subsided to 40 to 60 animals similar to Fig. 1. The two islands were ecologically similar and no satisfactory explanation has been provided for the different population responses.

IV. MULTIPLE STABLE STATES

So far, I have considered only simple models, though they have adequately mimicked many real life situations. But there are more complex hypotheses that must be considered.

A population at equilibrium is one that persists by fluctuating between certain boundaries. It possesses the characteristic of moving away from the boundary when the boundary is approached in the same sense as Holling's (1973) "domain of attraction" with its boundary. Taking the animal population in Fig. 1 and 2, the boundaries *X* and *Y* must include the two starting populations since both of these move towards some intermediate level (Fig. 4). Notice that the boundaries are far wider than the normal range of random fluctuations shown by the population.

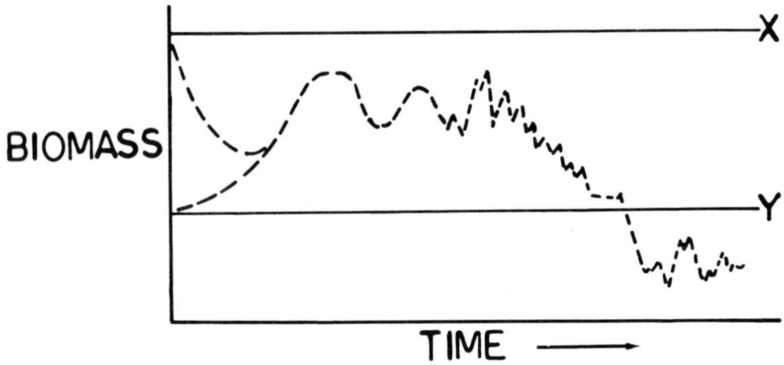

Fig. 4. The population or biomass region within which a herbivore population will return to the same equilibrium position. If the population is forced outside the boundaries (X, Y) of this region, then it will settle to a new equilibrium.

If the population is forced over either boundary, then the system moves to a new state. In Fig. 4 the population crosses the lower boundary and then continues to move away from the boundary, eventually settling at some new lower level. At least two situations could result in a new state: If a herbivore became too scarce, the vegetation may change in species composition so that there is permanently less food for the animals. Examples of this can be found in some South African reserves; severe reduction or even elimination of grazers in the early part of this century allowed the vegetation to change from grassland to dense scrub. Subsequently, grazers were not able to reestablish themselves (or remained at very low density) because the vegetation was unsuitable, and human intervention by bush clearing has been necessary for them to recross the boundary. A similar situation appears to have occurred in the Serengeti region of Tanzania when thickets grew up following the decline in grazers through rinderpest in the last century (Sinclair, 1979).

A second scenario that could result from a low herbivore population involves predators. The effect of predators can be inversely density dependent, that is they take an increasing proportion of the prey as the prey population declines. This causes the herbivore population to become even scarcer. Eventually the herbivores stabilize when there are so few of them that the predators cannot catch them or turn their attention to other species. This is illustrated in Fig. 5 where the herbivore population is at equilibrium when N_{t+1}/N_t is

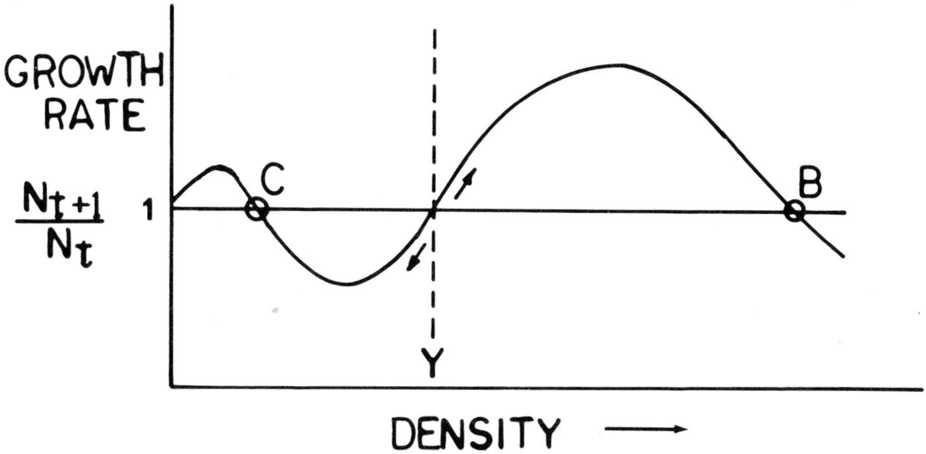

Fig. 5. The ratio of population size in one year relative to the year before, plotted against density. Left of Y the population moves to C; right of Y it moves to B.

unity. The two stable points are at *B* and *C*, determined by food supply and predators respectively, and the boundary is at *Y*. If density is forced lower than *Y*, predators take over and continue the decline to *C*.

An example of two stable points determined by food and predators is seen in the wildebeest of Kruger Park, South Africa (Smuts, 1978; Walker and Noy-Meir, 1979). The population was originally regulated by food but managers decided there were too many wildebeest and initiated culling. At about the same time, an increase in rainfall created long grass which the wildebeest did not like. They responded by breaking up into small groups and staying on the isolated short grass areas; the many small groups thus made them more vulnerable to predation. When seven years of culling ceased in 1972, instead of the expected increase in population, the decrease continued at least for the next five years because predators had taken over. In terms of Fig. 5, the culling had reduced the density past the boundary *Y*, while the increased rainfall had aggravated the situation by shifting the boundary further to the right.

What happens when an animal population increases above the *X* boundary is less well known but a possible scenario is suggested in the elephant/tree/grass complex in Kidepo Park, Uganda (Field and Ross, 1976). A high density of elephants has supposedly changed the previous *Acacia* woodland to an open long grass vegetation. The increased amount of long grass allowed more extensive and severe fires which then prevented regeneration of the trees. Hence, the area at the time reported was locked into a new vegetation state. The elephant population was not going to extinction though it may have been reduced, since elephants were able to live on long grass. In terms of Fig. 5, the vegetation had been reduced below its lower boundary and thus became dominated, not only by predators but by fire as well.

I have given the examples to show that these simple models can be related to real situations. Perhaps more importantly, apparently diverse and unrelated events may be understood in terms of a few rules, as illustrated by the models. The models now let us define "overabundance."

V. OVERABUNDANCE

Since it is feasible for natural systems to have more than one stable state, a shift from one to another cannot be considered a detrimental event; for example, a low herbivore population regulated by predators and a high one regulated by food are both possible. Almost certainly, a particular ecosystem will switch back and forth between these depending on other parameters, such as weather, altering the position of the boundary. Furthermore, it is known that in some systems the progress of vegetation succession can reduce the lower domain of attraction until it disappears (i.e., boundary *Y* in Fig. 5 moves far to the left) and the population appears to erupt suddenly (Peterman *et al.*, 1979; Peterman, 1980).

The causes of many herbivore increases in natural areas are largely unknown and biologists have been tempted to guess at simple, facile explanations without good evidence. The point is, an increase to a high population never seen before is *not* evidence that things are going wrong and that the population is overabundant. For the same reasons, it is not valid to draw conclusions from related declines in vegetation.

When can we conclude that populations are overabundant? The situation depicted in Fig. 3 resulted from a herbivore forcing the vegetation across a boundary below which extinction occurred. In general, one can say that a population is overabundant if it crosses into a domain of extinction.

There are problems, however, even with this concept. In Fig. 1, the initial decline in vegetation (which does not cross any boundary) is accompanied by changes in species composition. Palatable and, hence, vulnerable plants are initially common, but by the time equilibrium is reached they are rare or extinct. Meanwhile other species, originally absent, become common. In North American range management parlance, these plant types are termed "decreasers" and "invaders" respectively (Stoddart *et al.*, 1975). Range managers have misguidedly used these plant criteria to determine whether herbivores were overabundant or not by declaring arbitrarily that a high proportion of "decreasers" was good while a high proportion of "invaders" was bad (Caughley, 1976b). Unfortunately, this led to management decisions that there were always too many animals (the reason for 30 years of elk culling in Yellowstone). These criteria result in the ridiculous conclusion that the only good herbivore population is one vanishingly small.

In the progression toward equilibrium, some of the plants go extinct. Do we try to save all the species? If we do, we fall into the range managers' trap. If we do not, we recognize that systems are dynamic -- species change in abundance and some go extinct. If this is so, then where does the baobab tree of Africa fit in? If it is a decreaser (unlikely), we must allow it to disappear or become rare, and if it is an invader, then we need not worry. In short, we cannot necessarily decree a herbivore population as overabundant because it reduces one or two plant species in an array of food types.

VI. ADAPTIVE MANAGEMENT

It is possible to develop predictive models based upon the relatively small changes already observed in a particular natural area. These models may predict whether the end result of these changes leads to extinction or not. But the models are necessarily extrapolations beyond our region of knowledge and it would be dangerous to initiate an extensive management program of culling, for example, without first testing whether these models are correct. As mentioned earlier, the predictions of irreversible population disaster and desertification (associated with elephant, elk, and many others) are models or hypotheses, if somewhat unrigorous, in the same sense as are mathematical predictions.

What we need to know for accurate prediction of overabundance is:

(1) The location of extinction boundaries.

(2) How these boundaries shift with different parameters in the ecosystem.

In time, it may be possible to develop predictive models, but with our current knowledge, the only feasible way is to allow the perturbation experiments to take place and observe from these the behavior of the system -- in particular, the location of the boundary and the processes involved (competition, predation, dispersal). This approach has been termed "adaptive management" (e.g, Holling, 1978).

Perturbations may either be natural or induced. There are usually many unintentional man-made perturbations. Ideally, one should allow perturbations in a small system or part of a larger one in the hope that one can apply the results to a larger system, but this may not always be possible. We have learned some important things already. From the Kruger wildebeest experience, for example, we learned there are two stability domains and that culling moved the population too close to the lower boundary so that it was vulnerable to unpredictable shifts of the boundary. In other words culling reduced the resilience of the population to withstand normal environmental perturbations.

VII. CONCLUSIONS

This necessarily brief and superficial discussion of current ideas of ecosystem dynamics is presented to show that stability characteristics determining equilibrium, or carrying capacity, are complex and in general beyond our present knowledge of real ecosystems. Yet we need to know these characteristics before we can say whether populations are "overabundant" or not. I have suggested that we make use of methods allowing or inducing perturbations as the best approach to finding out how a system works and whether populations are overabundant.

REFERENCES

Caughley, G. (1976a). Plant-herbivore systems. *In* "Theoretical Ecology" (R. M. May, ed.) Saunders, Philadelphia, Pennsylvania.

Caughley, G. (1976b). Wildlife management and the dynamics of ungulate populations. *In* "Applied Biology" Vol. I. (T. H. Coaker, ed.), pp. 183-246. Academic Press, New York.

Field, C. R. and Ross, I. C. (1976). The savanna ecology of Kidepo Valley National Park. II. Feeding ecology of elephant and giraffe. *East Afr. Wildl. J. 14,* 1-16.

Holling, C. S. (1973). Resilience and stability of ecological systems. *Annu. Rev. Ecol. Syst. 4,* 1-23.

Holling, C. S. ed. (1978). "Adaptive Environmental Assessment and Management." Wiley, New York

Peterman, R. M. (1980). Influence of ecosystem structure and perturbation history on recovery processes. *In* "The Recovery Process in Damaged Ecosystems" (J. Cairns, ed.), pp. 125-139. Ann Arbor Science, Ann Arbor.

Peterman, R. M., Clark, W. C., and Holling, C. S. (1979). The dynamics of resilience: shifting stability domains in fish and insect systems. *In* "Population Dynamics" (R. M. Anderson, B. D. Turner, and L. R. Taylor, eds.) *Symp. Brit. Ecol. Soc. 20,* 321-341. Blackwell, Oxford.

Scheffer, V. B. (1951). The rise and fall of a reindeer herd. *Sci. Mon. 73,* 356-362.

Sinclair, A. R. E. (1979). Dynamics of the Serengeti ecosystem: process and pattern. *In* "Serengeti, Dynamics of an Ecosystem" (A. R. E. Sinclair and M. Norton-Griffiths, eds.) Univ. of Chicago Press, Chicago, Illinois.

Smuts, G. L. (1978). Interrelations between predators, prey and their environment. *BioScience 28,* 316-320.

Stoddart, L. A., Smith, A. D., and Box, T. (1975). "Range Management, Third Edition." McGraw-Hill, New York.

Walker, B. H. and Noy-Meir, I. (1979). Stability and resilience of savanna ecosystems. *In* "Symposium on Dynamic Changes in Savannah Ecosystems." S.C.O.P.E., Pretoria.

THE DEBATE ABOUT THE VICUNA POPULATION IN PAMPA GALERAS RESERVE

K. C. Otte
R. K. Hofmann

Proyecto Vicuna
Nazca
Peru [1]

I. INTRODUCTION

The purpose of the Pampa Galeras Vicuna Project has been widely misunderstood, particularly by certain conservation bodies which have stated that the Project is slaughtering vicuna in order to exploit its wool for the luxury market. The principal problem of the vicuna *(Vicugna vicugna* Molinia 1782) is reconciliation between the conservation of a rare species and the needs of the human population where it occurs. The problem is not one of local overabundance of an endangered species in a badly designed national park but is an artificial one related to conflicts with human interests. It may be classified as a transition between the Class 2 and Class 3 of Caughley's (this volume) overpopulation categories, or perhaps even Class 5. [2]

Eventually, the problem is how to compensate the local people for their agreement to tolerate vicuna on their land and allow this species to share the grazing with their domestic stock. Such economic incentives for conservation are the only ones likely to be accepted in developing countries.

The high value of vicuna wool stimulated excessive poaching; by the middle of this century the species had been greatly reduced in numbers and was in danger of extinction. Throughout history and during the days of the Inca Empire the vicuna has

[1]Address to which requests for reprints should be sent: *Apartado 84, Nazca, Peru.*

[2]See Workshop Report.

ISBN 0-12-385280-3

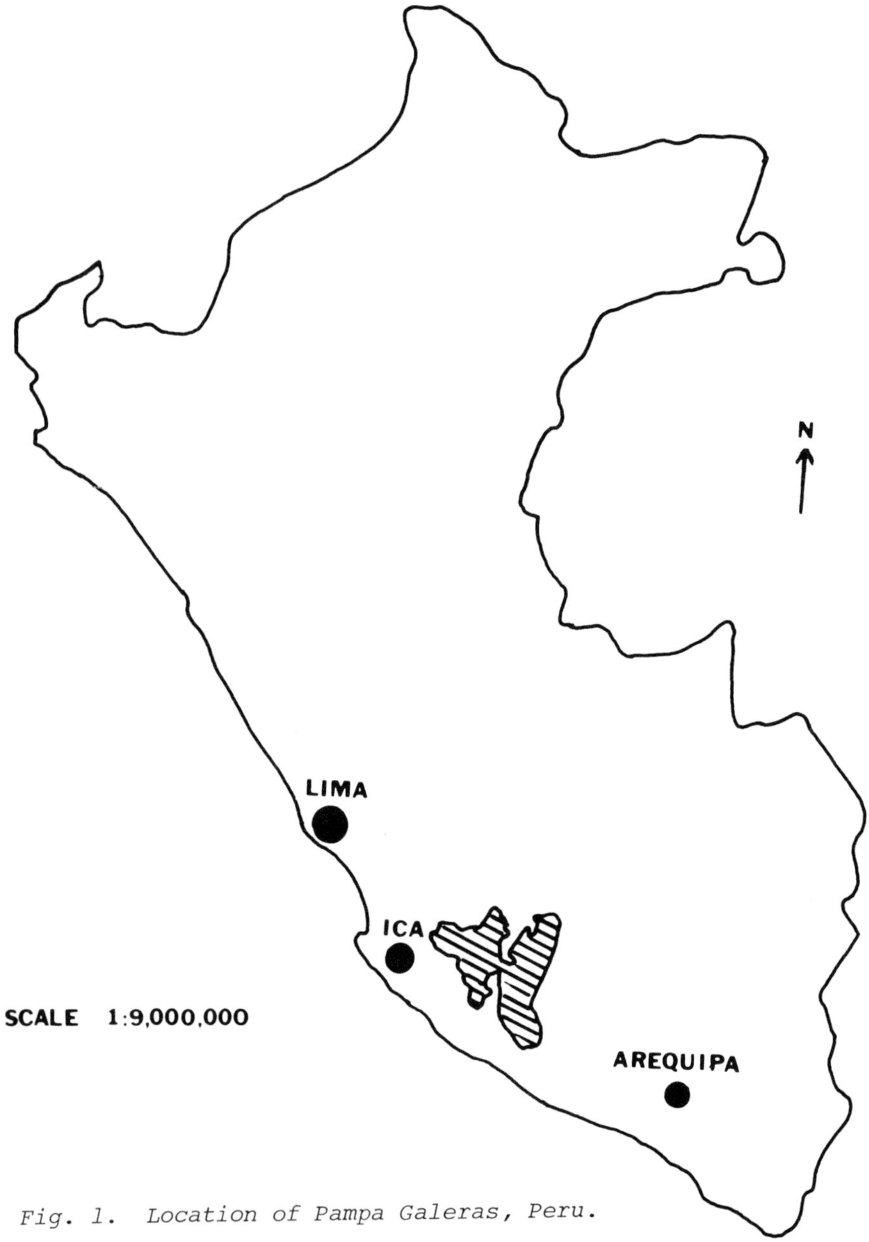

Fig. 1. Location of Pampa Galeras, Peru.

been fully protected by law, but the protection was not enforced. In order to save the species a reserve, Reserva Nacional de Vicunas de Pampa Galeras, was established in 1967 in the High Andes (Fig. 1). Originally, the Reserve was only 6,500 ha in area and contained approximately 1,200 vicuna, but it has since been expanded to 520,000 ha and it is intended that the eventual size will be about 1,000,000 ha. With the expansion of the Reserve, larger numbers of vicuna became protected and with the control of poaching, the population expanded until in 1980 numbers were in excess of 48,000. The reserve is not devoted only to vicuna; large numbers of domestic stock -- cattle, sheep, horses, donkeys, llamas, and alpacas -- share the range.

In 1972 the Project for the Rational Utilization of the Vicuna was established to manage the population for the mutual benefit of the species and the local people. The work is carried out by a team forming part of the German Technical Cooperation Program for Peru. The objectives of the Project include the exploitation of the vicuna through its meat, wool, and hides for the benefit of the local farmers and the repopulation of the Andes with vicuna throughout its former range in order to increase the economic potential of marginal areas through the capture and sale of living animals, sport hunting, and tourism. Such traditional ways of using wildlife would also be suitable for other species living in the Andes, such as deer, guanaco, viscacha, puma, and waterfowl. Exploitation needs to be on a sustained yield basis and part of the revenue, possibly through a tax system, should be plowed back into the Project to finance policing, management, and research.

II. BIOLOGICAL BACKGROUND: THE PUNA AND THE VICUNA

The most productive regions in the Andes are already intensively settled and used for cultivation or ranching. The remaining wilderness areas known as the puna, are marginal and have been mainly unproductive or exploited by unsuitable land use techniques. The principal livelihood of the Andean peasants of the puna is the keeping of exotic livestock on a semi-nomadic or transhumance basis.

A. *The Puna*

The puna is a grassy steppe characterized by tussocks of "ichu" grasses such as *Festuca, Stipa* and *Calamagrostis* species intermingled with "tola" dwarf bushes and shrubs (e.g., *Lepidophyllum, Senecio* and *Gynoxis* ssp.) During the rainy season or in permanently wet regions, a lawn-like sward develops from the ichu and tola vegetation on which most of the herbivores graze. The puna in Peru extends over some 6,500,000 ha at an average altitude of 4,200 m, varying between 3,800 and 4,800 m. On a land use basis, about 40% of the area is suitable only for vicuna, 30% for mixed vicuna and domestic stock, and 30% preferentially for use by domestic livestock only.

The puna is generally cold and windy. The mean annual temperature is 7.9°C and the mean annual precipitation is 463 mm, varying from 300 mm near the Pacific coast to 800 mm further east. The rain falls mainly during the months of December to April in the form of violent rains and hail storms. There are occasional snowfalls at night. Details of the annual fluctuations in rainfall and temperature at Pampa Galeras are given in Fig. 2.

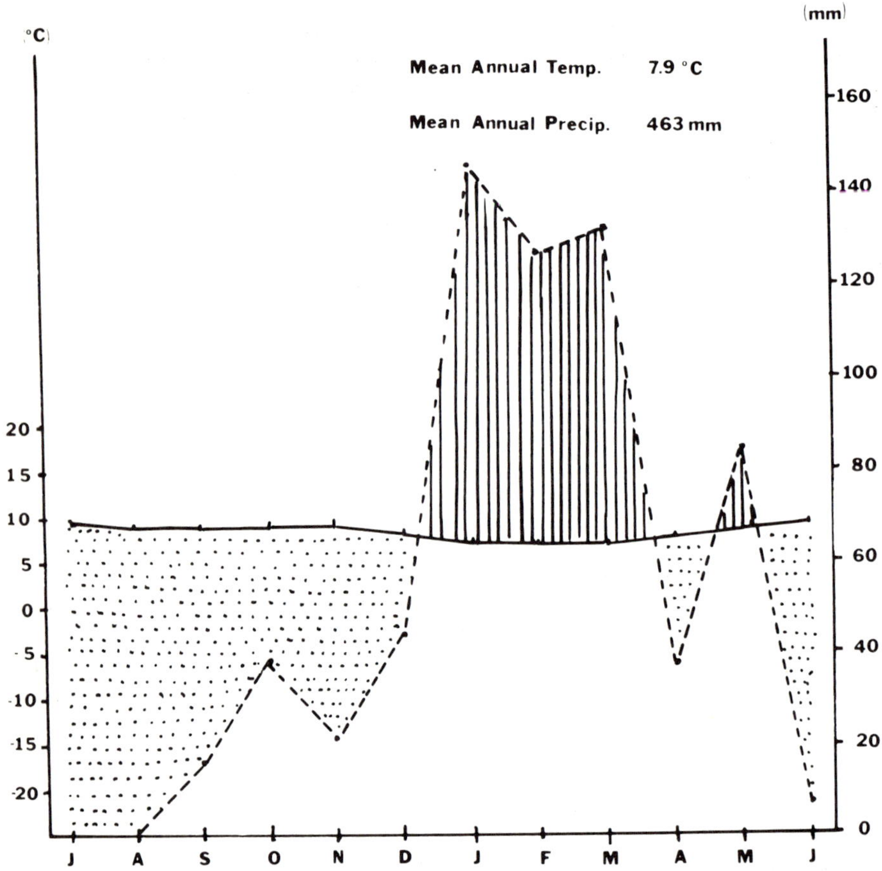

Fig. 2. Climatodiagram of Pampa Galeras (altitude -- 3,995 m), showing annual temperature and precipitation fluctuations, July 1966 to June 1977.

B. The Vicuna

The principal value of the vicuna lies in its wool, which is the finest known from any mammal. The average diameter of the wool fiber is 12.0 μm (range 10.0 to 13.5 μm) and the follicular density is very high at 7.9 follicles/mm^2.

The vicuna is the smallest species of the tylopods or camels and shows a number of features, besides its wool, which adapt it to its puna habitat. The lower incisors are unique among living artiodactyls in being continuously growing with open roots and enamel only on the labial sides, rather like the incisors of rodents. The teeth enable the vicuna to bite off the tough vegetation without uprooting the plants as often happens with domestic stock. Vicuna are also able to graze the very short grasses and prostrate herbs much closer to the ground than can domestic animals and so make further use of the pasture after the domestic herds have left. Another anatomical feature by which the vicuna is adapted to the puna is the soft pad on the third phalange of each digit. These pads ensure that the vicuna does not damage the soil surface by causing erosion as it walks over the puna (unlike sheep and cattle whose hard hooves break up the surface).

The calf is born at an advanced stage of development with a birthweight equal to about 15% (14.5% to 16.5%) of the mother's weight. It is hardy and begins to take vegetation when one month old.

The social life is based on a territorial system with three categories: territorial males with their harems of females, bachelor groups, and solitary animals (Koford, 1957; Franklin, 1976). The family unit is the most stable group consisting of a territorial male and varying number of females and calves. About 65% to 68% of the population are found in such family groups. Of these groups, some 75% to 80% are found on the better pastures and occupy their territories all year round. The remaining family groups are divided between two other categories -- the migratory and marginal territorial groups -- which are unstable. These groups are smaller in size and form an intermediate stage which would later take over the prime territorial sites.

The bachelor group is made up of a changing number of subadult to adult males and does not hold a territory. Bachelor groups, which have no leaders and are constantly changing in composition, form 26% to 28% of the population. They live in peripheral habitats and form a reservoir from which territorial males are recruited. From a management point of view, they represent an important sector of the population that can be exploited to increase the sustainable yield.

The remaining vicuna, comprising 5% to 6% of the total population, live solitary roaming lives and are mainly senile animals.

III. MANAGEMENT PROBLEMS IN PAMPA GALERAS

Because of low temperatures and scanty precipitation, humus is formed very slowly with soil low in microorganisms. Nitrogen and phosphorus are available only in small amounts in the somewhat acidic soils. The resulting low productivity, coupled with the wide diurnal fluctuation in temperature and severe night frosts, makes the community-owned puna unsuitable for agricultural development and extensive stock rearing is the most appropriate form of land use. The higher numbers of grazing animals have led to problems of erosion through overgrazing.

Plant species which indicate overgrazing, such as *Opuntia, Senecio, Lepidophyllum,* and *Pycnophyllum* ssp., are apparent everywhere. Necessary remedial action includes control of indiscriminate burning, regulation of herbivore stocking rates, pasture management with rotational grazing to leave some standing hay, and the reestablishment of some native "queunua" forests *(Polilepis* spp.)

The question arises whether the vicuna is one of these species from harsh environments which undergoes a series of population booms and crashes. With such a species it is impossible to achieve a state of balance in population numbers. There is some evidence that a stable limit cycle (Caughley, 1976) is a likely outcome between the vicuna and the puna.

A. Evidence of Overpopulation

Evidence of overpopulation comes from the identification of detrimental changes in the environment which could lead to the breakdown of the ecosystem. Since it is unlikely that any single measure will suffice to diagnose overpopulation, a number of criteria must be used. In the case of the vicuna these include the degree of soil erosion, the state of the vegetation, the condition of the herbivore, diseases, and the relationship with other species.

1. Soil Erosion. Soil in the Andes is shallow and can easily be eroded. One heavy rainstorm can wash off surface soil that has taken twenty years to accumulate. Gully and sheet erosion is presently occurring in the Pampa Galeras Reserve leading to loss of nutrients, spread of gravel, and a reduction in vegetation. This loss can be correlated with an accelerating increase in the vicuna population ($\bar{r}$ = 0.192).

2. *State of the Vegetation*. A theoretical analysis of the dynamics of plant growth and grazing predicts that the interaction between the vegetation and an efficient herbivore such as the vicuna may stabilize at several levels and reduction in the vegetation may not be reversible (Noy-Meir, this volume). The management implication is that although the herbivores, both vicuna and domestic stock, may give a high sustained yield, there is a high risk of a population crash. A knowledge of the interaction between plants and herbivores is necessary before overgrazing can be defined. Overgrazing occurs when the animal removes plant tissue faster than it can be replaced by photosynthesis. When this occurs, the plant draws on its root reserves leading eventually to the regression of the roots and death of the plant. The problem is exacerbated with the vicuna because the growing season for the vegetation is short, about four to five months per year, and the period during which photosynthesis can occur each day is only six to eight hours because of the low night temperatures. In the nongrowing season, food reserves are transferred to the roots leaving lignified tissue above ground which may not contain the 4% protein necessary to provide most ruminants with an adequate diet.

The simplest method for determining the quality of the rangeland is to select a key plant species that is palatable and abundant and to estimate the level of grazing on it. One such plant in Pampa Galeras is the "quishna grass" *(Calamagrostis vicunarum)*, which ten years ago was one of the most abundant forage plants for vicuna (Tovar, 1973). Today its frequency is low, being present on only 69.5% of the earlier sites and at a medium frequency of 9.8% (Hoces, personal communication 1980).

Another technique is to compare subjective classifications of the rangeland. Five years ago, 25% of the pasture sites were classified as "good," 65% as "fair," and only 10% as "poor." In 1979, 63.3% of the sites were classed as "poor" and 34.7% as "very poor." In 1980, the range had deteriorated further with 9% "poor" and 91% "very poor." By 1980, the relative vigor of the most important forage species diminished to 54.8% of its 1979 value. The basal ground cover by vegetation declined from 28% in 1976 to 11% in 1979 and net aerial primary production fell 43% from 405 kg dry matter/ha/yr to 231 kg/ha/yr in 1980.

The impoverished condition of the pasture in 1980 has resulted from a series of dry years (1977 to 1980) with only 49% of the average rainfall and a concommitant rise in the density of the herbivores (which decreased in 1980 as a result of the cropping of vicuna and removal of domestic stock). On the basis of the widely accepted criteria that the dry weight forage required for one animal unit each month on rangeland of average quality equals the live weight of the animal and that no more than 50% of the annual net aerial primary production on poor pasture should be taken by herbivores, it is estimated that the Pampa Galeras rangeland is overstocked by a factor of four to six. This places the vicuna overpopulation into Class 5 of Caughley's categories.[1]

3. The Condition of the Vicuna. The increase in vicuna numbers has probably resulted from the protection given to it against poachers and the control of feral dogs. Cases are well known of herbivore populations increasing until they damage their habitat after the removal of predators. The vicuna population has been counted annually by coordinated ground counts and the results are given in Fig. 3. This shows that the population is doubling every 3.6 years, principally because of the improved survival of newborn calves. Previously, human poachers and feral dogs accounted for some 40% of the neonate mortality with bad weather, causing pneumonia or bacillas diarrhea, accounting for 35% to 40% of the deaths. The remaining mortality was due to wild predators such as foxes, pumas, and condors.

Evidence that the vicuna population is too high comes from a number of sources.

a. Physiological condition. The Kidney Fat Index (KFI) has proved to be a useful measure of physiological condition, although there are some anomalies for cervids such as those reported by Dauphiné (1975). Measurements of KFI are now underway with vicuna, but the species deposits very little visceral fat apart from the characteristic sternal fat on each side of the sternum. The physiological condition of vicuna, therefore, is best reflected by its subcutaneous fat and body weight. Studies on bone marrow fat have not yet been much applied, but could be worthwhile.

Changes in fat reserves can occur in animals without marked changes in their appearance, and this is particularly likely with vicuna whose long woolly coat hides the body contours.

Observations of the vicuna have shown a markedly diminished fat deposit on the sternum, particularly towards the end of the year (October to January). There has also been a decline in

[1] *See Workshop Report.*

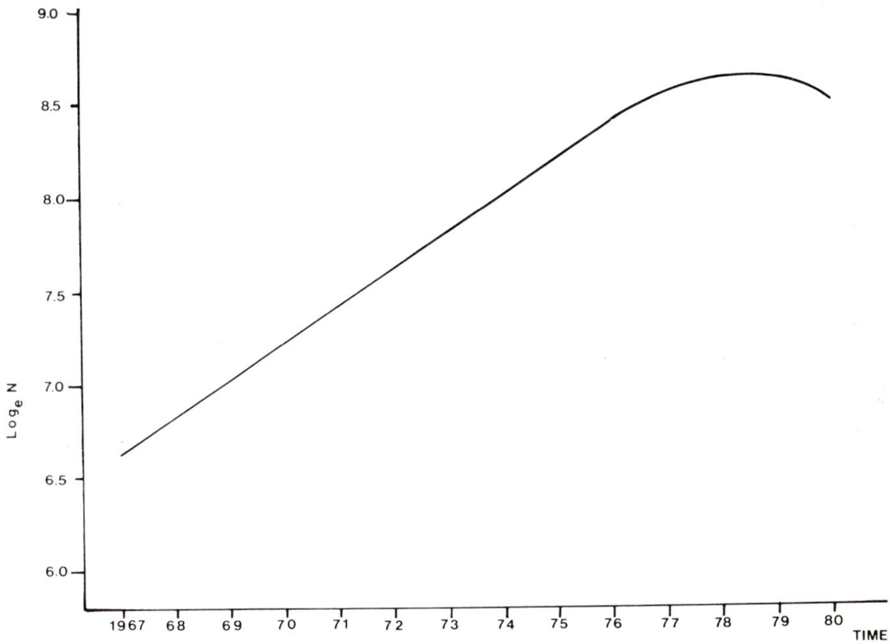

Fig. 3. Population growth of vicuna in Pampa Galeras.

the growth rate of the calves during their first and second years on account of food shortage. Body weights of males two years of age have declined from an average of 30.8 kg in 1977 to 27.9 kg in 1979. During the same period, the overall length of the body declined from 155 cm to 150 cm. Even greater reductions were recorded with younger vicuna.

b. Demographic vigor. Demographic vigor can be measured by the rate at which a population increases. It can be expressed as the survival/fecundity rate of increase (r_s) but this is a difficult factor to measure and calculation of the regression coefficient s over a period of time represents a more suitable technique.

While the Pampa Galeras vicuna population increased at the average doubling time of 3.6 years (r = 0.192; finite rate of increase = 1.21) between the years 1967 and 1978, there has recently been a pronounced change and the vicuna are now decreasing with numbers halving every 5.4 years ($\bar{r}$ = -0.128; finite rate of decrease = 1.14) (Fig. 4).

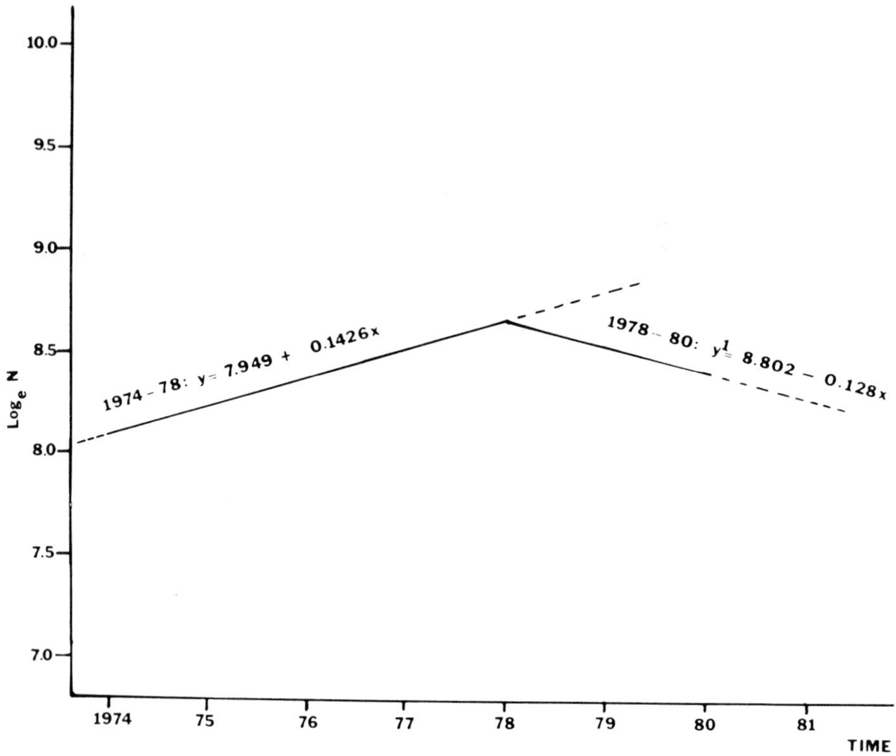

Fig. 4. Vicuna population growth turning point.

The decrease in the vicuna population has probably resulted from factors commonly found in other ungulates under conditions of poor nutrition -- increase in juvenile mortality, increase in age of first reproduction in females, decline in fecundity, and increase in adult mortality. An examination of the vicuna population shows that these factors are operative and could result in a population crash. Franklin (1976) reported that in 1969, 85% of the adult females (two years and older) produced calves and 76% of these still had a calf at the end of the year; the figures for 1980 were a pregnancy rate of 58%, 24% of females produced a calf of which only 27.5% were still alive in October. Because of the culling program currently being carried out at Pampa Galeras, it is difficult to assess the adult mortality but this is estimated at 15% to 20%.

c. Behavioral aspects. Several population studies have indicated that loss in condition is reflected in changes in some types of behavior which may be broadly grouped into two categories -- social interactions and feeding strategies.

Watson and Moss (1970) discussed dominance, spacing behavior, and aggression in relation to population limitation in vertebrates and concluded that behavior leads to limitation of a population under certain conditions. These include a large nonbreeding sector in the population, the ability of these nonbreeders to reproduce if the breeding animals are removed, and that breeding animals must not use up completely some necessary resource such as food, nest sites, or space. All these conditions are met to a greater or lesser extent in the vicuna. The bachelor groups form the nonbreeding sector and they are physiologically capable of reproduction, so the first two conditions are satisfied. The third condition is partly satisfied. Overcrowding of the vicuna population is indicated by a reduction in the mean territory size to some 22% of its former value (from 50.4 ha in 1970 to 11.3 ha in 1979), while the mean size of the family group remained more or less unchanged (6.5 in 1970 and 6.0 in 1979). The level of aggression increased particularly between territorial and bachelor males. There are examples in which two or more factors may limit the population (Watson and Moss, 1970), e.g., territoriality and dominance or food. This is especially likely to be the case with vicuna as far as social behavior and food shortage are concerned.

Feeding strategies are taken to include all aspects of the animal's feeding behavior as well as quantity and quality of the food and time spent feeding. When food becomes limiting for any reason, plant species of high forage quality are reduced and often eliminated (Klein, 1970). Changes in the feeding habits of the vicuna in Pampa Galeras have been found indicating both range degradation and overpopulation. The animals have extended their feeding periods by 15% to 20% over the past three or four years, especially during the late evening hours, and have continued feeding after sunset on moonlight nights. This change has been caused by a reduction of 56% in their preferred food plants during the last three years (Hoces, personal communication). The vicuna have now turned to ichu grass and even tola bushes which were hitherto practically untouched but which now form up to 40% of the diet.

4. Relationships between Vicuna and Other Mammalian Herbivores. The effects of overpopulation apply to other species besides vicuna, but in the High Andes the indigenous vicuna is at a competitive advantage over introduced domestic stock. Local ranchers do not allow their animals to die but sell them when the carrying capacity is lowered. General herd reductions on Pampa Galeras started in 1977, and herd size reached its lowest level in 1979 when only 16.8% of the original stocks remained.

5. *Predators and Disease.* Predator control of foxes and feral dogs has prevented their numbers rising with the increase in vicuna. On the other hand, there has been an increase in parasitic infestation rising in 1979 to three times its 1976 level. The organisms concerned are mainly ticks (*Amblioma parvitarsum*), lice (*Microthoracius minor*), and intestinal worms (*Nemotodirus* spp.) Even a few cases of mange (*Sarcoptes auquenidae*) have recently been diagnosed, a disease not recorded in Pampa Galeras over the past twenty years.

IV. MANAGEMENT OPTIONS

Management decisions must ensure that actions do not have undesirable side effects, of which there is a danger in marginal lands such as the High Andes. A balance must be found between the space allowed for an endangered species and the immediate socioeconomic needs. Hence the first essential step is to define the objectives for a particular area, as has been done here earlier in this chapter. Decisions usually have to be made with incomplete knowledge of the natural system and of the possible consequences of management. It is, therefore, desirable to plan alternative management strategies and to err on the side of caution. The best solution is for the wildlife manager to assume that the worst will happen and then to attempt to minimize the maximum possible loss. Such an approach may be called a minimax strategy. Each management project should be regarded as an experiment and, therefore, should include provision for monitoring the effects of management action.

The first management option is to do nothing and to let nature take its course, but this is often not an adequate response. The second option is to exploit the population for its maximum sustained yield. This approach is easier in ecosystems where events can be reasonably well predicted, but many environments, such as the Andes, are subject to unpredictable fluctuations which inevitably reproduce stochastic elements into any management plan. In such cases it is unlikely that there is a unique maximum sustained yield. Nevertheless, a minimum sustained yield may be a suitable concept, especially if precautions are taken in its calculation. The raw problem lies in the uncertainty over the resilience of the population to cropping and to changes in its age or sex structure. Judicious manipulation can substantially increase growth rate and hence the yield, as can changing the sex ratio by selectively cropping males in polygynous species like the vicuna.

Sustained yield cropping, therefore, must be based on management plans that are constantly reviewed and adjusted as necessary through a feedback system involving research, management, and monitoring.

An alternative is adaptive management in which provision is made for the protection of suitably large examples of natural ecosystems. Ecosystems can reach equilibrium only if they are sufficiently large, so it is better to maintain a few large protected areas than many small ones of the same overall size. Small perturbations in the equilibrium of the ecosystem can be allowed to run their course. Such an approach, of course, is suitable for the management of national parks and reserves. Although minimum interference should be an important principle, a policy of "do nothing" is as bad as overreacting to events.

V. MONITORING/RESEARCH

Effective management requires information which can be supplied through monitoring of such factors as the size, structure, and range of the populations, their food and cover requirements, and the adequacy of the various habitats to support wildlife. It is also necessary to follow the biological changes that occur as a result of management action and to record any trends that may lead to the establishment of new equilibria.

There is a danger that information will be collected for its own sake and not applied in management. The consequence then is the same as a policy of noninterference. There is also the problem that newly appointed local staff will wish to demonstrate their independence by interfering in management plans without considering the knowledge built up by former managers. This is what has happened over the past two years in the Pampa Galeras Reserve, where decisions made by the newly installed Executive Director of the Vicuna Management Project are endangering the survival of the species.

VI. CONCLUSIONS AND RECOMMENDATIONS

Management decisions concerning Pampas Galeras must take into account three principal facts. First, the vicuna is still considered an endangered species; second, its habitat is the delicately balanced and ecologically vulnerable Andean puna;

third, the main grazing areas are communally-owned land. Hence management must render some economic benefit to the Andean farmers without endangering either the vicuna or its environment. Culling, in its various forms, and enlarging the management area are the two principal solutions.

A. Culling

Culling is the selective removal of animals from a population either through shooting or by translocation. The main arguments for culling, given in the preceding pages, are that the Project is supposed to produce some revenue for the local people and lack of effective management has resulted in an overpopulation of vicuna.

The overriding requirement for the success of any management plan is adequate law enforcement, and this has been the main reason for the rise in vicuna numbers. But now, with the new Peruvian direction of the Project, poaching has proliferated, introducing an unpredictable element into vicuna management. Unless this can be stopped, management plans cannot be put into effect and vicuna numbers will decline.

Enlargement of the management area has always been, and remains, a principal objective of the Project. The "artificial" overabundance of vicuna in the Pampa Galeras, now requiring emergency action, has been created by the administrative inefficiency of the Peruvian direction, exacerbated by three consecutive years of drought.

B. Not Culling

The consequences of not culling vicuna in the Pampa Galeras will be detrimental to the environment and to the economy of the local people through a reduction in the available grazing. Without culling, there will be reduced carrying capacity and a decrease in the diversity of both animal and plant species. If this is allowed to continue unchecked, there could be a total loss of vegetative cover.

Factors which may limit the population naturally fall into four categories: inter-specific competition, natural enemies, environmental heterogeneity, and genetic feedback mechanisms. The first two factors, including poachers, are already operating; those of the third are coming into play. But the fourth, genetic consequences, will require several generations in which to develop.

Starvation will not automatically intervene as a convenient safety valve to remove a population surplus without harming the survivors. By the time starvation begins to cause mortality, the resources will have been overexploited. Genetic fitness tends to deteriorate when population density is high and resources scarce.

Special care is necessary in these mountainous rangelands where plant cover must be adequate enough to prevent soil erosion. Carrying capacity of the puna is lower than that of similar areas of flat land.

C. Recommendations

Culling through translocation and cropping is recommended as the best solution but, because of the problems already mentioned, certain precautions are necessary. Because of the evidence that many ecological systems have multiple equilibrium states, culling should aim at keeping the population well away from the boundaries of equilibrium systems. As the latter are at present unknown, it is recommended that experimental culling should be practiced with deliberate overharvesting of stocks in some areas to create vacuum or buffer zones but leaving other areas completely untouched. This recommendation was made to the Peruvian government some years ago but, unfortunately, so far there has been no response.

1. Translocation. The capture of wild vicuna (apart from selling a few to zoos, etc.) would make sense only if they were sheared and released but, because of the marked territoriality of the species, such operations would be extremely costly and justifiable on sociological grounds only if job opportunities were created for local people.

Translocation from areas of high density to ecologically more suitable regions is a feasible solution only if the numbers involved are manageable and suitable resettlement areas are within reasonable distances of 50 to 150 km (i.e., within a day's drive in a truck) from Pampa Galeras. The only region to which translocation over a greater distance would be justified is the newly established National Reserve of Aguada Blanca near Arequipa.

2. Cropping. Several wildlife studies have demonstrated that cropping rapidly reverses habitat changes caused by overgrazing. Grasses, which are *r*-selected, recover more quickly than *K*-selected bushes or trees to reduction in herbivore pressure. The object in the Pampa Galeras Project is to determine the characteristic return time of the system to its former level. This time will usually increase as harvesting takes the population to and beyond the maximum sustainable yield level (Beddington, 1979). Estimates indicate that for vicuna/puna systems, return times are of the order of 5 to 15 years, according to the level of reduction, available forage resources, etc. All cropping models for vicuna must take into account the domestic livestock which share their range.

Methods have been developed in Pampa Galeras which allow the quick and humane shooting of vicuna in large numbers and which permit the hygienic processing of the carcasses. The shooting of complete family units prevents social disturbance and reduces the population without much altering its structure. A further advantage is that reproductive rates and other demographic data can be maintained.

VII. SUMMARY

The case of the vicuna in Pampa Galeras, Peru, has been described in this chapter, and it is suggested that local overabundance has led to undesirable changes in the vicuna population and to overgrazing with a consequent loss of soil and nutrients through erosion.

The vicuna live on privately-owned community land where domestic animals share the range. Utilization of the vicuna is necessary to solve the problem of overabundance and to promote rural development. The Project was intended to be an animal production operation and it is only through the lack of effective management that the population of vicuna in the Pampa Galeras has undergone a natural eruption leading to several density dependent responses.

The situation should not immediately lead to catastrophic consequences for the survival of the species itself, but it definitely demands some culling operations, which should include translocation and cropping at different levels of intensity. At the same time control areas should be set aside where vicuna are not controlled in any way.

REFERENCES

Beddington, J. (1979). Harvesting and population dynamics. *In* "Population Dynamics" (R. M. Anderson, B. D. Turner, and L. R. Taylor, eds.), *Symp. Brit. Ecol. Soc. 20*, 321-341. Blackwell, Oxford.

Caughley, G. (1976). The elephant problem -- an alternative hypothesis. *E. Afr. Wildl. J. 14*, 265-283.
Dauphiné, T. C. (1975). Kidney weight fluctuations affecting the kidney fat index in caribou. *J. Wildl. Manage. 39*, 379-386.
Franklin, W. L. (1976). Socioecology of the vicugna. Ph.D. Thesis, Utah State University.
Klein, D. (1970). Food selection by North American deer and their response to overutilization of preferred plant species. *In* "Animal Populations in Relation to Their Food Resources" (A. Watson, ed.) Blackwell, Oxford.
Koford, C. (1957). The vicuna and the puna. *Ecol. Monogr. 27*, 153-219.
Tovar, O. (1973). Comunidades vegetales de la Reserva Nacional de vicunas de Pampa Galeras, Ayacucho, Peru. *Publ. Mus. Hist. Nat. Javier Prado Ser. B 27.*
Watson, A. R. and Moss, R. (1970). Dominance, spacing behavior, and aggression in relation to population limitation in vertebrates. *In* "Animal Populations in Relation to Their Food Resources" (A. Watson, ed.) Blackwell, Oxford.

THE VICUNA OF THE PAMPA GALERAS NATIONAL RESERVE THE CONSERVATION ISSUE

S. K. Eltringham

Department of Applied Biology
University of Cambridge
Cambridge
England

W. J. Jordan

People's Trust for Endangered Species
Guildford, Surrey
England

I. INTRODUCTION

The Pampa Galeras Reserve, which lies in the Department of Ayacucho, Peru, was established in 1964 when the population of vicuna (*Vicugna vicugna)* in the 6,500 ha area was believed to be under 1,000. As a result of protection, vicuna numbers increased until by 1979 the population size was reported to be 43,471 in a greatly expanded reserve (which now spreads over 577,750 ha). This population represents about 80% of the estimated total number of vicuna in the world. The Pampa Galeras Reserve is not devoted exclusively to vicuna -- cattle, sheep, llamas, alpacas, horses, and donkeys are also kept on the rangeland. The Project for the Rational Utilization of Vicuna was established in 1965 by the Peruvian Ministry of Agriculture. The policy of the Project is set out by Hofmann and Otte (1977); once numbers have built up to a suitable level, the principal intention is to round up the vicuna every two years, shear them, and release them back into the wild. The profits from the venture, which would be under state control, would be used to improve facilities

ISBN 0-12-385280-3

for the local people. Mention is also made of the possibility of translocating vicuna to other suitable regions. In optimally stocked areas some game cropping is contemplated, yielding further income through hides and meat. It is not proposed to domesticate vicuna or to interbreed them with llamas or alpacas.

II. CROPPING

The Project, therefore, envisages the Reserve as an area where vicuna numbers should be built up (through the control of illegal hunting) so the population can be exploited commercially by methods that include cropping. This in itself is not necessarily against conservation interests provided certain safeguards are observed. In the first place the productivity of the population must be known so that a sustained yield may be calculated. Although there are various mathematical models from which yields can be estimated, the results obtained from them are only approximate and a certain degree of trial and error is necessary. The best policy is to err on the side of caution and to monitor the population regularly during cropping so that signs of overexploitation can be recognized. For a cropping operation, therefore, it is necessary to know the size of the exploited population with some accuracy.

In the case of the vicuna it should be pointed out that the species is listed as vulnerable in the International Union for the Conservation of Nature and Natural Resources (IUCN) Red Data Book and consequently great care is necessary to ensure that the animals are not overexploited. Another important caveat is that the vicuna is listed on Appendix I of the Convention on International Trade in Endangered Species of Wild Fauna and Flora (CITES) which prohibits trade in its products. No income can legally be acquired through the exploitation of vicuna until 1989 when the present moratorium expires. Any skins or wool obtained before then will presumably be stored, and the objective of providing funds for local people will not be immediately realized.

A further worry is that a cropping program, or even a shearing program, increases the risk of illegal killing. Once legal trade in an animal product is established, it is not too difficult for unscrupulous persons to channel their own illicit goods into the system. This has certainly been the case in the ivory trade, for instance. A cropping operation in the Pampa Galeras Reserve might well be operated properly and illegal hunting controlled, but other, far smaller, populations of vicuna would be at risk not only in Peru but also in the other South American countries where the species is found. Some of these populations are so small they could easily become extinct if exploited.

Finally, there is always the risk in a cropping program that economic factors will take precedence over ecological interests. In order to meet financial targets, it may be necessary to take a yield in excess of that which the wildlife population can sustain. The history of the whaling industry does not inspire confidence that ecological considerations will always prevail in wildlife exploitation.

Our conclusion is that although cropping of vicuna could be unexceptional on conservation grounds provided rigid control is exercised, it is fraught with dangers that make it inadvisable. We believe that shearing would be a far safer form of exploitation and probably more lucrative in the long run. We are uncertain of the effect of shearing on the survival of vicuna in the very cold environment, but we assume that research into this problem has been or will be carried out.

A. Is the Population Large Enough to Crop?

There is no doubt that vicuna have recovered very considerably in the Pampa Galeras Reserve and adjacent management zones. According to Brack (1980), the population increased from 1,000 in 1967 to 43,471 in 1979. However, it transpires that most of this increase is artificial and due to periodic inclusion of fresh regions into the census areas. Nevertheless, a population of over 40,000 would be large enough to crop. Various conservation bodies, however, challenged the accuracy of the count and the International Fund for Animal Welfare (IFAW) asked S. K. Eltringham to carry out an aerial survey. This was arranged through the Ministry of Agriculture in Peru and was carried out in April 1980 with the cooperation of the Air Ministry in a Peruvian Air Force plane. The task was a formidable one since an aerial count of an animal population had never been made at that altitude (about 4,000 m) before, and the usual light aircraft used in aerial counting could not operate at that height. Although far from ideal for this type of activity, a Twin Otter was used to complete the census. The technique used was the standard one of a systematic, stratified sample covering 7.4% of the area with sampling intensities ranging from 5.4% to 13.7% depending on the density of the vicuna. The total recorded was 11,489 ± 2,689 in the whole management zone. This total is certainly too low, for it is well known that aerial counting of animals the size of a vicuna considerably underestimates numbers. It is usual to correct underestimates from simultaneous ground counts, but these could not be made in the present case and there is no other reliable way of checking the total. Errors of up to 50% have been recorded with African antelopes. However, if the total vicuna population given by Brack is correct, the accuracy of this aerial count is as low as 26.4% (or more probably 22.1% since his figures refer to 1979 and if the population had continued to rise at the rate indicated, the 1980 population, with which the aerial total should be compared, would be about 52,000).

The aerial total has been dismissed as useless, but it serves to cast doubt on the ground totals given by Brack. Those counts are made by rangers on foot or horseback who count the animals over a period of some six weeks in September and October. We have not been able to assess the accuracy of the technique, but clearly there is the probability of some margin of error. Further misgivings arise when one takes a closer look at the data in Brack's report. There are some discrepancies between the totals given in his Table II and those that can be derived from his Tables IV and VI. There are also several mathematical errors in the total columns of Table VI. In addition the figures have to be interpreted with care because of the new regions that have been included in the census areas. Allowance can be made for these new regions after 1976 when separate totals were recorded for the individual management zones. If three of the added regions (Chalhuamayo, Palloa, and Soras) are discounted, the 1976 total of 24,448 vicuna (from Brack's Tables IV and VI) had risen to 40,136 in 1979. Details are given in Table I of this chapter.

Our analysis shows annual rises within the Zona de Influencia as high as 57.5% with an average annual increase of 40.7%. Even within the Zona Nuclear, where the population decreased 13.7% between 1978 and 1979, there was a 31.7% increase in the previous year. These increases are extremely high for an ungulate species breeding once per year and producing a single calf. Mortality would need to be extremely low before such rates were possible. For instance, in the Serengeti National Park in Tanzania, where buffalo and wildebeest are undergoing eruptions, the annual increases are not nearly as high. Sinclair (1973) shows an increase in buffalo from 24,500 in 1958 to 58,659 in 1972 -- an annual increment of 6.4%. Within this period, wildebeest have increased from 263,362 in 1961 to 692,777 in 1971 -- a rate of 10.2%/annum.

Examples of population increases in temperate zone ungulates give a similar picture. An increase in deer from 4,000 in 1905 to 100,000 in 1924, following elimination of predators on the Kaibab Plateau in Arizona, represents an annual rate of 18.5%. We can cite the increase of 25 reindeer (released on St. Paul Island in the Pribilofs in 1911) to a total of 2,046 in 1938, an annual increment of 17.7% (Scheffer, 1951), as an example of a population eruption following the introduction of an ungulate species into a virgin habitat with abundant food.

TABLE I. Annual Totals of Vicuna in the Pampa Galeras Reserve and Neighboring Regions[a]

	1976		1977		1978		1979
Zona Nuclear							
Total	*17388*		*18006*		*23716*		*20475*
Annual difference		*+ 618*		*+ 5710*		*- 3241*	
Percentage difference		*+ 3.6%*		*+ 31.7%*		*- 13.7%*	
Annual increment = 5.6%							
Zona de Influencia							
Total	*7060*		*9614*		*12486*		*19661*
Annual difference		*+ 2554*		*+ 2872*		*+ 7175*	
Percentage difference		*+ 36.2%*		*+ 29.9%*		*+ 57.5%*	
Annual increment = 40.7%							
Both Zones							
Total	*24448*		*27620*		*36202*		*40136*
Annual difference		*+ 3172*		*+ 8582*		*+ 3934*	
Percentage difference		*+ 13.0%*		*+ 31.1%*		*+ 10.9%*	
Annual increment = 18.0%							

[a]*According to corrected data taken from Brack (1980).*

Because of the improbability of the reported increases in vicuna numbers, Brack's data have been subjected to a computer analysis. This was carried out by J. R. Beddington of the University of York. Beddington wrote a computer program that modeled the vicuna population and varied certain factors to see what would be required to achieve the increases recorded by ground counts. Assumptions made in the model are:

(1) The population had been increasing at an annual rate of 15% prior to 1976.

TABLE II. *A Computer Model of the Vicuna Population of the Pampa Galeras National Reserve and Surrounding Regions*[a]

Computed 1979 total	Pregnancy rate percentage	Adult survival percentage	Calf survival percentage
44350	100%	95%	90%
41645	100%	95%	80%
40563	80%	95%	95%
40293	100%	95%	75%
----------	----------	----------	----------
39481	80%	95%	90%
39217	60%	99%	95%
37588	100%	95%	65%
37317	80%	95%	80%
37115	100%	99%	50%
33530	100%	95%	50%
32177	90%	95%	50%
31488	100%	90%	60%
29138	100%	90%	50%

[a]Assumptions are a population rise of 15% prior to 1976, age at first calf of three years, annual reproduction with a single calf each year, and a life span of 13 years. The 1976 population is assumed to have been 24,750 and the population variables have been manipulated to see what is necessary to achieve the reported 1979 total of 40,136. The combination of factors that have produced a figure in excess of this value are shown above the dashed line.

(2) A female produces her first calf at three years of age (suggested by Koford, 1957).

(3) One calf is produced each year.

(4) The life span of vicuna is thirteen years.

Factors varied were the proportion of females that become pregnant each year, the calf mortality, and the adult mortality. The results are given in Table II.

Using the higher of Brack's figures for 1976, a total of 40,136 vicuna could have been reached by 1979 only if the values for the population variables had been improbably high. In all cases, adult survival would have had to have been at least 95%, while calf survival could not have dropped below 75%. If the proportion of pregnant females falls from 100% to 80%, both calf and adult survival must be at least 95%.

These figures are scarcely credible since Koford (1957) reports that calf mortality is about 50% after the first two years of life. Data in Brack's report support this since it can be calculated from details on the population structure that the proportion of females with young each year varied from 33.0% in 1979 to 55.7% in 1976. The time of year when the counts were made is not shown in Brack's report, but presumably further deaths could have occurred; consequently, these figures represent the minimum mortality. Table II shows that with a calf mortality, of 50%, the computed 1979 total is nowhere near 40,000 even with a 100% pregnancy rate and 90% adult survival. Clearly, the possibility that the reported rise between 1976 and 1979 was due to reproductive increase is so remote it may be discounted.

Closer examination throws further doubt on Brack's figures. The annual increment from reproduction cannot be higher than the number of females present in the previous year in a species that produces only one calf annually, yet in the Zona de Influencia Brack's data (after correction) shows such discrepancies. Thus, between 1978 and 1979, the annual increment exceeded the number of females present in 1978 by 1,532; between 1977 and 1978 the increment exceeded the 1977 total of females by 1,272. Brack has stated that the recorded increases are not due entirely to reproduction -- immigration and the addition of new count zones have also contributed. The latter factor has been allowed for in this analysis and need not concern us further. Brack gives no evidence for the existence of immigration (e.g., no animals were marked or followed), and it is difficult to see how immigrants could be recognized and distinguished from residents. One might also wonder from where the immigrants came and why there has been a progressively increasing tendency to migrate. If Brack's figures are correct, the minimum number of immigrants entering was 1,272 in 1978 and 1,532 in 1979, assuming that all females, including yearlings, bred and that all calves survived to maturity. If a more realistic figure of 50% calf survival is assumed, the necessary numbers of immigrants rise to 3,303 and 4,404. The comparable figure for 1977 is 1,124. In the absence of any evidence, it is difficult to accept that such massive immigration has taken place.

In view of the many inconsistencies, it must be concluded that the ground count results given by Brack (1980) are unreliable and should not be used in estimating the maximum sustained yield that may be taken from the population. Before further cropping is undertaken, the number of vicuna should be checked by a different technique. The aerial count reported here has shown that aerial survey is feasible in the High Andes and, with certain refinements, is likely to produce reliable figures. Future aerial counts should be made in an aircraft fitted with a radar altimeter and in one which allows the observer to be in closer contact with the pilot than was possible in the Twin Otter. Conversion factors must be obtained by making ground/air comparisons in each of the density zones.

B. Is Culling Necessary for Ecological Reasons?

There remains the possibility that vicuna numbers need to be reduced because of damage being caused to vegetation. Brack (1980) states that this is the case in the Zona Nuclear and recommends drastic action to prevent a major catastrophe in 1980. He advocates reducing the population in the zone by 10,000 through translocation of 4,000 vicuna to other regions of Peru and culling by shooting of 6,000 animals. The cull of 6,000 would consist of 3,000 bachelor males and 3,000 animals from family groups (which consist of a male and approximately 4 adult females). Brack states that the cull of 3,000 from family groups is necessary in order to finance the project; however, if conservation groups wish to save these 3,000 vicuna, a sum of $300,000 must be paid for the capture and translocation program as well as to recompense the local people. He advocates, therefore, the removal of only 7,000 vicuna for ecological reasons since, if the necessary ransom is paid, the other 3,000 would be spared. The proposal to shoot several thousand of a vulnerable species in order to finance the translocation of others is not one that would appeal to many conservationists and is a dubious management principle. Surely it would be possible to raise the necessary finances without resorting to crude ultimata of this nature. No proposals were made by Brack to reduce the number of vicuna in the Zona de Influcencia.

Evidence of overpopulation that would justify the culling of a protected species must be convincing and should be sought in the state of the vegetation and in the condition of the animals themselves. The proposal to cull vicuna raised international protests as well as opposition from conservation bodies, notably PRODENA, within Peru. As a result, IUCN and the World Wildlife Fund (WWF) dispatched a consultant, Archie Mossman, to Peru to report on the matter. He spent June 26 to

29, 1979 in the Pampa Galeras Reserve. His visit was followed in August by that of Major Ian Grimwood, an IUCN consultant, and Charles de Haes, Director-General of WWF, both of whom conferred with the Peruvian authorities in Lima although they did not visit the Reserve. Having considered the reports of their representatives, IUCN and WWF gave their full support to the cull (IUCN, 1979). Mossman maintained that the rangeland was in a very bad state over the whole area and that its condition was declining. However, he did not go far from the Reserve Headquarters during his short visit because of altitude sickness. He did not carry out a proper survey of the vegetation nor did he attempt to count the animals. Nevertheless, he considered that Brack's population estimates were probably valid.

In view of the continuing controversy, IFAW asked W. J. Jordan to visit the Pampa Galeras to report, as a veterinary surgeon, on the status of the vicuna and to evaluate the feasibility of translocation. His visit was made in November 1979. No evidence was found to indicate that the vicuna were in any way in bad condition. Large numbers were carefully observed at close range and all appeared to be fit and healthy, raising their heads rapidly on approach and regarding the observer with bright eyes and erect ears. Movements of the body were deliberate, rapid, and alert. The animals moved with no signs of weakness, lifting their limbs and taking steps in a purposeful manner. Animals weak from starvation often stumble but despite the uneven terrain, which contains many tussocks and large stones, the vicuna walked quickly and smoothly. There was no tendency to sway slightly as is usually the case with weak animals. All the vicuna observed were feeding and appeared to have no difficulty in locating food. No carcasses were seen, despite the assertion by Brack that large numbers had died, nor was there any sign of the predicted outbreak of disease.

These observations give no support to Brack's contention, (repeated in the IUCN report) that the vicuna in the Pampa Galeras Reserve are in danger of starvation and that large numbers need to be culled for their own good.

We are not competent to comment on the condition of the rangeland vegetation. It is reported to be declining in cover, and although this may well be so, it does not necessarily follow that the reduction is a result of grazing pressure from vicuna. This species feeds by biting the hard vegetation with its sharp, open-rooted incisors, it does not pull plants up by the roots. It walks on pads and does not have sharp hooves to cut up the sward. Any damage to the vegetation is more likely to be caused by the domestic sheep, goats, and cattle which share the range. It must also be borne in mind that the region has experienced several years of drought and any reduction in the vegetation could be a direct result of lowered rainfall. Despite the drought, all springs examined during the visit made in November were flowing although it was the end of the dry season.

Our conclusion, therefore, is that no evidence exists to suggest that vicuna need to be culled for their own good. The fact that numbers have increased or that vegetation has declined is not necessarily indicative of overpopulation. It is more likely that the vicuna population is still rising toward its carrying capacity. The only evidence that the carrying capacity may have been reached is the decline in numbers recorded by Brack (1980) between 1978 and 1979 in the Zona Nuclear. He cites this as a reason for culling vicuna, but it could more reasonably be interpreted as a sign that the vicuna are regulating their own numbers and the population will stabilize without management interference. In any case, it would be unwise to take action based on one year's observation before it is confirmed that a trend is occurring.

It is likely that the fear that rising numbers equals habitat damage stems from African experience with elephants. Indeed, we have been informed by correspondents that failure to crop vicuna could result in a Tsavo situation. (Tsavo National Park in Kenya suffered severe damage from elephants in the 1960's.) But vicuna are not elephants, if only in that they lack the latter's bulk and destructive feeding habits. We know of no case in which a natural population of an ungulate the size of the vicuna has caused serious habitat damage through excessive numbers. Temporary imbalances are soon corrected by reduced recruitment, increased mortality, or by emigration. Brack claims that the latter option is not open to vicuna in the Zona Nuclear because the region is isolated from the surrounding puna, except to the north, by deep valleys or mountainous slopes. It is not clear why the vicuna could not migrate through the northern route but in any case, we found no indication, during either the aerial count or the ground visit, that the region was hemmed in by impassable natural barriers. This was also the impression of Jan Oelofse (personal communication) who flew over much of the region in a helicopter in 1979.

C. *Translocation*

The alternative to culling by shooting is translocation of vicuna to other regions. IUCN (1979) dismisses this possibility on the grounds of expense, but the costs do not appear to have been accurately assessed. The International Fund for Animal Welfare has offered to finance a translocation program and believes that it could be done economically.

IUCN also raised the problem of biological hazards involved in translocation (e.g., the likely disruption of family groups), and it has been suggested that such transfers would be cruel and cause numerous deaths. During the November visit, this aspect was discussed with Oelofse, an expert in animal capture from South Africa, who was in Peru to assess the feasibility of translocating vicuna. Oelofse found that a family group could readily be shepherded by helicopter and driven into a trap. We also found that a family group could be herded on foot. The animals showed no inclination to stampede when approached but merely moved away a little distance and then stopped. Some captive vicuna at the Reserve Headquarters were closely observed. The pens in which they were held were each about the size of a tennis court. On being approached, the animals moved to the far fence but did not hurl themselves at the wire nor try to jump over or scramble underneath. They could easily be driven into a corral or onto a transport vehicle without need to tranquilize. Oelofse observed that the animals were not disposed to jump over obstacles; on the contrary, they tried to burrow underneath. Thus, high fencing would not be needed to enclose vicuna as would be necessary for impala, for example. He considered that vicuna could be transported safely in darkened trucks with a small chink of light entering at floor level. This would encourage the animals to go down on their knees in an effort to get out; they would, therefore, avoid being thrown about as the truck traveled over rough terrain.

The best time of year for moving vicuna would be around May when the females would not be heavily pregnant and the young (born in January or February) would be large enough to avoid being trampled underfoot in the trucks. Experience with stress in captured animals in other situations suggests it would be better to move the vicuna immediately after capture -- not to hold them for several weeks as has been the practice in some operations of this kind.

We understand that an experimental translocation of vicuna from Pampa Galeras to a reserve at Aguada Blanca was successfully carried out in late November 1979.

III. CONCLUSIONS

A. Cropping

We accept that it would be possible to crop vicuna for meat and hides, but we do not believe that this is a wise policy for reasons given above. A successful cropping operation requires more information than is presently available to assess the maximum sustained yield. In view of the doubts cast on the numbers of vicuna present in the Reserve, estimates of the population size should be obtained by at least two different techniques. If an income has to be generated from the vicuna, we recommend the population should be exploited by shearing.

B. *Culling*

We are not convinced that culling is necessary. It has yet to be shown that the recorded vegetational changes represent a trend and are not simply part of a normal cycle. Even if the changes are permanent and deleterious, it has not been established that vicuna are responsible. Until more evidence implicating the vicuna has been gathered, we do not believe conservationists can support a cull. Should it be established beyond reasonable doubt that too many vicuna are present, we feel the best solution is to translocate the surplus. In any case, it makes good sense to build up small populations of vicuna elsewhere in the species' range by translocation in order to avoid a situation where "too many eggs are in one basket."

ACKNOWLEDGMENTS

Our traveling and out-of-pocket expenses were defrayed by the International Fund for Animal Welfare. The aircraft and crew for the aerial survey were provided by General Herman Boluarte of the Peruvian Air Force by kind permission of the Minister for Air, General Luis Arias Graziani. The aircraft was piloted by Captain Armando Aquayo with Lt. Armando Abenzur as copilot. Free accommodation was provided for the survey team at the Paracas Institute near Pisco. Sr. Felipe Benavides, President of PRODENA, made most of the arrangements for our visits and took part in the aerial observations. We are most grateful to Dr. John Beddington of the University of York, for his computer analysis of the ground counts. Other persons who provided invaluable help included Dr. Antonio Brack, Ing. Luis Cueto, Ing. Enrique Felices, General Carlos Gamarra (Minister of Agriculture), General Ernesto Fernandes Lañas, Mr. Ian MacPhail, Mr. Tony Morrison, Mr. Jan Oelofse, Commandant Miguel Riglos, Commander Ernesto de Romaña (ADC to the President), and General Augusto Romero.

REFERENCES

Brack, A. (1980). Situacion actual de la poblacion de vicunas en Pampa Galeras y zonas alendanas y recomendaciones para su manejo. Unpublished manuscript. Proyecto Especial Utilizacion Racional de la Vicuna, Ministerio de Agricultura y Alimentacion, Peru. 15 pp.

Hofmann, R. and Otte, K. (1977). Utilization of vicugnas in Peru. German Agency for Technical Co-operation, Eschbom. 48 pp.

International Union for the Conservation of Nature and Natural Resources (1979). Vicuna cull backed by IUCN and WWF. *Int. Union Conserv. Nat. Nat. Resour. Bull. 10*, 69.

Koford, C. B. (1957). The vicuna and the puna. *Ecol. Monogr. 27*, 153-219.

Scheffer, V. B. (1951). The rise and fall of a reindeer herd. *Sci. Mon. 73*, 356-362.

Sinclair, A. R. E. (1973). Population increases of buffalo and wildebeest in the Serengeti. *East Afr. Wildl. J. 11*, 93-107.

THE ELEPHANT SEAL

Burney J. Le Boeuf

Crown College
University of California
Santa Cruz, California

I. INTRODUCTION

Northern elephant seals have been expanding their breeding range and increasing in number every year since sealers reduced the world population to less than 100 individuals during the 1880's. At present, these seals breed on 12 islands or island groups from central Baja California, Mexico, to the Farallon Islands near San Francisco in central California (Fig. 1). This is virtually the same range the species occupied before exploitation began in the last century. Present population size is estimated to be over 60,000 animals (Le Boeuf, 1977). No other large mammal has made such a remarkable recovery from near extinction. However, the population continues to increase and crowding is evident in certain parts of the range. From 1968 to 1981, my students and I monitored dispersal of individuals within the population and fluctuations in the size of colonies. This facilitated the study of colony dynamics on Año Nuevo Island, a rookery we studied intensively. Early in the study period, Año Nuevo was a small new colony of elephant seals with low female density in harems. Pup production increased annually. By 1980, the colony had become very large and crowded. This chapter will document the increase in density on this rookery, report observations made during the colony's growth, and address problems inherent in studying changes that occur with increasing density.

ISBN 0-12-385280-3

BREEDING BEGAN	PUPS PRODUCED IN 1977	ESTIMATED TOTAL ANIMALS	
1972	104	457	FARALLON IS.
1961	750	3200	AÑO NUEVO I.
1930's	3891	17,696	SAN MIGUEL I.
1950's	74	322	SANTA BARBARA I.
1940's	693	2960	SAN NICOLAS I.
1977	2	9	SAN CLEMENTE I.
1950's	49	224	LOS CORONADOS
1977	1	3	I. SAN MARTÍN
-	6922	28,304	I. DE GUADALUPE
1930's	1590	7,016	IS. SAN BENITO
1960's	121	572	I. CEDROS
1977	12	64	I. NATIVIDAD
TOTALS	14,009	60,827	

Fig. 1. A 1977 summary of northern elephant seal breeding locations, the approximate time that breeding began, pup production, and estimated total population size. Adapted from Le Boeuf (1977), Le Boeuf and Mate (1978), and Le Boeuf and Bonnell (1980).

II. POPULATION GROWTH

The population of northern elephant seals more than quadrupled in the last 21 years. A plot of total northern elephant seals censused periodically since 1890 reveals that the population is still in the exponential phase of growth (Fig. 2). The curve fits the equation $Y = ae^{bx}$ with $a = 29.35$ and $b = 0.09$. At the colony level, four new island rookeries were established during the 1970's and one breeding site was established on the continental mainland.

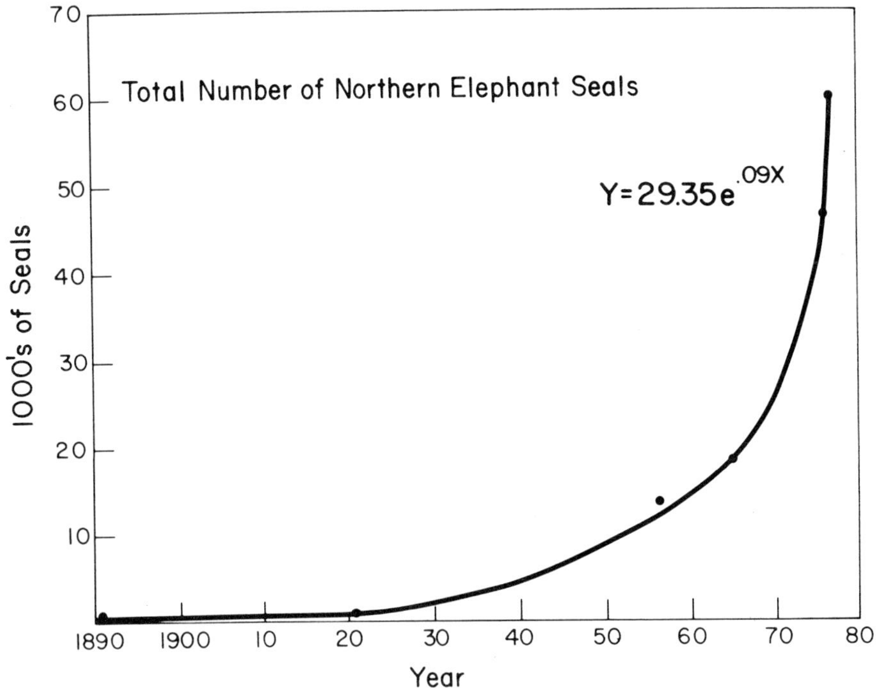

Fig. 2. Exponental growth of northern elephant seals since the nadir of the population in 1890.

A. Colony Changes Associated with Increasing Density.

During the recovery phase of the elephant seal population, some rookeries reached their carrying capacity and sent off immigrants to form new colonies. Año Nuevo Island was colonized in 1961 by immigrants from southern California rookeries. Pup production rose annually so that by 1980 over 1,200 pups were born on the two beaches on this tiny 8-acre island. By the early 1970's signs of crowding began to appear. First, the pup mortality rate began to increase (Le Boeuf and Briggs, 1977). Fig. 3 shows that the increase was gradual at first, but the pup mortality rate began to skyrocket in bad storm years so that in 1978, 40% of the pups born died before reaching weaning age. Storms did not kill pups directly so much as indirectly by increasing density. Secondly, some Año Nuevo-born seals began establishing new breeding sites in the vicinity of Año Nuevo Island. In 1972 breeding began on Southeast Farallon Island, 89 km north of Año Nuevo (Le Boeuf *et al.*, 1974). Eight years later, pup production was up to 250 pups

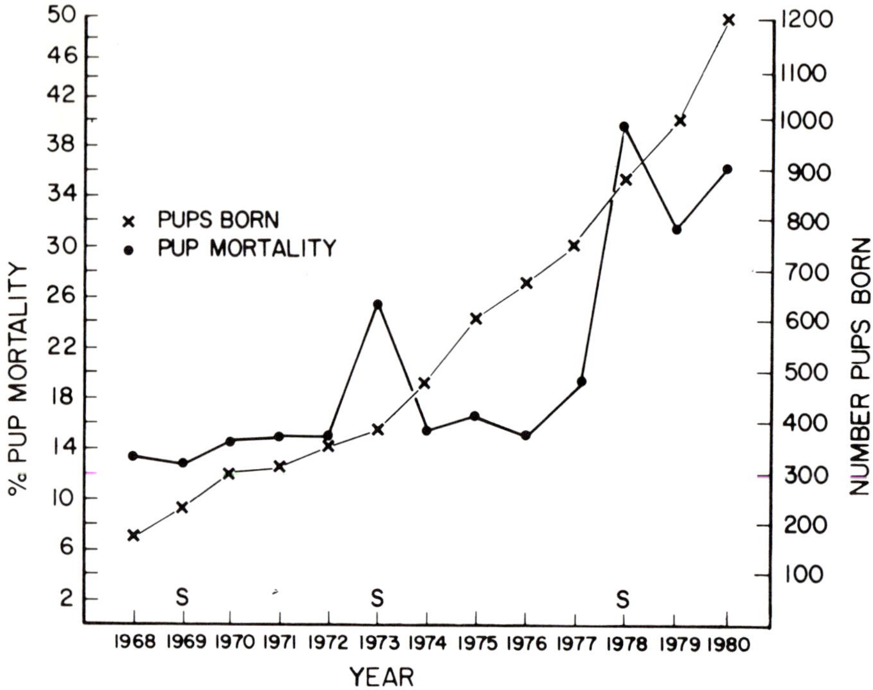

Fig. 3. Variation in pup mortality and annual pup production on Año Nuevo Island during the period 1968 to 1980. The mortality rate is calculated from pups that died prior to weaning. S denotes storms which struck the rookery at peak season when the maximum number of females and pups were present. Adapted from Le Boeuf and Briggs (1977) and Reiter et al. *(1981).*

per year. Seventy-six percent of the tagged seals that gave birth here during the period 1972 to 1978 were born on Año Nuevo Island. Ninety-two percent were young females, five years of age or less. Similarly, the male pioneers were also young and Año Nuevo-born. In 1975 breeding began on the mainland less than 1 km across the channel from Año Nuevo Island. Pup production increased annually; 150 pups were born here in 1980. Eighty-seven percent of the females were Año Nuevo-born and very young.

B. Causes of Pup Mortality and Emigration of Young Females

Young females began to breed in new locations because they could not compete effectively with older females in the high density conditions that prevailed (and continues to prevail) on

Año Nuevo Island (Fig. 4). A study of reproductive success as a function of female age (Reiter *et al.*, 1981) revealed that the probability of a female producing a pup that survives to weaning increases with age (Table I). Young females were least successful under high density conditions. The low reproductive success of young females was due in large part to their inferiority to older females in size, dominance, and maternal experience. Young females gave birth at a poor time for rearing pups, at peak season or later, and in the poorest location, the periphery of the harem where conditions led to a high probability of mother/pup separation and pup injury.

The reproductive success of young females that gave birth in the low-density harem on the Año Nuevo mainland (Fig. 5) or on Southeast Farallon Island was higher than that of their counterparts who remained on Año Nuevo Island and gave birth in the crowded harem (Table II).

C. *Other Indicators of Crowding*

As the number of females in a harem increases, the area unoccupied by the females spreads out because the aggressiveness of females tends to maintain inter-individual spacing. When a beach occupied by a harem becomes fully utilized in this way, the arrival of additional animals leads to an increase in density. On the Point Harem on Año Nuevo Island, a stage was reached where landing became so difficult that females began to breed in marginal areas, formerly unoccupied beaches with a rocky rather than a sandy substrate. Thus, a fully utilized beach and use of marginal areas is a sign of crowding.

High density in a harem is indicated by a high frequency of mother/pup separation, the presence of many orphaned pups, a high frequency of milk stealing, disruption of nursing, trampling of pups by males, female biting of pups, and finally, high pup mortality. Several females may gang up on an orphan and in a frenzy, bite it repeatedly and kill it. These chaotic conditions -- exacerbated further by bad weather -- make for a high level of arousal and bickering among females. This would appear to cause them to expend their limited fat reserve (their only source of food and water for themselves and their pups because females do not eat or drink while nursing) in activities other than nursing. Consequently, I predict lower weaning weights for pups nursed in high-density harems as opposed to low-density harems, as well as a shorter nursing period for females (the female's nursing period and stay on land is limited by her fat store).

In crowded harems the age composition of males is heavily weighted to adult bulls. For example, on Pilot Rock Beach at Isla de Guadalupe, which was extremely crowded at peak season

Fig. 4. Part of a crowded harem on Año Nuevo Island during the 1980 breeding season.

Fig. 5. A small low-density harem of young females.

TABLE I. *The Percentage of Females in Various Age Categories That Produced Pups Which Survived to Weaning Age on Año Nuevo Island*[a]

Breeding season	Age										
	3	4	5	6	7	8	9	10	11	12	13
1976	25 (4)	65 (17)	47 (17)	100 (4)	86 (14)	100 (2)	-	-	-	-	-
1977	36 (11)	50 (10)	73 (11)	88 (8)	100 (3)	100 (15)	-	100 (1)	100 (1)	-	-
1978	0 (4)	13 (16)	0 (1)	20 (5)	30 (10)	50 (2)	30 (10)	0 (1)	100 (1)	100 (1)	100 (1)
1979	13 (16)	25 (8)	40 (10)	33 (3)	100 (1)	67 (3)	100 (3)	100 (5)	-	100 (1)	-
Totals	20 (35)	39 (51)	51 (39)	65 (20)	68 (28)	91 (22)	46 (13)	86 (7)	100 (2)	100 (2)	100 (1)

[a]N is in parenthesis. From Reiter et al. (1981).

Table II. The Proportion and Percent of Young Females, Five Years of Age or Less, That Produced Pups Which Survived to Weaning Age in Good Condition and Were Weaned by Their Own Mothers[a]

Breeding season	*Año Nuevo Island Point Harem*	*Southeast Farallon Island*	*Año Nuevo mainland*
1976	*15/42 = 47%*	*7/8 = 88%*	*0/1 = 0*
1977	*13/27 = 48%*	*7/12 = 58%*	*2/4 = 50%*
1978	*1/19 = 5%*	*8/17 = 47%*	*10/12 = 83%*
1979	*6/31 = 19%*	*Not available*	*10/17 = 59%*

[a]*From Reiter* et al. *(1981).*

in 1973, 60% of the males on the beach were full grown and over eight years of age. In contrast, at Año Nuevo Island, then a low-density peripheral rookery, only 29% of the males were adults; 55% of the colony was composed of pubertal males six years of age or less (Le Boeuf, 1974). As the number of animals breeding on Año Nuevo Island increased and crowding set in, the age composition of males has approached the proportions observed earlier at Isla de Guadalupe. I suspect that a similar change in age composition occurs among females.

Although there are relatively few young males near crowded harems, compared to older males, the former can copulate more frequently in this setting than in lower-density harems because it is easier for them to sneak into a chaotic crowded harem undetected than into a small low-density one. In the latter, one adult male usually prevents all males from approaching the females and he alone does all of the mating. Thus, copulations by young pubertal males are a sign that the harem is crowded. In the vicinity of crowded rookeries it is common to observe single males hauling out on isolated beaches or coves. During the last ten years, subadult males have been observed frequently on mainland beaches north and south of Año Nuevo Island.

D. *Immigration and Recruitment During Crowding*

The Año Nuevo colony has grown through immigration and recruitment. Every year new females and males from southern rookeries have bred here (Fig. 6). Elephant seals are site tenacious; once they breed in a certain place, over 70% of them return to the same place to breed. Elephant seals are also

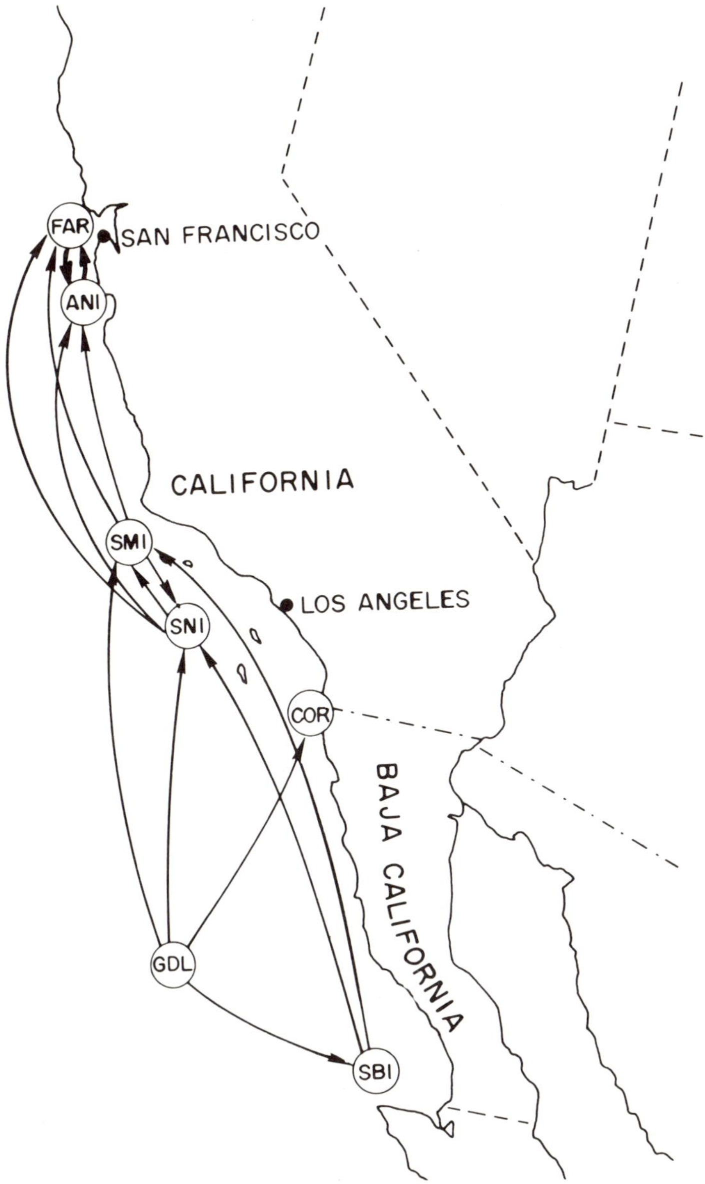

Fig. 6. The major routes of dispersion and immigration of northern elephant seals based on tagging studies of pups conducted during the 1970's. The following abbreviations are used: FAR = Southeast Farallon Island, ANI = Año Nuevo Island, SMI = San Miguel Island, SNI = San Nicolas Island, COR = Los Coronados, GPL = Isla de Guadalupe, and SBI = Islas San Benito.

philopatric; over 70% that survive to breeding age return to give birth at their birthplace. It is interesting that immigration of animals from southern rookeries to Año Nuevo did not stop once the latter began to show signs of crowding. Rather, young females from the southern rookeries (San Miguel and San Nicolas Islands) began to appear on Año Nuevo Island as well as on the new breeding areas (Southeast Farallon Island and the Año Nuevo mainland).

III. METHODOLOGICAL PROBLEMS

What I have summarized thus far about signals and symptoms of crowding on a rookery must be liberally salted before ingestion. There are serious problems in studying and measuring changes with increasing density in field studies on long-lived pinnipeds. First, the changes occur over a number of years. During this time there may be changes in personnel, fluctuation in motivation by the observers, and changing priorities in research (not to mention lapses of funding and difficulties in acquiring the necessary permits). One does not usually set out with the single-minded purpose of studying changes that occur with density over the course of a 14-year longitudinal study. Second, there is the problem that one does not know what to look for. Because of the time factor, the changes are slight, subtle, and easily overlooked. One must collect data consistently through the years so that appropriate statistical procedures can be applied. Finally, there is the most serious problem of all -- the difficulty of measuring some of the most important variables increases with density! Assessing the annual pup mortality rate is a good example. When the Año Nuevo Island colony was small, we removed each dead pup to avoid the bias of it being buried or washed away and thus underestimating the count or overestimating by counting one corpse twice. Early counts were not so much estimates as absolute values of pup mortality. But as the harems grew in area and number, it was no longer possible to approach the dead pups, let alone pull them out. It became necessary to estimate more and more using various methods applied by various observers over the years. The irony is that at a time when it was most important to know precisely what was happening to the pup mortality rate, i.e., at high densities, the measure was most suspect. Thus, the natural inclination of the scientist to be conservative in the face of uncertainty results in softening our conclusions about the effect of density.

IV. PROSPECT

During the last few years, the number of elephant seals breeding on Año Nuevo Island and the Año Nuevo mainland has increased to such a point that elephant seals have become a tourist attraction in the state park and a common sight on other beaches in the area. An increasing number of seal/people confrontations have occurred, some with good endings and others with disastrous consequences (usually for the seals). Given the present rate of increase in the seal population, one can anticipate conflicts between seals and people in the future. The conflict is not likely to be over food but over beach space. Before this time comes we must ponder the question: How far will we allow a species like the northern elephant seal to come back from near extinction? How many individuals will we tolerate on mainland beaches before people begin to consider the comeback-success species of the last few decades a cancerous growth in the present decade, a threat to people, a menace?

REFERENCES

Le Boeuf, B. J. (1974). Male-male competition and reproductive success in elephant seals. *Am. Zool. 14*, 163-176.

Le Boeuf, B. J. (1977). Back from extinction? *Pac. Discovery 30*, 1-9.

Le Boeuf, B. J. and Bonnell, M. L. (1980). Pinnipeds of the California Islands: Abundance and distribution. *In* "The California Islands: Proceedings of a Multidisciplinary Symposium" (D. Power, ed.) Santa Barbara Museum of Natural History, Santa Barbara.

Le Boeuf, B. J. and Briggs, K. T. (1977). The cost of living in a seal harem. *Mammalia 41*, 167-195.

Le Boeuf, B. J. and Mate, B. R. (1978). Elephant seals colonize additional Mexican and California islands. *J. Mammal 59*, 621-622.

Le Boeuf, B. J., Ainley, D. G., and Lewis, T. J. (1974). Elephant seals on the Farallones: population structure on an incipient breeding colony. *J. Mammal. 55*, 370-385.

Reiter, J., Panken, K. J., and Le Boeuf, B. J. (1981). Female competition and reproductive success in northern elephant seals. *Anim. Behav. 29*, 670-687.

THE PACIFIC WALRUS

G. Carleton Ray

Department of Environmental Sciences
University of Virginia
Charlottesville, Virginia

I. INTRODUCTION

The objectives of this workshop were directed more toward large, terrestrial mammals -- especially herbivores -- than toward such marine carnivores as the walrus *(Odobenus rosmarus)*. Nevertheless, the walrus is a benthic "grazer," to which the problem of "overabundance"[1] similarly applies. By this I mean that walrus feed upon largely sessile organisms, subject to local depletion in a manner similar to the plants on which terrestrial grazers rely. The analogy of the walrus' grazing to grazing by herbivores becomes more real, however, by virtue of the recent concern that this species may be approaching its carrying capacity and that its population may soon overreach this level and "crash" in response to depletion of its food supply.

In this brief discussion, I will focus on our state-of-the-art knowledge of the Pacific walrus population and its relationship to environmental factors which are useful in defining carrying capacity (K). I will make the point that although our knowledge is scanty, it may be good enough for predictive purposes in a very general way. Some theoretical considerations follow regarding the possible interpretations

[1] *I use the term "overabundance" only because it was used frequently by the workshop. Its obvious nonscientific connotations are similar to other ambiguously loaded words, such as "crash," which have quite different meanings in different contexts.*

ISBN 0-12-385280-3

of over (and under) abundance; the question is, does it make any real, long-term, ecological or biological difference whether or not the walrus population exceeds *K*? However, it is clear that the matter does not stop there. I cannot restrict myself entirely to scientific issues, despite the fact that they form the focus for this volume. Thus, I will consider "critical habitat" as defined by scientific and social factors. Society's perceptions about population trends are especially important when dealing with a species upon which subsistence hunting and a cottage industry (ivory carving) depends. Societies which depend heavily upon a resource, as Eskimos do on the walrus, cannot permit a "crash" to occur; such societies frequently argue for increased economic yield by hunting. I will, finally, draw some conclusions, the principal one of which is that the walrus' case is an excellent example of the need for integrated management on a system (both social and eco-) basis.

I also note a fundamental issue with regard to marine mammals in general. Neither the population modelers nor the oceanographers have much considered marine mammals from ecological or behavioral perspectives. Thus, we are able to make only crude guesses about marine mammal population responses to either depletion or to *K*-equilibrium levels of abundance. Yet, upon these, scientific notions of overabundance ultimately depend.

II. THE PACIFIC WALRUS POPULATION

Overabundance relates to the numbers of animals in a population, the age structure of the population, the life history and behavior of the species, ecological relationships, and ecosystem dynamics. Equally, it relates to the population's relationship to *K*. What do we know of these things in the case of the Pacific walrus?

Fay (1957) estimated the original size of the Pacific walrus population to be at least 200,000 animals. Severe depletion over a hundred-year period resulted in low points during the late 1800's and again in the mid-1950's. The last commercial harvests by the Russians occurred during the late 1950's (Krylov, 1968). From 1958 to 1975 several aerial surveys have suggested that the population has at least doubled during the past two or three decades. Present estimates suggest recovery to the original population size of at least 200,000 animals (Fay, in press, and personal communication). It is possible that a trend in population numbers, perhaps even a rate of population increase, could be established with the aid of these surveys. However, the surveys occurred at various times of the year and used various methods; further, they only peripherally considered behavioral factors such as hauling-out patterns or ecological ones such as ice dynamics or food supply. This is, in part, because the surveys until very lately were based upon little prior knowledge of where the walrus concentrations were located, the degree of clumping of animals, their haul-out patterns, and such ecological considerations as sea ice habitat relationships.

Furthermore, the ability to sight walrus is subject to very wide variance. For example, Ray and Wartzok (1980, and in preparation) found that on three successive days in April, one group numbered 450, 1,200, and 9,500 animals. The twenty-fold increase in numbers resulted from continuous hauling-out of animals that were always in the vicinity, during increasingly calm and warm weather. We have similar data for other groups as well. Such dramatic changes in numbers of sightable, i.e., hauled-out, animals appear to be typical, as land and ship based observations by a number of other investigators have also shown (e.g., Estes and Gilbert, 1978). The conclusion is that surveys which do not actually measure such variance are able to show only very generalized trends in population.

Nevertheless, other sorts of evidence do indicate, perhaps even more strongly than numerical assessments can, that the walrus population of the Bering and Chukchi Seas has increased in the last two or three decades. The population now occupies all of its historic range (Fay, in press). It appears to have changed its location of feeding (Fay *et al.*, 1977) so that it now exploits recently unused areas. There is also evidence for declining physical condition and increasing natural mortality (Fay and Kelly, 1980), and suggestions that productivity of the population may also be declining (Fay, 1981). These latter two points are, in fact, evidence of population response to *K*-levels of abundance.

All of this evidence, taken together, plus certain other recent findings about productivity and population structure, indicate that the Pacific walrus population reached low ebbs twice in the past 100 years as a consequence of human exploitation and may have recovered to original numbers after each decline. Now, the population appears to ride a crest in numbers, which we may assume to be near *K*-level. However, we do not now have, nor are we likely to obtain in the near future, sufficient data to predict the timing or magnitude of any population change which may be the result of oscillations about *K*-equilibria, except in a general way.

An equally significant matter concerns the level of *K* itself. Here, we are on even less secure ground. Two factors no doubt play significant roles: food and sea ice. Stoker's (1978) work is the only comprehensive account to date of the distribution and diversity of walrus food in the Bering and

Chukchi Seas. Nevertheless, little information exists on the natural productivity of walrus food there under grazing pressure. There is only a hypothetical prediction of the potential impact of a proposed clam fishery on walrus food (Stoker, 1977) and an estimate of possible impact by walrus on their food supply (Fay *et al.*, 1977; Fay and Lowry, 1981); none of these studies can provide more than crude predictions about the impact of walrus on their food supply, although this matter is of obvious importance in defining overabundance.

Burns *et al.* (1980), Wartzok and Ray (1980), and Ray and Wartzok (1980) have investigated the sea ice habitat of the Pacific walrus. These studies outline some of the adaptations of walrus to ice for hauling-out and yield a preliminary appreciation of preferential selection of certain ice densities and types for this purpose. It is premature to quantify these results as a basis for guessing at carrying capacity, but it is important that further work be done along this line. Ice distribution may be an even more important determinant of walrus distribution than food since walrus require ice of a certain type for rest and reproduction. Because of this requirement, sea ice distribution may determine impact on walrus food supply in time and space even more than the walrus' independent choice of feeding areas.

Obviously, ecosystem-oriented research of a type necessary to suggest K-equilibrium levels for walrus lies mostly in the future. The truth is that indicative but incomplete data form the basis for much of our understanding of walrus population dynamics and ecosystem relationships. Although difficult management decisions cannot await detailed evidence, the present situation clearly, over the long term, will not suffice. As Fig. 1 indicates, when questions about P, K, and r can only be answered within very large margins, then there exists little opportunity to predict either K-equilibrium levels or the effects of increased take on either the population or its environment; nor can management options be satisfactorily formed from either the ecological or socio-economic point of view.

III. THEORETICAL CONSIDERATIONS

Caughley (1979) poses the question, what is carrying capacity? He warns that our use of the term is far from clear: "We tend to think of carrying capacity largely as a density of animals, but we could equally think of it as a characteristic density and composition of vegetation." His well-taken point is that K represents an equilibrium between animal populations and what they feed upon, not merely a

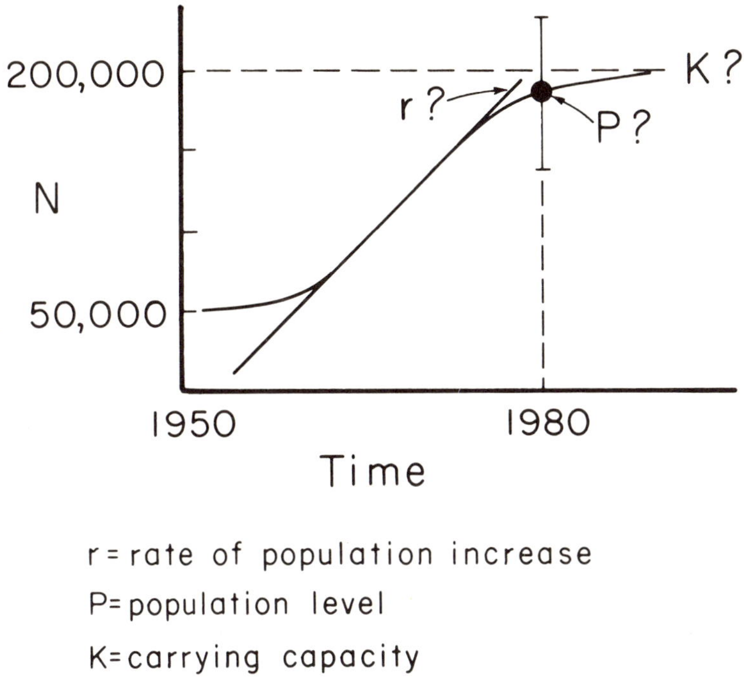

Fig. 1. Questions regarding r, P, and K can only be answered within large boundaries. Therefore, predictions about the relationship of the Pacific walrus population cannot accurately be made. Indirect environmental and population indicators may provide better evidence that this population is near K-level than numerical assessments.

population relative to some static, hypothetical environmental level. It is, of course, necessary to keep in mind that such notions as carrying capacity, *K*-selection, and the like, are just that -- theoretical notions in urgent need of testing for such elusive animals as marine mammals. For example, Estes (1979) and Fowler (1981) speak of marine mammals as *K*-selected in general. However, Ray (1981) illustrates a possible range of *r* to *K*-selection strategies for marine mammals, following Southwood's (1976) observation that there are *r*-*K* continua within taxa. *K*-selection theory is important in the design of management actions because the more *K*-selected a species is, the more it is adapted to *K* levels and the less likely its population is to suffer a dramatic and permanent decline as a result of its own perturbations; an instance of the latter is that of the caribou on St. Matthew Island, Alaska (Klein, 1968). For *K*-selected species, predictions of such a crash are on thin theoretical grounds in the terms in which society usually perceives a crash -- a catastrophic event with permanent effect. As an argument for culling, one could point out that some *K*-selected species, such as elephants, do exceed *K* due to environmental or other perturbations. But it is more to the point to perceive the nature of the perturbations, their effects on the population, and the time scales of the overabundance/crash cycles, which may be very long, approaching decades or even centuries.

Carrying capacity is no more static in time or space than populations are. *K* oscillates as *K*-selected populations do; it is the equilibrium between these two factors that is central even more than whether or not the population is in danger of overshooting *K*. Other chapters in this volume consider this essential point. The implication for management is obvious, but exceedingly difficult to assess in the case of marine mammals; that is, management should be reoriented to ecosystems if management of species is to be successful. To the terrestrially-oriented, such a statement may seem obvious. But to those dealing with the sea, in particular fisheries (under the roof of which marine mammals generally lie), the ecosystem approach is largely considered to be futuristic.

With regard to the Pacific walrus specifically, it is likely that the population may be approaching -- or be at or even beyond -- what may be popularly considered carrying capacity in the older definition of that concept. Put in other terms, it may be a dynamic equilibrium at which level both *K* and the population are bound to oscillate in response to one another -- to what extent it is impossible to say. All that is needed for this presumption is, first, that the walrus is relatively *K*-selected and, second, that its environment remains relatively intact. The first presumption is, I think, true (Ray, 1981). The second is also probably true, since there is no significant fishery for walrus food, little environmental pollution, and few deleterious results of human activities yet as far as this species is concerned. However, this situation will almost certainly not prevail in the face of intensifying development of minerals, oil and gas, and fisheries in the Bering and Chukchi Seas, making all the more urgent the need for ecosystem-oriented management.

It is fair to ask, even on theoretical grounds, what the consequences would be if walrus exceeded *K* or, contrarily, were to be kept at less than *K* levels by management techniques in order to increase yield. The obvious consequence of the first would be a decline in numbers by means of adaptive mechanisms in response to *K*-equilibrium. Presumably, this ecologic "crash" is "normal" and is not to be made equivalent to "crashes" of *r*-selected species or those which occur as a result of perturbations of the environment through human interference; in these cases, severe habitat damage is involved. The second condition is more difficult to predict. Reducing a population is a means to increase yield; generally, it is not much concerned with ecologic consequences, since yield models are rarely ecological in nature, at least for fisheries. However, population/environmental feedbacks could serve to modify conclusions about the ecological effects of walrus. That is, there is a possibility that walrus may enhance the productivity of their own ecosystems. Walrus are benthic grazers. Reducing them could conceivably result in loss of nutrients and lesser overall ecosystem productivity because their bioturbation may have a significant impact on the release of nutrients "locked" in sediments (Ray, 1973). Thus, their feeding behavior could result in transport of nutrients from bottom to surface and from place to place. This raises the interesting prospect that if walrus were reduced in numbers, so might be the carrying capacity of the ecosystem for them and other species due to the possibility of nutrient loss or decrease in nutrient cycling efficiency. Oceanographers and population dynamicists have yet to grapple with problems of this sort for large marine organisms.

IV. A METHOD FOR ASSESSMENT

Obviously, diverse data must be integrated in order to gain an understanding of the many biological and ecological relationships involved in overabundance. Also, obviously, overabundance relates to social values and perspectives, as many other chapters in this volume testify. For marine mammals, such values are exceedingly varied; they range from conflicts between marine mammals and fishermen; through dependency of some subsistence cultures on marine mammals, to economic gain from marine mammal products, and finally to the "thou shalt not kill" of absolutist protectionism. Few animals illustrate such a range of perspectives as well as marine mammals, as recent history so adequately illustrates.

Ray *et al.* (1979) approached the problem of integrating such diverse factors by considering "critical habitat" from biological, ecological, and socio-economic points of view. The Pacific walrus was used for illustrative purposes in our study. The method could be applied to other species for which a suitable array of data exist. For the Pacific walrus, such data do exist, but cannot yet yield answers to the problem of overabundance; rather, the critical habitat approach illuminates complex relationships and indicates times and places for limited management actions and for priority data needs.

The results of critical habitat analysis are:

(1) An indication of where and when the population may be most biologically vulnerable.

(2) The identification of ecological processes upon which both the population and its habitat depend.

(3) The identification of past, present, and potential conflicts with human activities which may affect the population and/or its environment.

It is the synthesis of these matters that must guide management. Thus, it is my belief that planning for the conservation and management of marine mammals, such as the Pacific walrus, would benefit from critical habitat analysis. This is true for the simple reason that it is essential that we learn to predict how and where, perhaps even when, species are most vulnerable. Population dynamics models, which currently dominate marine mammal management in general, cannot yield such insights on their own. Obviously, assessment of all the relevant variables for critical habitat analysis is not a simple task, nor are sufficient data usually available. This makes even more urgent the use of a method for organizing and evaluating whatever information is available so that relatively informed judgments can be made and research needs identified.

Following this logic leads me to the inevitable conclusion that assessment surveys, as well as other forms of population research, should reemphasize the gathering of environmental data. The need for numerical assessments on which to base management regulations -- such as a permitted take level -- usually overwhelms the need to gather other kinds of equally needed data, for example on habitat relationships. The walrus' case is a good example of this need.

V. CONCLUSIONS

The objectives of this workshop, enumerated elsewhere in this volume, will now be considered:

(1) For Pacific walrus, predictions about population responses to management have not gone far beyond the probability that the population is, in all likelihood, expanding and may be at original levels. But there is clearly not enough knowledge about the ecosystem to make predictions concerning carrying capacity.

(2) With further research about the walrus' feeding strategies and ice dynamics relationships, predictions might be possible about the environmental effects of high population levels.

(3) Examination of feeding pressure on benthic communities offers hope of developing diagnostic characteristics for early stages of the effects of population increase. Other possibilities relate to the animal's physical condition, population structure, survivorship of age classes, and productivity. But little might be gained by too much dependence on any one set of data alone, either populational or environmental. As Caughley (1979) points out, both the population and its environment must be considered together; consideration of either without the other may lead to serious error.

(4) Options for management of Pacific walrus populations are limited at present to hunting, mostly by subsistence hunters and, secondarily, by commercial means. However, both tend to be highly selective. Thus, one must be wary of the hunt as a method of culling. A more satisfactory method might be culling of entire social groups as has been suggested for sperm whales and elephants, in order to reduce, among other things, the socially disruptive effects of principally killing productive adults.

Will the walrus population "crash?" Alaska natives have expressed the fear that this may be so (Fay, 1978). Why have they expressed this fear? Notions of carrying capacity expounded by scientists and managers, in my opinion have led to popularized notions of this concept that are now quite out of date. The very word "capacity" suggests a container of sharply defined content (e.g., "Milk: Contents, one quart"). Further, the physical condition of walrus may be poorer than of specimens taken 10 to 20 years ago (Fay and Kelly, 1980), and the productivity of the population may be lower as well (Fay, 1981, and personal communication). These findings suggest to segments of the public that the population is in "stress." These public perceptions are in sharp contrast to a scientific view that walrus may be responding to their environment in a highly adaptive way. Clearly, the problem which must be faced is making clear statements not subject to public misinterpretation. Once again citing Caughley (1979): "For purposes of management we would be better served by an idea of carrying capacity that gave equal stress to the densities of the plants and the animals" (that is, to both population levels and environmental matters). How can this be put so as not to reinforce older, static beliefs of carrying capacity as a warning point where populations will inevitably become of poor condition and of low productivity?

An ecosystem approach to overabundance accepts as "normal" that *K*-selected species can and do exceed a so-called *K*-levels. Thus, populations will decline and environments will also become depleted. However, it is the extent and permanence that is in question. This related to the degree of depletion which the population may have suffered in prior times, the time scale of environmental responses, and the species' adaptive strategies within the *r*-*K* continuum. Even a *K*-selected species, if depleted to the extent of the blue whale, to less than one twentieth of former numbers, may theoretically react on the road to recovery much like an introduced species. The caribou of St. Matthew is, of course, more *r*-selected *and* introduced and, for some segments of the public, its crash has become a model. The walrus may have been depleted to one fourth of original numbers and, I think, is not equivalent to either blue whales or caribou. I would suspect that any walrus "crash" would be modest, perhaps not even numerically detectable by present survey techniques. This is not to say that we should ignore the possibility of a population decline and habitat damage.

A resolution to diversity of views about this matter might involve the "adaptive management" techniques of which Holling (1977) has spoken or the "critical habitat" technique described above. The use of such techniques should serve to distinguish more clearly between scientific or ecological objectives and economic or social objectives for management by asking: How can we maximize benefits from the ecosystem and for society as a whole? That is, how can we manage walrus to maintain healthy population/environment equilibria and also obtain a reasonable yield?

Whether one perceives a population as overabundant or not does indeed relate to what one wants out of the system -- yield or ecologic equilibria. That is, if the Pacific walrus population does crash, where is the harm done? Is this a "normal" situation en route to a stable equilibrium, or "bad" because it reduced yield and damages habitat? The answer is not easily apparent, of course, but it could be partly in the eye of the beholder.

Obviously, management decisions relating to overabundance are not susceptible to solution from any one point of view. A proviso is that biology and ecology should be the "underpinnings" for social and economic matters, particularly for marine species and their ecosystems where our abilities to repair damage done by mismanagement or perturbation is minimal. In this regard, one social consequence of increased culling, in the interest of yield and on the basis that no harm to the system will ensue, is that increased yields soon become habit to the point of social need, or even of addiction.

It has been the usual course of history that we have erred on the side of incaution. One hopes that history will not repeat itself in the case of the walrus.

ACKNOWLEDGMENTS

I extend special thanks to F. H. Fay of the University of Alaska, and also to A. M. Johnson of the U. S. Fish and Wildlife Service and Douglas Wartzok of The John Hopkins University for their comments on the manuscript. This acknowledgment should not be construed to mean they endorse what I have to say. Healthy controversy is, in fact, required so that such difficult matters as "alternatives to culling" can reach their own equilibrium levels.

REFERENCES

Burns, J. J., Shapiro, L. H., and Fay, F. H. (1980). The relationships of marine mammal distributions, densities, and activities to sea ice conditions. Final report to the Alaska Outer Continental Shelf Energy Assessment Project, June. 172 pp.

Caughley, G. (1979). What is this thing called carrying capacity? *In* "North American Elk: Ecology, Behavior, and Management" (M. S. Boyce and L. D. Hayden-Wing, eds.), pp. 2-8. Univ. of Wyoming Press, Laramie.

Estes, J. A. (1979). Exploitation of marine mammals; *r*-selection of *K*-strategists? *J. Fish. Res. Board Canada 36*, 1009-1017

Estes, J. A. and Gilbert, J. R. (1978). Evaluation of an aerial survey of Pacific walrus. *(Odobenus rosmarus divergens)*. *J. Fish. Res. Board Canada 35*, 1130-1140.

Fay, F. H. (1957). History and present status of the Pacific walrus population. *Trans. North Am. Wildl. Conf. 22*, 431-443.

Fay, F. H. (1978). Morbidity and mortality of marine mammals. Environmental Assessment of the Alaska Continental Shelf, Quart. Repts. Prin. Investigators, Oct.-Dec., Vol I. NOAA, Env. Res. Labs. Boulder, Colorado, pp. 3-15.

Fay, F. H. (1981). Modern populations, migrations, demography, trophics, and historical status of the Pacific walrus. Annual report to the Alaska Outer Continental Shelf Energy Assessment Project, June. 40 pp.

Fay, F. H. (In press). "Ecology and Biology of the Pacific Walrus." North American Fauna, U. S. Fish and Wildlife Service, Washington, D. C.

Fay, F. H., Feder, H. M., and Stoker, S. W. (1977). An estimate of the impact of the Pacific walrus population on its food resources in the Bering Sea. *U.S.N.T.I.S. PB Rep. 273-505*, 38 pp. Springfield, Virginia.

Fay, F. H. and Kelly, B. P. (1980). Mass natural mortality of walrus (*Odobenus rosmarus*) at St. Lawrence Island, Bering Sea, autumn 1978. *Arctic 33*, 226-245.

Fay, F. H. and Lowry, L. F. (1981). Seasonal use and feeding habits of walruses in the proposed Bristol Bay clam fishery area. Final draft report to the North Pacific Fisheries Management Council, Anchorage, Alaska. 60 pp.

Fowler, C. W. (1981). Density dependence as related to life history strategy. *Ecology 62(3)*, 602-610.

Holling, C. S. ed. (1977). "Adaptive Environmental Assessment and Management." Wiley Interscience, New York.

Klein, D. R. (1968). The introduction, increase, and crash of reindeer on St. Matthew Island. *J. Wildl. Manage. 32*, 350-367.

Krylov, V. I. (1968). Present condition of Pacific walrus stocks and prospects for their rational exploitation. *In* "Lastonozie Servernoi Chasti Tikhogo Okeana" (V. A. Arsen'ev and K. I. Panin, eds.), pp. 185-200. Pischevaya Promyshlennost', Moscow.

Ray, G. C. (1973). Underwater observation increases understanding of marine mammals. *Mar. Technol. J. 7*, 16-20.

Ray, G. C. (1981). The role of large organisms. *In* "Analysis of Marine Ecosystems" (A. R. Longhurst, ed.), pp. 397-413. Academic Press, New York.

Ray, G. C., Salm, R. V., and Dobbin, J. A. (1979). Systems analysis mapping; an approach towards identifying critical habitats of marine mammals. *U.S.N.T.I.S. PB Rep. 80-111594*, 1-16 and plates 1-6. Springfield, Virginia.

Ray, G. C. and Wartzok, D. (1980). Remote sensing of marine mammals in Beringia: the results of the Besmex (Bering Sea Marine Mammal Experiment). Final report NAS2-9300 to NASA, Ames Research Center, April. 77 pp.

Ray, G. C. and Wartzok, D. (In preparation). Remote sensing of marine mammals: the results of Besmex (Bering Sea Marine Mammal Experiment). *NASA Spec. Publ.*

Southwood, T. R. E. (1976). Bionomic strategies and population parameters. *In* "Theoretical Ecology: Principles and Applications" (R. M. May, ed.), pp. 26-48. Saunders, Philadelphia, Pennsylvania.

Stoker, S. W. (1977). Report on a subtidal commercial clam fishery proposed for the Bering Sea. Final report. Marine Mammal Commission, Washington, D. C. 33 pp.
Stoker, S. W. (1978). Benthic invertebrate macrofauna of the eastern continental shelf of the Bering and Chukchi Seas. Ph.D. Thesis. Univ. Microfilms International, Ann Arbor, Michigan.
Wartzok, D. and Ray, G. C. (1980). The hauling-out behavior of the Pacific walrus. *U.S.N.T.I.S. PB Rep. 80-192578*, 1-46. Springfield, Virginia.

THE WORKSHOP REPORT

In essence the reason for this workshop was concern about what has been perceived as overabundance in populations of endangered or vulnerable species. Rigorously defined this means too many animals. But who says there are too many animals in a population and by what criteria is that view reached? As the workshop progressed it became clear that one paper had presented categories of overpopulation which provided an acceptable framework for discussion. This was Caughley's seminal paper, simply entitled "Overpopulation."

Caughley recognized four classes of overpopulation:

Class 1 The animals threaten human life or livelihood. The species in question threatens human life or health, or threatens livelihood, for example by attacking domestic livestock.

Class 2 The animals depress the densities of favored species. This class reflects a value judgment; one species may threaten to drive another to extinction locally.

Class 3 Too many animals for their own good. This has usually to do with some aspect of the physical condition or breeding success of the members of a population.

Class 4 The system of plants and animals is off its equilibrium. This class involves basic ecological principles, data, and interpretation. It was agreed that this class was of essential concern to the workshop (the only one in Caughley's view), although other classes were of great importance.

In the discussion about diagnostic criteria a fifth class was formulated, namely:

Class 5 There are too many animals for the "system's good." This class was formulated to direct attention to hidden changes -- as for example the loss of essential

ISBN 0-12-385280-3

minerals from the soil over an extended period of time. While many participants felt that these kinds of phenomena could be considered special cases of Class 4, the use of this classification is illustrated in Caughley's chapter and its value was not questioned.

What is particularly important for classification is the fact that many problems have been presented as belonging to Class 4 when, in reality, they belong to Class 2 or 3. In this way action to "cure" what was claimed to be "overpopulation" was given the sanction of ecological necessity. Much of the discussion of case histories and diagnostics centered around these points.

T. ECOLOGICAL CONTEXT

Another type of classification provided a useful framework for reference when considering how tractable or intractable problems of local abundance might be. This classification related to the feeding ecology, size, and life span of the key species in a system; a possible classification had been set out in the preliminary paper circulated by the steering committee, and is fully presented in the chapter by Laws.

The feeding ecology of animals has a fundamental influence on the nature of their interactions with their habitats. In considering the types of situations, the problems they could lead to, and possible action to resolve them, the following broad dichotomous classification might be adopted:

(1) Environment -- terrestrial or marine.

(2) Large or small mammals.

(a) Short-lived/low body weight/high intrinsic rate of increase.

(b) Long-lived/high body weight/low intrinsic rate of increase.

(3) Lifespan of prey and/or productivity of the food resource.

(a) Short-lived/high productivity per unit biomass; typical of grasses and plants of the field layer (terrestrial) and of marine plankton.

(b) Long-lived/low productivity per unit biomass; typical of shrubs and trees (terrestrial) and marine benthos, particularly the larger vertebrate prey.

The temporal relations between the lifespan and productivity or rate of increase of the food and its animal consumer strongly influences the nature of the interaction and its reversibility. For example, wildebeest and vicuna exemplify the combination of 2a and 3a, reindeer 2a and 3b, elephant 2b and 3b, white rhino 2b and 3a, and so on. Each of these examples involves different processes and time scales. The numbers of impala can increase very rapidly (20%/year), overshoot the carrying capacity, and overgraze grasslands, but this process is rapidly reversible (providing topsoil is not lost). Elephants increase slowly (a maximum of 7%/year); under natural conditions with the opportunity of emigration they may not overshoot the carrying capacity, but where their range is restricted, their impact may become extreme. Downward adjustment of their reproductive rate is slow, and as a result of this time lag and low natural mortality rate, the standing stock biomass of elephant is reduced slowly and the standing crop of woody vegetation may be drastically reduced. The rate of regeneration of trees (e.g., baobabs) is slow and in a national park that is effectively an "island" surrounded by human cultivation some species of trees may become extinct. Between these two extremes there is a broad spectrum of examples.

II. CASE STUDIES

Case studies were presented in the form of contributed papers by a number of participants. (See chapters by Hanks *et al.*, Walker, Bell, Cumming, Klein, Owen-Smith, Harwood and Lavigne, Estes, Cobb, and Chapman.) Table I distills the essential characteristics of the problems encountered in management of these species. Assignment to one or more of Caughley's classes of overabundance was attempted in order to differentiate social or political problems from ecological ones. Brief summations of the case studies follow.

In South Africa many of the problems are seen as a consequence of the small size of the protected areas, including national parks. (Of the 46 reserves listed by Hanks *et al.*, only Kruger National Park is larger than 500 km^2.) Also, reserves frequently are isolated, being fenced and surrounded by farmland.

The current population of 2,000 white rhinos in Umfolozi Game Reserve in Natal has been maintained by the relocation of over 2,000 animals during the last two decades. However, rhinos may be culled by shooting in the near future because of changes in composition of vegetation and the increased potential for soil erosion.

TABLE I. Essential Characteristics of Case Studies

Species	Country/Region	Cause/Problem
South African wildlife (37 species)	Parks and reserves in South Africa	Small size and isolation of parks
Elephant Buffalo Warthog Zebra Waterbuck Wildebeest Impala	Parks and reserves in Zimbabwe	Increase in human population and livestock
Deer	North America	Human activities; removal of habitat
White rhino	Natal, South Africa	Population increase of species
Humpback whales	Northwest Atlantic	Reduced food source (capelin)
Grey seals	Farne Islands	Concentrated breeding areas
Walrus	North Pacific	Population increase of species
Sea otters	California	Population increase of species
Leopard	Africa	Fur trade; nuisance to humans

Diagnostic characteristic	*Method of control*	*Caughley's class of overabundance*
Animal and veld condition; loss of topsoil	*Killing; translocation*	*Class 3 or 4 (possibly Class 2)*
Soil erosion	*Killing; translocation*	*Class 4 (possibly Class 2)*
Deterioration of vegetation; condition of deer	*Hunting*	*Class 3 (possibly Class 4 on islands)*
Change in vegetation; soil erosion	*Translocation; shooting proposed*	*Class 4 (possibly Class 2)*
Disruption of human fishing activities	*Scare devices; attempts to free entangled whales*	*Class 1 (perceived as Class 3)*
Soil erosion; high mortality of pups	*Shooting; preventing females coming ashore*	*Class 2 (perceived as Class 3)*
High mortality; poor condition	*None; some native hunting*	*Class 3 (perceived as Class 4)*
Changes in size structure of invertebrates and in species composition of algae	*None; shooting and translocation proposed*	*Class 1*
Presence of leopards	*Shooting of nuisance animals; translocation; captivity*	*Class 1*

The problems of overabundance in Zimbabwe are linked with the history of human occupation, stretching back 2,000 years to the original introduction of livestock into the country. Subsequent events of importance include exploitation of elephants for ivory, the effects of the 1890 rinderpest epidemic, attempts at tsetse eradication, and a fourteenfold increase in the human population in the last 80 years.

Problems involving deer in North America appear to reflect the nature of their habitat and their behavior. Habitat is often extensive and rather uniform. The deer are nonterritorial and locally high densities may not be reduced by emigration. These problems are intensified by human activities -- cutting of forests, removal of habitat, competition from livestock, removal of natural predators, and hunting.

Four species of marine mammals were considered in the case studies. The population of Northwest Atlantic humpback whales, which were formerly heavily exploited, has been protected since 1955. Whales feeding close inshore may become entangled in fixed fishing nets, damaging the nets, and possibly dying. The principal food species (capelin) has recently been dramatically reduced in offshore areas; other food species in inshore waters may have increased.

On the Farne Islands, breeding grey seals have increased the rate of soil erosion from islands with fragile soil caps. This erosion has little effect on the seals but may make the islands unsuitable for breeding puffins and physically unattractive to people.

The North Pacific walrus population has increased over the last few decades and now occupies much of its former, eighteenth century range. (However, population estimates from aerial surveys can be grossly unreliable.)

Many sea otter populations have recovered from being heavily exploited in the nineteenth century. Sea otters can dramatically alter species composition of marine communities to the detriment of shellfisheries, but possibly to the benefit of other industries.

The single terrestrial carnivore discussed under case studies was the leopard. This species is both valuable (because of its skin) and an apparent nuisance to humans (because it eats domestic animals and, occasionally, people). The leopard is representative of the nuisance level (Caughley's Class 1) of many other large terrestrial carnivores.

III. THEORETICAL CONSIDERATIONS

Discussion concentrated on Caughley's Class 4 of overabundance which covered animal populations that had or might become more numerous than the density appropriate to their long-term equilibrium. Such a situation typically arose in one of three ways: A herbivore population started from a low initial density and if unharvested crashed and then converged to equilibrium, or due to migration the initial density was higher than that at equilibrium, or a predator was removed from the system.

Caughley expressed the belief that management could take one of two alternative paths. One, involving culling, held the population below its long-term equilibrium. The other, involving no culling, permitted the population to approach its equilibrium without interference.

Discussion on these options, and particularly the option of no culling, centered on one central point. This point was, in the course of an irruption there might be such damage to the system that the long-term equilibrium density a herbivore might have attained would be reduced -- for example, if soil erosion reduced plant productivity. An extreme case of such a problem would be where the long-term equilibrium for the system was so depressed the herbivore species concerned became locally extinct.

Caughley viewed such dire developments as unlikely and maintained there were few if any cases where such disasters had occurred. Other participants in the workshop believed that restriction of annual migration by fencing of national parks made it more probable that herbivore populations, artificially constrained in this way, would reduce the productivity of the vegetation. Some examples of drastic changes in habitat type had been observed in a number of domestic grazing systems. It was not agreed whether such examples could act as pointers in the management of national parks.

Walker presented a modeling analysis of a savanna ecosystem. This system permitted a number of possible equilibrium states affecting the relative abundance of woody and grassy vegetation. The central message from this analysis was that some oscillation in herbivore density actually increased the resilience of the equilibrium of the model system. Walker noted that many management systems were aimed at reducing variability both temporally and spatially and argued that this could well involve unfortunate and unintended consequences for the system's ability to withstand environmental changes.

Walker highlighted the interaction in the Kruger National Park between lions and wildebeest as an example of a system with multiple equilibria. Here climatic change altered the effectiveness of predation by lions and foraging by wildebeest so that in wet seasons wildebeest populations were particularly vulnerable to significant and continued reduction. Discussion about this presentation focused on the need to quantify the relationships and conduct experimental testing of the models.

Sinclair's presentation concentrated upon the importance of recognizing the difference between fact, opinion, and hypothesis. He emphasized the necessity to test models used in management and to conduct, where possible, disparate management regimes in different areas. This latter procedure was aimed at obtaining information about the system in the most efficient way.

The overall burden of Sinclair's presentation was that the fallibility of scientific advice given to management must be recognized.

Owen-Smith followed up his case study of rhino population dynamics with the presentation of a management plan which involved removal of numbers of this species from the Umfolozi Game Reserve. Justification for culling was the fear that the long-term equilibrium might otherwise be a low rhino density following severe erosion of the soil. Argument focused on the untested nature of this assumption, and there was no agreement on its likelihood. Further discussion on practicalities indicated a sympathy with the current management of this highly endangered species -- including culling -- because extinction was seen as a possible consequence of nonintervention.

Noy-Meir's presentation on parallels from the husbandry of domestic livestock was seen to be highly relevant by the workshop participants. One of the implications for management is that herbivore density which may theoretically give maximum sustained animal production involves a high risk of a crash. A lower density, giving a lower but safer level of production, is usually preferable. However, the choice of the "optimal" herbivore density (in domestic grazing systems) is not entirely determined by ecological factors but also strongly influenced by economic parameters and social factors (tenure and ownership, and the objectives of the manager).

IV. DIAGNOSTIC CHARACTERISTICS

Dr. Daniel B. Botkin of the University of California at Santa Barbara acted as chairman of the workshop session on diagnostic characteristics. In his presentation Botkin responded to the question of how to tell if a population of animals is overabundant , by citing the point at which detrimental effects on the vegetation and soil begin. The soil starts to degrade and the chemical nutrients necessary for the sustenance of life are lost through erosion. Certain measures of soil conditions are crucial to assess the status of an ecosystem and the question of whether any individual animal population or set of animal populations is overabundant. (The scientific basis for these assertions has been published elsewhere.)

It is possible that soil processes provide a sensitive indicator of deterioration before a catastrophic crash in the target animal population. This conjecture needs to be demonstrated for any particular reserve and Botkin recommended that soil analysis be included in monitoring programs for reserves in which large mammals, particularly herbivorous ones, are important. Soil nutritional status is of fundamental importance because the recovery of a soil once the chemical elements have been lost is a geological process requiring an extremely long time and involving mechanisms such as weathering of rock, transport of new materials by air and water, and the biological fixation of nitrogen. Thus protection of the soil's nutrient status should be of prime concern. It has been shown in many parks and in many studies of wildlife that animals are limited by some nutritional quality in their food rather than by the total abundance of food. So a prime problem for terrestrial herbivores is to get enough vegetation of high nutritional quality. The vegetation, in turn, depends on soil conditions (or the amount of grazing and other kinds of disturbance) and on rainfall as determinants of the nutritional quality of the leaves and shoots. The quality of the edible parts of plants is also affected by competition with other plants. An ecosystem that supports an abundance of mammals must have a rapid cycling of nutritional elements. This can take place only when there is moderate grazing which removes the nutrients from the plants and stimulates new shoot growth; the nutrients, in turn, must be rapidly released from the animals' feces and urine and made available to the roots of the vegetation. Under light grazing -- or with no grazing -- many plants will translocate nutrients from shoots to underground organs or produce stems which are of relatively low nutritional quality. Under moderate grazing, plants are stimulated to take up nutrients to produce young shoots and leaves which are of high nutritional quality. Thus, moderate levels of disturbance lead to higher nutritional quality in the vegetation -- a more rapid cycling of the chemical nutrients necessary for life and a higher carrying capacity for large herbivorous mammals. A study of soils in Africa indicates the percent base saturation of the soils is controlled by biological activity in areas of moderate to high rainfall (rainfall greater than 700 or 800 mm/year and less than 1,500 mm/year). In this range, overabundance or underabundance of large mammals can lead to conditions in which nutrients are more readily lost from the soil by processes of fluvial erosion. A system that had maintained a high abundance of large mammals, because of a high nutrient status and a rapid cycling of chemical nutrients, could shift to a lower level of soil fertility that is only capable of maintaining a much smaller biomass. Botkin suggested there may be several regions of stability of soil nutrient status in which high abundance of large mammals could be sustained. Once the soil nutrients have been lost, the carrying capacity for large mammals is reduced to a much lower but also stable region.

In summary, Botkin considered what characteristics should receive attention in attempts to diagnose overabundance in each of Caughley's four classes. He originated the thought that a fifth class might be added -- "there are too many animals for the *system's* good." By adding this fifth class, attention would be directed to subtle changes in the soil; however, most of the workshop participants felt this was just a special case of the system being off its equilibrium, and so was already encompassed in Caughley's Class 4.

Caughley's Class 1 comprises conflict in which the abundance of certain wild mammals is perceived as a direct threat to human life or livelihood. Diagnosis in the remaining Classes 2, 3, and 4, which all involve biological processes, would need to be directed to the following categories:

(1) The biological characteristics of the target species.
(2) The biotic characteristics of associated species including competitors, predators, parasites, and prey.
(3) Characteristics of the community, e.g., species diversity.
(4) Abiotic measures of the ecosystem, e.g., chemical elements and energy sources in air, water, soils, and sediments.

It was stressed that monitoring only the target species involves the risk of missing early warning signs, thereby delaying the detection of Class 4 "overabundance" noted above. Botkin speculated that ecosystems in terrestrial environments which appear to be in equilibrium are "leaky" and may run down over time due to the loss of chemical elements and that past history, in particular the effects of rare events, should not be neglected in this discussion.

What are scientists actually trying to diagnose? It was agreed that in general they are attempting to identify detrimental changes which could lead to the probable breakdown of an ecological system, where breakdown implies the extinction of the target species.

The question remained whether it was possible to predict which changes are reversible and which are irreversible. For example, Bell's definition of a national park[1] is unambiguous, but not all of the workship participants were convinced that his was an appropriate definition of the function of a national park. There were a variety of views as to what actually constitutes system breakdown, and clearly value judgments are involved in the diagnostic process. There are many examples where wildlife managers thought something had to be done to prevent system change or breakdown and culling programs were initiated; but, unfortunately, there are few examples where controlled experiments were conducted and this has limited what we have learned in the process. The workshop session concluded that much research is still required to enable diagnosis of overabundance in an ecological context.

Diagnostic characteristics relating to two species -- the northern elephant seal and the vicuna -- were discussed in this session.

A. *Northern Elephant Seal*

Le Boeuf reviewed the case history of the northern elephant seal, a population which has undergone a rapid increase from fewer than 100 animals to more than 60,000 at present. Scientists have monitored dispersal of animals which has accompanied the increase in population size. At high densities there has been an increase in pup mortality on breeding beaches. Other indications of crowding include fully used beaches, use of marginal beaches, non-harem bulls mating with females in harems, and additional mortality to pups through trampling or being killed by females. An apparent increase in the white shark population has been noted, and a potential increase in conflicts between elephant seals and humans. However, none of these changes imply any breakdown in the system. The limiting factor in the case of northern elephant seals appears to be available space for reproduction and such feedback is almost without time lags.

[1]*"The purpose ... is to preserve in perpetuity representative examples of the plant and animal communities ..." (Bell, this volume).*

B. Vicuna in Peru

Otte outlined the case of the vicuna on the Pampa Galeras (the location of the National Vicuna Reserve) in Peru. He suggested the status of this species could be diagnosed as a plausible example of local overabundance which is leading to changes in biological parameters of the vicuna population, overgrazing of natural grasslands, and loss of soil and nutrients through erosion.

Worldwide, the vicuna is still classified as an endangered species. It is found in Peru on privately-owned and community lands (including the area of the Reserve, where the human population also maintains domestic animals). Otte suggested that the long-planned program for the utilization of vicuna was also an appropriate solution to the perceived problems of overabundance. Vicuna wool is one of the finest animal fibers known, and part of the resulting financial benefit could be used for the translocation of surplus vicunas to other conservation areas.

Eltringham discussed the use of aerial surveys as a diagnostic tool to monitor changes in density over time. He pointed out that aerial counts, even if inaccurate, can at least be used as indices to reveal changes in numbers or trends in abundance over time. He also noted that changes in density may be more important than changes in absolute numbers.

In the case of the vicuna, Eltringham felt it was not clear whether the population was being cropped to produce revenue for the local residents as compensation for allowing the vicuna to increase in numbers, or culled to overcome a perceived problem of local abundance. Reference was made, in connection with the second alternative, to a paper by A. Brack (Executive Director of the Vicuna Management Project) which argued that culling was necessary to avoid a major catastrophe, not only to vicuna but also to other ungulates including domestic species and to the local people.

It appeared to the workshop participants that the vicuna operation in Peru was related to animal production for wool, skins, and meat, and therefore was not relevant to our discussion. The vicuna have, with the cessation of predation, apparently increased in abundance, and have exhibited several density-dependent responses. There was no evidence that this would lead to catastrophic effects and no unequivocal evidence has been presented to show that culling is necessary to prevent "overpopulation."

C. *General Discussion*

From available evidence few generalizations can be made at present about the diagnosis of overabundance. Past measures of overabundance have been ineffective because of lack of data and uncertainty about which questions should be addressed. It was suggested that more consideration be given to system properties than to the target species alone. Manipulation of systems should be done in a way which maximizes the amount of information gained from the experiment, i.e., informative management to maintain future options. In terrestrial ecosystems more consideration should be given to soils and nutrient flow.

Two views emerged about the necessity of culling. One view suggested there was no evidence that culling was necessary to prevent system breakdown. The other view stressed the importance of time scale, suggesting that culling might be necessary in the case of long-lived herbivores such as the elephant and white rhino.

V. METHODS OF CONTROL

Four papers were referred to in this discussion. Bell's paper distinguished between technical and aesthetic problems in culling and he argued that the basis for deciding what outcome is appropriate is usually aesthetic. Bell pointed out that the habitat used by elephants in Kasungu National Park apparently can exist at one of three equilibria: woodland, scrub, or grassland. Woodland can only persist at very low elephant densities. Scrub occurs at moderate densities, where it persists as a browse-supply system with high productivity. At very high elephant densities, scrub is opened to fire and grassland develops. All three habitats are desired and necessary, and their persistence requires the maintenance of different densities of elephants in different areas. Cumming argued that this situation was not applicable in Zimbabwe, and Walker emphasized that the relation between elephant and plant association is variable.

Chapman's paper outlined the history of exploitation of the northern fur seal and identified the single management objective as maximum sustainable yield (MSY). Following the exploitation of females in the 1950's, the population has not recovered according to predictions. Among other overexploited populations of otariids, some have recovered and others have not. It was concluded that the behavior of complex systems is difficult to predict and that exploitation of the female component may result in a drastic population decline.

A paper by Smuts[1] described how culling in Natal game reserves is based on defined management objectives together with an assessment of the abundance and physiological condition of each species. There is thought to be no practical alternative to killing and live capture in order to meet these objectives. For the purpose of translocation, it is imperative that there be a suitable release site, potential problems are considered, and the social structure of the herd is taken into account. Removal of young females is the most effective means of population control, but this is not possible for all species (e.g., elephant).

In his paper Owen-Smith advocated the use of "vacuum zones" for culling. He argued the following points:

(1) There is a greater need to cull large species.

(2) Those species that can affect the structure of the vegetation need to be culled.

(3) The need to cull depends on the size of the reserve.

(4) Amplitude of population fluctuations depends on the amount of spatial heterogeneity.

(5) Wide-ranging species and species dependent on dispersal for population regulation are most in need of culling.

The objective of culling is to prevent the system from crossing a boundary into instability. The concern of park managers should be how best to implement this in space and time, so culling is carried out with a mimimum number of removals, with aesthetic considerations in mind, and with the objective of gaining information. By designating vacuum areas, more precision in culling is possible:

(1) A buffer against population fluctuation is provided.

(2) Core areas remain unaffected.

(3) Disruption to natural selection is minimized.

(4) Spatial heterogeneity is enhanced.

(5) Culled animals can be easily removed with appropriate physical planning.

Noy-Meir discussed the use of artificial water holes as an alternative to the culling of abundant herbivores. The main objectives of artificial water holes are to attract animals into unused areas and to stop the expansion of bare areas around existing water points. He concluded that the addition

[1]*Incorporated into chapter entitled "Management of Locally Abundant Mammals -- The South African Experience."*

of water points will increase population and increase the quality of the animals. However, it may also increase the amount of bare areas unless the number of herbivores at each watering point is carefully controlled. This is an easy solution to relieve local overabundance in the short term, but in the long view it is likely to be counterproductive except in certain special cases.

Cobb discussed alternatives to culling endangered large carnivores. As an example he selected leopards in Meru National Park. Translocation has not been a suitable option since these animals seem to have a strong tendency toward homing. Translocation to zoos is not feasible because most large carnivores breed easily in zoos and wild ones are not required, and wild animals are expensive to move. Botkin suggested there are stronger density-dependent behavioral regulations on population growth in carnivores than herbivores. Large carnivores are considered overabundant most often for social rather than ecological reasons.

VI. SYNTHESIS BY DISCUSSION GROUPS

On the last two days of the workshop, participants divided into smaller discussion groups to attempt a synthesis of ideas. Reports were prepared on specific topics:

(1) Points of agreement.
(2) Time/space considerations.
(3) Diagnostic characteristics.
(4) Has intervention in protected areas achieved its aims?

These reports were presented to the entire workshop when it again assembled and a consensus reached or opposing viewpoints identified. Much debate arose concerning topics 2, 3, and 4 and the issues have been summarized here after further circulation among the participants.

A. Points of Agreement

(1) An essential first step in the management of an area is to identify the objectives for that area (e.g., to maintain a species, to maintain a diversity of species, to maintain diversity of a system, to maintain ecological processes, to provide an ecological control area) recognizing there may be objectives that are mutually incompatible or unattainable.

(2) Caughley's four classes, as amplified by discussion, provide a broad framework. Most scientific considerations relate to Class 4 and can be tackled mainly by the application of ecological principles and by experimentation. Problems under the other classes raise social, political, and economic questions and involve value judgments, as well as posing ecological questions.

(3) It is recognized that in any situation there may be, in theory, several possible outcomes ranging from disaster to attainment of an acceptable equilibrium. The approach adopted by management should recognize this and, ideally, in order to keep options open for the future, at least two types of experiments should be set up in subcompartments within an area. These should comprise:

(a) A variety of manipulations in time and space.

(b) Leaving the system alone for a long period and monitoring changes to test different hypotheses about its natural behavior.

The first class of experiments will yield results on a time scale of years; the second class may require a much longer span of time measured in decades. Usually managers are unable to wait for the longer periods for results before making decisions, but if such experiments are not undertaken, the answer will never be known.

(4) Areas differ in the relationships they encompass between space, time, climate, soils, flora, and fauna, and there are dangers in extrapolating from one area to another. Sometimes it is preferable to carry out active adaptive management considering each case separately, in the light of current theory and evidence. If part of a system is unique, the options open to managers are more limited.

(5) The history, spatial diversity, and occurrence of rare events in an area dictate the approach to this kind of experimental management.

(6) When making decisions under conditions of uncertainty, it is necessary to assign probabilities to the results of adopting alternative management strategies, and usually to err on the side of caution.

(7) Constraints are imposed on the management of a system by the size of the area under consideration and the behavior of the species involved (e.g., territorial, migratory, or if resident, home range size).

(8) There may be less risk of disaster in adopting a policy of nonintervention where small, rapidly-responding species of animals exploiting rapidly-renewable vegetational components are concerned than for large, slowly-responding species which exploit vegetational components where recovery involves long time lags and effects upon other species of animals.

(9) Because more can be learned from perturbations than from equilibrium states, use should be made of perturbations, either natural (e.g., fire, drought) or uncontrolled (e.g., poaching, human population pressure leading to compression) as well as experimental manipulation of systems (e.g., fire, predator control, removal of herbivores, fencing, exclosure plots, provision of artificial water points) to provide a sound basis for future predictions. If spatial compartmentalization is not feasible (e.g., for migratory species) temporal comparisons should be made (e.g., use natural events, such as wildebeest or caribou recovery from low densities or change in migration routes). Density-dependent responses can be studied in the same population over time or in different populations in space (e.g., southern elephant seals in exploited and unexploited populations).

(10) Every opportunity should be taken to maximize the information from management since active adaptive management is a powerful research tool. Although one objective of park management should be to minimize interference, a policy of "do nothing" could be as bad as overreacting to events. There is a pressing need to ask the right questions and to define key experiments to answer these questions (e.g., the effect on base saturation of soils of removing animal and plant biomass).

(11) The provision and distribution of artificial water points is a complicated question. While water points may provide a temporary solution and postpone the need to cull to relieve overabundance, they can often make the problem worse in the long term. Used in a careful and sophisticated way, they may be useful.

(12) Political or economic reasons for culling should not be obscured by irrelevant scientific excuses.

(13) Scientific evidence is unlikely ever to be so strong that areas of uncertainty and disagreement are eliminated; general rules will not invariably apply.

B. Time/Space Considerations

A working group discussed the following general problem: In what situation (if any) would the unrestricted growth of a mammalian population pose a significant threat to its own survival? The typical pattern of mammalian population growth is an overshoot of the population size that should be reached at equilibrium with subsequent damped oscillations to that equilibrium. This has been documented for many ungulate populations; the theoretical framework to explain it is that of delayed density dependence. Furthermore, theoretical considerations do lead to the recognition that such a population overshoot may produce changes in the productivity of the ecosystem, via physical damage such as soil erosion or by several grazing mechanisms which result in an alternate and less productive equilibrium state. Noy-Meir reviews such systems, and in his earlier work he has indicated that some range management practices aim at avoiding this phenomenon. In the first set of conditions, where a less productive equilibrium state is established, the physical change is only reversible on a very long time scale. In the second case, where management has intervened, judicious manipulation of the system may reverse the process and the more productive equilibrium be reattained relatively simply.

The working group could not reach agreement on a system of classification which might be used to indicate what systems, if any, were most susceptible to such phenomena beyond the obvious one of poor, easily erodible soil. However, it was felt that large mammalian herbivores feeding on long-lived, slow-growing plant species might be the most susceptible. A significant number of the group believed that all such problems were only of academic interest and did not occur in practice. However, it was clear that there must be a lower limit to the size of a reserve which could support a population and that smaller reserves would be more susceptible to the problems of overshoot.

The underlying mechanisms which may be responsible for some instances of apparent overabundance are the interactions of herbivores with patches of vegetation that are differentially grazed. Movement of herbivores to other patches following overgrazing of one patch facilitates recovery. Such systems are similar to some that have been investigated theoretically and in laboratory experiments with arthropods in which a critical size is necessary for the stable continuation of these systems. If this were indeed the case, it presents interesting problems for the choice of size of nature reserves.

C. Diagnostic Characteristics (Draft Guidelines for Diagnosis)

If it is claimed that a certain animal population is overabundant and the removal of excess animals is necessary, what reaction should an advisor make? First, it must be made clear that it is not possible to give a checklist of specific actions which will apply in general to every ecosystem, to every population, and to every management goal. Ecology cannot as yet provide generalizations of this kind. Many ecologists believe that the science of ecology will *never* afford such broad generalizations. Lacking a checklist, what is possible? One can attempt to establish a procedure so that, given a specified management goal, a series of pertinent questions is posed, the answers to which will allow the manager of a particular habitat to determine whether or not the claim of overabundance is valid. Such questions might be:

(1) Is there a trend of increasing population size or density accompanied by a decrease in resources, a change in the habitat, or a decrease in the status of associated populations? (There resides in this the problem of separating trend from normal climate-induced fluctuations.)

(2) If there is a trend, is it likely to slow down toward a stable state (or region) or continue into an eruption/crash process? (Effectiveness and time lag of density-dependent process in the population and its resources will be key factors.)

(3) If an eruption/crash is expected, is it likely to settle down to a stable state (or region) similar to the previous one or result in extinctions or other major changes in the ecosystem?

There are four major categories of pertinent measurements to be made in relation to such questions. The first category of these measurements comprises biological characteristics of the species of prime interest; e.g., abundance, changes in abundance, birth rate, death rate, growth rate, and age structure or size structure of populations. The second category comprises the biological characteristics of associated species. (These can be divided into trophic levels such as competitors, predators, and parasites, and resources which are the foods of the animals and the soil nutrients for the plants.) The third category consists of measures which express the integration of an entire biological community (an example is species diversity). The fourth category consists of abiotic measures -- that is those parts of the ecosystem which provide storage and transport of chemical elements and of energy among organisms -- the fluid media, air and water, and the sediments.

A manager needs to determine which of these categories are most relevant to his problem. In the past, management and conservation authorities have tended to concentrate on only the first category. A common, but erroneous, focus on measurements in the first category may provide information too late to be useful. An analogy would be a doctor monitoring the numbers of bacteria in a patient's blood, but not measuring his body temperature. While the abundance of the bacteria might be the cause of the disease, it will not be evident to a doctor what abundance will produce disease in any particular patient. On the other hand, the temperature, which is symptomatic but not causal, is in this case the crucial measure of interest. A manager's concern should be, therefore, to determine what is the appropriate ecological symptomology in addition to the appropriate ecological cause and effect. The symptomology will often include measures that move beyond the biological characteristics of the target species. Further, it is unlikely that any single measure will be sufficient. Pairs or groups of measures will be necessary in almost all cases. A manager should be advised to make several measures from more than one of the above categories.

Next, we sketch a framework for predicting the consequences of culling versus no culling. With respect to a potential problem species, the first step is to identify the essential subsystem, defining which areas and variables must be included and which can be ignored.

Within the subsystem, the specific consequences of overabundance need to be considered in relation to five suggested measures:

(1) The physical structure of the system must be considered by size (the smaller the area the more likely intervention will be necessary) and topography (the potential for erosion increases with increased average slope).

(2) The environment must be studied. Growth conditions (arid regions recover very slowly), soil type (deep sandy soils are less susceptible to increased herbivore pressure), and changes in soil nutrient status should all be considered.

(3) Some questions regarding habitat (particularly the vegetation) of the subsystem should be asked: What is the historical record of the vegetation? Has it changed drastically in the past? What is the potential degree of destruction and the potential recovery time? (For example, a woodland of 200 year old trees will be more difficult to regain than an annual grassland.) What is known of the dynamics of the subsystem? Is it likely the vegetation is always recoverable, or is it possible there are multiple stable states? If multiple stable states may occur -- what are they, where are the thresholds likely to be, and which factors will influence the position of the thresholds? When identifying the measurable attributes of the ecosystem, which are the most important in the system's ability to recover? (In the case of seed reserve in an annual grassland, for example, to what extent is vegetation regeneration sensitive to changes in soil surface characteristics or changes in hydrology?)

(4) Other components of the system (both animal and plant) must be examined to determine if (and how) they are adversely affected by overabundance of other species.

(5) In relation to the "problem species" itself, observed changes in demographic vigor of the population must be studied. (Are these changes due to a decrease in juvenile mortality? Decrease in age at first reproduction? Increase in fecundity? Decrease in adult mortality rate?) Along with demographic vigor, the physiological condition and behavioral attributes of the population need to be measured. Physiologically, if there is a local overabundance, how has this affected body conformation and condition as assessed by external appearance, deposited fat reserves, blood chemistry, and body growth? Behaviorally, to what extent has overabundance influenced rate and vigor of social interaction and feeding strategy (including qualitative and quantitative aspects of the diet)? Finally, assessment of the mechanisms of dispersal or immigration should be examined -- to what extent has local overabundance been caused by interference with dispersal?

Those measures that will be useful in a particular situation for particular management goals at this time can only be determined by experiment, so it is necessary for management to proceed in close coordination with scientific research. A manager should, therefore, attempt to carry out several treatments and designate a control area in which no "positive" management action is taken. If a manager feels it is necessary to respond to apparent overabundance by initiating removal, he should at least divide his managerial area into two units -- undertaking culling in one, and not culling in the other. If possible, a more extensive experiment should be designed. Obviously, this suggestion is made with the understanding that some unique reserves cannot be treated in this way. For example, if an area is the last remaining habitat of some species or representative example of some ecosystem, the only reasonable action perhaps is to preserve the entire area. However, a protected area can be put in one of three categories in terms of management objectives. Some habitats are small, fragile, and unique and demand immediate protection and action; the manager can only do what immediately appears to be most reasonable based on expert opinion and testimony. Other areas are large, obviously resilient, and will continue with little change over the manager's time horizon whether he does anything or not. The third situation concerns areas large enough to permit several kinds of manipulations simultaneously and in which it seems probable that unless some positive actions are taken an undesired catastrophic change in a population or in the entire system will occur. In such situations the manager will need to work in close cooperation with scientists to conduct well-designed experiments as part of the managerial procedure, in order better to understand the effects of managerial actions. As a regular facet of his job, the manager should also arrange to monitor the variables of interest.

Few areas established as reserves have any routine monitoring of crucial variables. This can lead to disaster, since if nothing is measured about the system there is no way to determine its state, and therefore no way of knowing whether it has changed in a significant way.

Sets of populations and ecosystems are characterized by threshold events, by breakpoints, and potentially by multiple stable regions. It is important for the manager to attempt to identify where the breakpoints lie that would direct the ecosystem to one or another stable region -- an attempt should be made to determine what conditions are essentially transitional between one region of one kind of stability and another region of another kind of stability. The manager should also be aware that variations are characteristic of natural systems. Eruptions and crashes in populations are not uncommon nor usually disastrous, the rule rather than the exception. Populations can vary widely in a terrestrial ecosystem as long as the ecosystem has the chemical elements necessary for life and does not lose them at a significant rate; i.e., at a rate greater than had occurred prior to the conditions that led to the claim of overabundance. The manager should also be aware of the distinction between preventive actions and corrective actions, analogous with the preventive and corrective medicines recognized by physicians. Obviously, it is preferable to seek preventive actions but in many cases (particularly at this time) only corrective ones can be taken. The above suggestions about coordinating actions to implement decisions about management within an experimental scientific context including monitoring, should help to make a greater emphasis on preventive management possible in the future.

The importance of the soil chemistry in a terrestrial ecosystem must be recognized. The manager should determine whether the system in question is vulnerable to the loss of soil nutrients. The fundamental task of management is to protect the physical and chemical characteristics of the substrate. It is also possible -- though not yet determined experimentally -- that monitoring of soil water for chemical ions could reveal changes that might in turn indicate a serious disruption of the system. For example, in an ecosystem where the target population is increasing or is in some apparently satisfactory steady state, revelation that the runoff contains a heavy load of chemical elements necessary for life (representing a considerable net loss of these elements from the soil) would be a warning of impending trouble. Being thus warned the manager should consider seriously whether the current abundance of the target species is having a negative effect on the soil, enhancing the nutrient loss, and calling for remedial action.

D. Has Intervention in Protected Areas Achieved Its Aims?

The very creation of protected areas involves some measure of intervention. Some ecological processes, most often predation or exploitation by man, are arrested, some processes are set in motion, while yet others are unaffected. Custodians of protected areas have to decide whether or not to arrest or modify change within the ecosystems under their care. A common expression of such change is an increase in populations of large mammals and a decrease, or at least change, in plant populations. Since much management has been directed toward manipulating large mammal populations, it is pertinent to ask whether the assumptions leading to culling have generally been supported or contradicted by subsequent experience. It is similarly pertinent to ask whether the assumptions leading to "no culling" have been vindicated.

The treatment of problems of overabundance involves three main stages:

(1) Definition of the problem.
(2) The action taken (including "no action").
(3) An assessment of the extent to which such action resulted in a predicted or expected outcome.

Ideally such information should be readily available. When we began to consider these stages at the workshop, however, it soon became clear that even in many well-documented cases such fundamental features as how the problem was initially preceived, what results were predicted, and what resulted from management action, were either not known or were in dispute. For these reasons the brief consideration of a large number of examples, such as we attempted in a short working session, was inadequate in several important respects.

To arrive at a basis for assessing the results of intervention (or nonintervention) in protected areas, we felt it desirable to examine in detail a few case histories. A broad framework for a case history evaluation was proposed as follows:

(1) Initial state of the system: Including such information as when protection began, the status of the population in question and of the ecosystem at that time, and the influences which changed when protection was imposed.

(2) Trends in the ecosystem: An examination of trends in the population or ecosystem which contributed to the problem as well as a consideration of such key influences as changes in human populations, climate, fire, and artificial water holes.

(3) Perception of the problem: When was the problem first perceived and how did these initial perceptions of the problem fit into the categories defined by Caughley? Did perception of the problem change with time?

(4) Action taken: What options were open to managers and which options were taken? Were any clear aims stated or predictions made about the results of management action?

(5) Current status: What is the present status of the ecosystem and how does this relate to 1 and 2 above?

(6) Evaluation of management action: Were the explicit or implicit aims of intervention realized fully, partly, or not at all?

The examples chosen, two from North America and two from Africa, reflect the experiences of workshop participants and seek to cover the following four main categories of problem and reaction:

(1) Nonintervention where an overabundance of moose was viewed as a threat to vegetation and an overabundance of wolves was viewed as a threat to moose, e.g., moose and wolves on Isle Royale.

(2) Culling where an overabundance of herbivores was viewed as a threat to the system, e.g., elk in Yellowstone National Park.

(3) Culling where an overabundance of large herbivores was viewed as a threat to vegetation and soil cover, e.g., elephant, buffalo, and zebra in the Deka catchment of Wankie National Park, Zimbabwe.

(4) Intervention followed by nonintervention where overabundance was viewed as a threat to vegetation, e.g., elephant and woodlands in Chizarira National Park, Zimbabwe.

1. Moose and Wolves in Isle Royale National Park, Michigan. (Compiled by David R. Klein from references at the end of this case history and personal communication from R. O. Peterson.)

a. Initial state. Isle Royale is a 544 km^2 island in northwestern Lake Superior which has been part of the U. S. national park system since 1940. The closest land is southern Ontario, 24 km to the north. Moose have been present on the island since the early 1900's, apparently swimming from the adjacent mainland; wolves became established there about 1948, probably crossing on the lake ice during the winter. Caribou were present in small numbers at least irregularly until about 1925, and the coyote was common after the turn of the century but disappeared when wolves became established. The disappearance of caribou coincided with the decline of the species on the adjacent Canadian mainland, apparently at least partly as a result of extensive forest fires that destroyed much of their habitat but which stimulated the growth of deciduous shrubs and trees of value to moose. No other ungulates or large carnivores have been present on the island in recorded history.

b. Trends in the ecosystem. Prior to the establishment of wolves on Isle Royale the moose population underwent an irruption, reaching in excess of 1,000 animals by 1930. Population growth from the time of their establishment on the island followed the typical sigmoid curve. The high moose density was associated with decline of the most important food species from overbrowsing. Winter starvation caused a subsequent decline in moose numbers.

A fire in 1936 temporarily destroyed browse plants on a quarter of the island and the moose population reached a low of a few hundred animals at this time. By 1948 moose had again increased to an estimated 800 and once again a decline in browse plants was noted.

An attempt was made by park authorities to establish wolves on the island in 1952 partly to serve as a natural check on the moose population, but also to create a protected wolf population. Although wolves had apparently already gained access to the island naturally, this was not confirmed at the time, and four zoo-bred wolves were released. They failed to adopt a wild existence and three were subsequently removed.

The natural establishment of wolves on Isle Royale, although unanticipated by park authorities, was nevertheless a welcome addition to the natural fauna of the island. Wolves and moose apparently reached a dynamic equilibrium on the island. Usually the wolves hunted in a primary pack of 10 to 16 animals with one or rarely two smaller subsidiary packs. Total numbers normally did not exceed 20 during the 1960's. Moose numbers during this period stabilized around 600 animals in late winter with fluctuations apparently associated with changes in forest succession following fire, which influenced production and availability of browse for moose.

Predation by wolves is the major source of moose mortality on Isle Royale. This has stimulated productivity in the moose and has apparently reduced the incidence of losses due to disease, parasitism, and starvation. In studies carried out in the 1960's it was found that wolves usually tested moose before attempting a kill; those vigorous individuals that effectively defended themselves were left alone.

c. Perception of the problem. Up until the early 1970's, wolves apparently controlled their own numbers through pack territoriality which limited the number of wolf packs that could exist on the island. Pack size is apparently limited by intragroup social hierarchies which normally limit breeding to one pair in each pack. In addition breeding took place only in the largest pack prior to the 1970's. Pup survival has been dependent upon the winter hunting effectiveness of the pack.

Apparently the wolves kept the moose herd within relative balance with its food resource until the early 1970's when the pattern of wolf/moose relationships changed markedly. At that time moose numbers were high (1,000 or more animals) and wolf numbers relatively low (less than 10) due to the exodus of a pack from the island. A series of winters of heavy snow beginning in 1969 coincided with the deterioration of browse plants from the heavy browsing pressure of the large moose population, with the result that weakened moose of all sexes and ages were particularly vulnerable to wolf predation. With the more easily available prey, wolves once again began to increase. Breeding began to occur in more than one pack and by 1975 wolf numbers had increased to an unprecedented 44 animals. By 1976 the ratio of wolves to moose had changed from one wolf per eighty moose in 1969 to one wolf per twenty moose. The increased wolf density in winter was apparently maintained through the easy availability of moose. In summer, however, the wolves showed pronounced adaptability in their food habits, consuming much larger proportions of beaver than at lower densities and even making significant use of fruit when it was available.

d. Action taken. The concept of nonintervention as a policy guiding management of wildlife populations on Isle Royale was a de facto policy from the time of the island's inception as a national park and the arrival of the wolf, a suitable predator of moose, seemed destined to lead to a more favorable balance between the moose and their food.

e. Current status. Rather surprisingly, the wolf population remained high up to 1980 when it reached a new peak of 50 animals. Equally surprising has been the resilience of the moose population in the presence of high wolf density, which after suppression during the deep snow years recovered to an estimated 650 to 700 moose in 1980. By early 1981, however, the wolf population had declined to an estimated 30 animals while the moose population remained at the 1980 level.

f. Evaluation of management action. As an example of nonintervention in the management of large mammals on wildlife reserves, Isle Royale appears to be "successful" in the anthropomorphic sense that it led first to a stable, and later, a dynamic equilibrium, with improved condition of the moose and minimization of overbrowsing of the habitat. It is of significance to note that during the first ten years of study of the wolf/moose relationship on Isle Royale, a relatively well-balanced equilibrium existed between these two species. This balance, however, was disturbed by a series of winters of heavy snows and the coincidence of low wolf numbers with high moose numbers when the severe winters began. The equilibrium of the system was stressed and it is presumed that moderately large fluctuations in both wolf and moose numbers may occur in the immediate future until a more stable equilibrium is achieved or the system is again stressed by climatic or other environmental extremes.

g. References.

Allen, D. L. (1979). "Wolves of Minong." Houghton, Boston, Massachusetts.

Mech, L. D. (1966). "The Wolves of Isle Royale." U. S. National Park Service Fauna Series 7, Washington, D. C.

2. Northern Yellowstone Elk. (Compiled by A. R. E. Sinclair from the description of this system given by D. B. Houston in the references listed at the end of this case history and by personal communication.)

a. Initial state. The initial conditions of the vegetation in the northern range of Yellowstone are known from a remarkable set of photographs taken between 1871 and 1890. They show a vegetation of conifer forests on the hills and grasslands in the valleys with aspen (*Populus tremuloides)* groves, while willow *(Salix* spp.) occurred in riverine areas.

b. Trends in the ecosystem. Based on the photographs, changes in the vegetation have been observed over the period 1870 to 1970. In particular many of the aspen stands have disappeared, conifer forest has spread more uniformly over the hillsides, and willow has declined in some areas.

Human influences through heavy livestock grazing in the early part of this century altered some of the grassland, but livestock were removed in the 1930's. A severe drought in the late 1930's apparently coincided with a reduction in grass/herb cover. This cover subsequently improved when higher rainfall returned. Willows also disappeared during the drought.

A further human impact occurred through reduction of forest fires. Prior to 1870, fires burned conifer forests at a frequency of one every 20 to 25 years, but subsequently the fires were reduced to a periodicity of 80-year intervals. Fires are necessary to promote regrowth of aspen, and fire reduction has resulted in the disappearance of aspen stands.

c. Perception of the problem. During the 1930's, certain winter range ridgetops were considered severely overgrazed by elk. Workers at that time assumed, without direct evidence, that such areas had developed in the previous 20 years, while elk numbers (around 12,000) remained approximately constant.

Similarly, elk were known to browse aspen, and the latter's disappearance was attributed to overbrowsing by elk. Willow was thought to have suffered the same fate.

d. Action taken. As a result of these conclusions, elk were culled from the early 1930's until 1968. Initially, only a few hundred were killed annually, but when no response was seen in any of the above vegetation criteria, severe culling (in the thousands) was imposed from 1955 to 1968 by which time only 3,000 to 5,000 animals remained.

e. Current status. In 1968 culling was stopped because it appeared to have had no effect on the vegetation; the elk population increased back to 12,000 by 1976 and has subsequently held steady (although some recent culling has lowered the level to about 11,000 animals). Again, no significant change in vegetation was observed.

f. Evaluation of management action. Primarily, culling was initiated upon the indirect evidence from vegetational changes that were *assumed* to be caused by elk overabundance. The lowered elk numbers resulting from culling, and their subsequent higher numbers, showed that elk were not the cause of the changes in the vegetation. Culling, therefore, was not successful in realizing management aims of increasing vegetation because the wrong criteria was used.

Recent changes in the management of fire, allowing more fires in the system, has seen the return of some aspen. The ridgetop grasslands have been shown to be stable (contrary to the implicit assumption that the overgrazed grasslands are unstable, leading to degradation). New evidence has shown that willows also disappeared in areas never used by elk, thus suggesting a climatic or edaphic cause for their reduction.

g. References.

Houston, D. B. (1973). Wildfires in northern Yellowstone National Park. *Ecology 54*, 1111-1117.

Houston, D. B. (1976). The northern Yellowstone elk, part III and IV. Vegetation and habitat relations. Research report. Yellowstone National Park. 444 pp.

Houston, D. B. (1978). The northern Yellowstone elk: winter distribution and management. Mimeo. Symposium on Elk Ecology and Management, Laramie, Wyoming. 21 pp.

3. Reclamation of the Deka Catchment in Wankie National Park. (Compiled by David H. M. Cumming.)

a. Initial state. The Deka catchment in the northwest corner of Wankie National Park comprises shallow rocky soils on basalts with deep black clay soils in valleys and depressions. The vegetation is an open tree/scrub savanna with a generally sparse grass cover on the shallow soils and medium to tall grasslands in the vleis (moist, grassy drainage depressions). At the turn of the century the area was divided into ranges, but cattle and game populations were low after the rinderpest of 1895. Much of the catchment later became a game sanctuary and its owner, H. G. Robins, made provision for its incorporation into Wankie National Park upon his death in 1939. A few artificial water points had been provided by Robins and, because cattle were grazed in the sanctuary, predators were controlled. His aim had been to build up the game population.

b. Trends in the ecosystem. Upon incorporation into the park, cattle were removed. A low level of predator control continued through to the late 1950's. Populations of elephant, buffalo, and zebra increased while comparisons with Robins' earlier records indicated that other species had remained fairly stable. Rainfall during the 1960's was lower than during the three previous decades. Early fire history and changes in fire frequency are not known.

c. Perception of the problem. The problem seems to have been first fully appreciated in the late 1960's as one of accelerated erosion highlighted by gully erosion and drying out of moist vleis. The catchment was then surveyed in some detail in 1972 by Robinson *et al.* (1973) who saw the problem as one of overstocking and inadequate fire control leading to an increase in bare ground, runoff, and consequent sheet and gully erosion. They emphasized that, without an improvement in grass cover and a reduction in runoff, the problem of gully erosion in the vleis could not be controlled.

d. Action taken. Recommendations by the survey team and soil conservation specialists included a 50% reduction in the populations of large grazers (elephant, buffalo, and zebra), a system of fire breaks providing for complete protection of the shallow soil areas, and controlled rotational burning of the vleis. At the time, the shallow soil areas could not have sustained a grass fire. The recommendations also involved a series of mechanical measures to arrest gully erosion and to restore vegetation and moisture levels in the formerly moist vleis.

In 1971 and 1972 a total of 848 elephant were culled in the Deka catchment area and a further 888 were removed during 1974 and 1976 to the south of the catchment. The effect of these culls was to maintain a dry season density of around 0.4 elephants/km^2 in the Robins area of the park. In 1974 and 1975 the number of buffalo removed was 935 and the number of zebra removed between 1974 and 1976 was 339. The recommended fire control and mechanical erosion control measures were implemented at the same time (Braithwaite, 1980).

e. Current status. By 1980 the catchment had shown a remarkable recovery, and grass cover was sufficiently well developed to warrant a burn to remove moribund grass. Gully erosion had been arrested and water tables in the moist vleis had been restored. The large mammal populations remained at lower levels than before the culling, and the level of grass utilization was generally low.

f. Evaluation of management action. The aims of the three-pronged management effort were realized. The recovery of grassland may have been assisted by a higher rainfall during the recovery period than had occurred during the previous decade. Mean annual rainfall for the 11 years prior to the 1972-1973 drought was 546 mm. During the drought years, the mean annual rainfall was 328 mm, while the six seasons following the drought showed a mean annual rainfall of 761 mm. The difference between the means is not, however, statistically significant. There is little doubt the reclamation of the vleis would not have occurred without the provision of mechanical structures. The effects of a substantial reduction in large herbivores, an absence of fires on shallow soils, and the effects of rainfall are, however, not readily separated. Robinson *et al.* (1973) argued that a reduction in bare ground and, hence, in runoff on the shallow soils together with a reduction in trampling in the vleis, was essential if gully erosion was to be arrested. While it could be argued that grass cover on the ridges and shallow soils may have recovered without a reduction in large herbivores, this argument is not easily supported for the recovery of the vleis.

g. References.

Braithwaite, P. G. (1980). Gully control in basalt soils. *Zimbabwe Agric. J.* 77, 253-258.

Robinson, J. C., Hill, J. R. C., and Rushworth, J. (1973). Deka River catchment survey. Unpublished Department on Conservation and Extension/Department of National Parks and Wildlife Report.

4. *Elephant and Woodlands in Chizarira National Park, Zimbabwe.* (Compiled by David H. M. Cumming.)

a. Initial state. The park falls within the Sebungwe region (see Cumming, this volume, Fig. 7) and encompasses most of the Chizarira range of hills which rise abruptly from the valley to the south of Lake Kariba. The undulating, and in parts highly-dissected, plateau supported a well-developed *Brachystegia/Julbernadia* woodland. Although well watered and used as a hunting ground by tribesmen before protection, it does not appear to have been settled or cultivated in recent times. The plateau was included in a nonhunting reserve in 1958. It became a game reserve in 1963 and a national park in 1975. It extends over an area of 1,910 km^2.

Elephant populations were at a low level in the 1950's (see Cumming, this volume) and aerial photography for 1951, 1959, and 1965 shows little change in woodland cover. Park policy stipulated that "particular attention will be directed toward preserving the existing balance between major physiognomic veld types."

b. Trends in the ecosystem. Resettlement of subsistence agriculturists from the Zambesi began in the late 1950's with the formation of Lake Kariba. They settled on the alluvial areas to the south, west, and north of the park. Tsetse control hunting intensified in the mid-1960's and provided further constraints on the elephant populations of the area. Changes in the woodland accelerated after 1965 and Thomson (1975) recorded tree felling rates of c. 20%/annum in 1972 and 1973. Aerial counts of elephant between 1968 and 1973 showed an upward trend, with a dry season population of at least 750 elephant in 1973. A correction for these counts suggests, however, that the population was about 1,800 animals. With the introduction of roads and fire breaks, early burns (which might have invaded from surrounding areas) were contained, but this resulted in an increase in the incidence of hot burns late in the dry season.

c. Perception of the problem. Damage to woodlands in Chizarira was reported in 1964 and 1965. During a reconnaissance of the plateau in 1966, I estimated that in parts some 20% of the trees had been damaged by elephant but noted that "evidence was inadequate to warrant control of elephant on the plateau." Following an aerial survey in September 1968, Kerr reported widespread damage to woodlands and recommended a substantial reduction in the elephant population as well as fire control. After the culling in 1972 and 1973 (plus estimates of the rate of tree loss) an offtake of 500 elephant was planned in 1974. This number was later reduced to 300, and then to the 104 animals which were removed. A reduced rate of woodland damage, measured in 1974, seems to have prompted this change in plan, together with a recommendation that only 50 animals should be removed in key areas during 1975. An increase in tree loss measured in 1975 was followed by recommendations to intensify fire control so as to promote regeneration of young trees and coppice growth. It was argued that the *Brachystegia boehmii* woodlands were at the beginning of a cycle of regeneration and that elephants should not be culled.

d. Action taken. The number of elephant culled in the park in 1972, 1973, and 1974 was 176, 122, and 104 respectively. Attempts to control and contain late dry season burns through a system of fire breaks were largely unsuccessful. At the end of 1976 the park headquarters was destroyed and options for management over the next three years were limited. In an effort to minimize fire damage to woody vegetation, a policy of early burning in *Brachystegia* woodlands was adopted in 1978.

e. Current status. The aerial survey of 1980 revealed a dry season population of c. 3,000 elephant in the park, giving a crude density of 1.5 elephant/km^2. The rate of woodland damage reported by Thomson (1975) continued and the *Brachystegia boehmii* woodland has all but been eliminated from the park. The only remaining woodlands on the plateau, *Brachystegia glaucescens* on steep rocky slopes and ridges, are now being attacked by elephant and have already been severely reduced. The status of grassland, and the question of sheet and gully erosion in an area very susceptible to erosion, has not yet been investigated. Field reports indicate that populations of large mammals other than elephant have declined; this is particularly true of sable antelope and tsessebe.

f. Evaluation of management action. The culling of elephant was not carried out at a sufficiently high level to arrest or alter the trend in woodland damage. The woodlands, consequently, have not been preserved as policy for the park intended. A contributing factor was an inadequate knowledge of the number of elephants involved. The alternative strategy of controlling fire also has not arrested the decline in woodland, and the prediction that the woodland was about to regenerate has not been supported even in areas protected from fire.

This case history provides an example of the changing perceptions of an overpopulation problem and of the failure of limited culling, coupled with changes to the fire regime. These changes in the fire regime probably served inadvertently to aggravate the reduction of woodland.

g. References.

Kerr, M. A. (1968). Aerial survey of Chizarira Game Reserve and proposed extension and the implications. Unpublished report. 6 pp.

Thomson, P. J. (1975). The role of elephants, fire and other agents in the decline of a *Brachystegia boehmii* woodland. *J. South Afr. Wildl. Manage. Assoc.* 5, 11-18.

5. *Summary.* The two North American examples have run on long enough to show that nonintervention in Isle Royale was justified and that culling elk in Yellowstone was unnecessary. The African examples suffer perhaps from too brief a perspective, particularly in the case of elephant and woodland, where time scales need to be longer than the twenty or so years which have elapsed since the problems were first perceived. Other pertinent examples in southern Africa which are treated in detail in preceding chapters of this volume include: white rhino and grasslands in Natal (Owen-Smith), elephant and woodlands in Malawi (Bell), elephant and woodlands in Zimbabwe (Cumming), and a range of case histories from South Africa (Hanks *et al.*)

VII. CONCLUSIONS

A certain population of a rare mammal (for instance the sea otter, the hippopotamus, or the vicuna) may be said to be overabundant. But is it? Or in what way are there too many animals?

The workshop identified four classes of overabundance. First, there are those numerous cases where wild animals threaten the livelihood or even the lives of local people. Leopards kill sheep and goats, and elephants raid crops. No delicate ecological balance is at risk and experts other than ecologists will have a variety of solutions.

Second, it may be held that "too many" animals of one species depress another favored species. For example, in many African national parks, hunting dogs were shot because they killed and disturbed the attractive herds of grazers. Elsewhere, one herbivore has gained ascendancy and depressed, by competition, the density of others.

Third, it may be argued that there are too many animals "for their own good." Deer held at low density by hunting tend to be fatter, healthier, and more fecund than unharvested stocks, something that hunters prefer. Conversely, wild herd animals, like wildebeest, when at high density may appear below top condition and yet be enduring no more than a natural cycle. Their condition, however, might worry the human visitor.

All of these first three classes derive from human perceptions of what is good or desirable. They do not reflect some inexorable ecological change.

The fourth class of overabundance is the one of central concern. It is that in which the animals and plants are out of equilibrium -- and a population of a certain valued mammal is above its equilibrium density. This is the crucial category because ecological changes of unpredictable outcome may be set in motion; one such outcome being the collapse and extinction of the animal population concerned. (It was emphasized that some of these ecological changes might be extremely subtle -- changes in nutrient status of the soil, for example -- and hence easily overlooked.)

This categorization identifies the problems about which conservationists need to worry; but the workshop emphasized that it is not always obvious to which category a problem belongs. Nor are scientists infallible arbiters. Many problems that have been put into Class 4 have proved to belong elsewhere, even in Class 1. This does not mean the problem will go away, but the worries will be of a social or political nature rather than ecological.

Turning to the central problem, the numbers of a large herbivore have reached disquieting proportions in a reserve or park originally set aside to protect them. The species is out of balance and some factors of its environment are deteriorating. Should some portion of the offending animals be killed? The workshop showed no inclination to jump to this conclusion.

Of course, culling need not mean killing. Animals can be translocated to another range, or park, or zoo, but such an escape from the dilemma is not likely to be available indefinitely.

How can culling by shooting be countenanced? The first consideration should be concerned with the diagnosis of the "overpopulation." The key factor will be change in the environment that is seen as deleterious. Dense elephant populations are associated with a severe loss of trees; white rhinos with overgrazing and soil erosion. Cause and effect must be established, however, and not assumed, and the means by which this deduction is reached, the evidence and research, must be set out. Misinterpretation can be made too easily. Elephants have been blamed for destroying trees that, in fact, were dying because of changes in soil salinity. If soil erosion is caused by hoofed animals, it needs to be clear which species is the culprit.

Many participants in the workshop were reluctant to accept the need to cull on the grounds that the long-term outcome of *not* culling remains unknown. If a system is out of equilibrium, then it might, if left alone, go through a cycle that would lead to complete recovery; but to allow time for this answer to emerge involves a risk which, at worst, could lead to the extinction of the animal being conserved.

Diagnosis of overabundance requires information on a diversity of biological processes which may be itemized under five headings. First are the biological characteristics of the target species itself; physiological condition, body conformation, and reproductive statistics are particularly indicative. Second, similar features of associated species -- including competitors, predators, parasites, and prey -- help to place the target species in the context of an ecological community. Third, of special importance in terrestrial systems, is the condition of the plant resources that sustain the herbivores by providing not only food but also shelter and the preferred habitat. Shallow water marine systems present parallel complications, while deep water systems present the special dimension of long-distance migration. Fourth, characteristics of the community (for example, species diversity) contribute to a more complete ecological assessment. Finally, abiotic measures of the ecosystem -- particularly of minerals and energy sources in air, water, soils, and sediments -- could be of special value in providing some early warning signs.

There is no shortcut to the assessment of these variables, but their careful documentation permits evaluation by the widest possible audience. The manager must first clearly define his goals and, ideally, he may be able to experiment with different procedures which should be compared with a control area in which no "positive" management action is taken. All the problems associated with the critical questions of diagnosis were extensively discussed at the workshop and the formulation of appropriate action was set against a theoretical background. The need to recognize several stable stages in a changing ecosystem was emphasized; the manager should attempt to determine what conditions are transitional between one kind of stability and another. This is an area of applied research in which the modeling of systems might offer considerable illumination.

The case histories presented showed that no formula was universally applicable. Examples of the apparent success of intervention or nonintervention were presented, as well as other examples where management goals, if defined, had not been achieved.

Methods of control proved to be a subject equally difficult to evaluate, because the philosophy of management largely determines how assiduously policies are carried out. The outstanding examples of active intervention are to be seen in South Africa; there culling is an essential method of achieving stated management goals. The search for ecologically sound, and humane, methods of culling continues, however, and the creation of vacuum zones was seen to have considerable merit. In this method of control animals are killed in a defined area only which, being depopulated, attracts excess animals that continuously move away from "overcrowded" areas. This would not be expected to work for all species. Some cautionary examples were available from exploited (as opposed to protected) populations of mammals. Attempts to achieve maximum sustained yields from fur seal populations, for example (which included exploiting females) had resulted in a failure of the population to follow the predicted course of recovery. Translocation, the use of artificial water points, and forcing habitat cycles out of phase were all discussed but, while each was seen to have some advantages, none was without apparently intractable problems. There was a strong implication that large reserves offer far greater opportunity for sympathetic management than small ones.

In defining common ground the participants agreed there should be:

(1) Definition of objectives by management.

(2) Correct identification of the problem -- social, political, or ecological.

(3) An attempt at an experimental approach.

(4) The adoption of active adaptive management in both space and time.

(5) An assignment of probabilities to possible alternative outcomes.

(6) Particular regard to key species and whether or not they are rapidly-responding species or ones with long time lags in their population changes.

(7) An endeavor to make use of, or at least to learn from, all perturbations -- both natural and artificially imposed.

(8) A refusal by researchers and managers to supply spurious scientific justification for a decision to cull which has been made for political, social, or financial reasons!

In addition to agreement on these pragmatic issues, it became abundantly clear and there was general agreement that theoretical concepts, and the models they engender, are of vital assistance in formulating the problems of management. Models also enable rapid assessment of the results of intervention.

The value of the workshop lay in the great diversity of views represented and not in the attainment of unanimity. Indeed, the problems at issue are so complex that unanimity could not be expected. But the value of vigorously shaking all accepted views and thinking things through again from the beginning did become apparent, and a consensus was reached on many issues. Even so, the deliberations can be regarded as only a beginning.

In the modern world wildlife must compete for land with a multitude of other uses. Wildlife conservation has powerful protagonists and popular support and is now developing a body of scientific theory that will enable conservation policies to be put into effect with commanding assurance instead of pious hope. It is our intention that this volume should promote the development of that theory.

INDEX

A

Abalone, 168
Abiotic measurements, 326, 335, 351
Acacia spp., 133, 148, 253
  tortillis, 111, 113
Acoustic survey, 155
Adaptive management, 255, 271, 312
African hunting dogs, 8, 349
Aguada Blanca National Reserve, 273, 287
Albert National Park, *see* Kivu National Park
Aleutian Islands, 168, 170, 171, 172, 173, 176, 193
Algae, 177, 218, 220, 239, 321
Amchitka Island, 174, 176, 177
Andes Mountains, 259–289
Año Nuevo, Island and mainland, 291–301
Artificial insemination, 189
Artificial watering points, 94, 102, 330, 331, 333, 339, 345, 352
Aspen, 8, 10, 343, 344
Attu Island, 173, 174

B

Baboon, 37, 44
Baikaiae plurijuga, 96
Baja California, 168, 171, 291
Bering
  Sea, 157, 159, 199, 305, 306, 308
  Strait, 159
Bison, 221
Blesbok, 32, 33, 35, 36, 37, 38, 39, 40, 44, 46, 49
Bontebok, 37, 39
Bothriochloa spp., 63
  insculpta, 134
Botswana, 45, 96, 99, 131
Brachystegia spp., 70, 72, 78, 81, 82, 83, 84, 87, 347
  boehmii, 111, 348
  glaucescens, 348
British grey seals, *see* Seals
Brownsman, *see* Farne Islands
Buffalo, 23, 31, 32, 37, 44, 46, 73, 75, 113, 114, 148, 218, 221, 226, 280, 320, 345, 346
Burning, *see* Fire
Bushbuck, 33, 35, 37

C

Calamagrostis spp., 261
  vicunarum, 265
California, 170, 178, 179, 320
  central, 169, 172, 291
  southern, 293
California Department of Fish and Game, 169
Cape Department of Nature and Environmental Conservation, 25, 29, 30, 31, 32, 36, 39, 40, 42, 43, 48
Capelin, 153, 154, 155, 156, 320, 322
Carbohydrates
  soluble, 78
  structural, 78
Caribou, *see* Reindeer
Carrying capacity, 9, 10, 40, 41, 84, 122, 133, 136, 143, 146, 156, 158, 168, 170, 173, 176, 181, 219, 227, 228, 247–257, 272, 273, 286, 293, 303, 306, 307, 308, 309, 310, 311, 319, 325
Caughley's classes of overabundance, 7, 8, 9, 10, 13, 14, 15, 16, 57, 120, 152, 156, 157, 161, 178, 242, 244, 247, 259, 266, 317, 318, 321, 323, 326, 332, 340, 349, 350

Cheetah, 73, 185
China, 15
Chirisa Wildlife Research Area, 109, 110, 112
Chizarira National Park, 75, 107, 109, 110, 111, 112, 347–349
Chukchi Sea, 157, 159, 305, 306, 308
Clams, 168
 fishery, 306
Classes of overabundance, *see* Caughley's classes of overabundance
Cod, 154
Colophospermum mopane, 96, 111
Combretum, 75, 96
Commander Islands, 193
Convention on International Trade in Endangered Species of Wild Fauna and Flora (CITES), 186, 187, 189, 278
Coronation Island, 123
Crash, *see* Eruption/crash
Critical habitat analysis, 309, 310, 312
Crop protection, 182, 228
Cymbopogon spp., 63
 plurinodis, 134

D

"Decreaser" plant species, 237, 241, 254, 255
Deer, 13, 119–127, 218, 248, 320, 321, 322, 350
 red, 221
 roe, 124, 125
Deforestation, 120, 224, 248, 322
Deka Catchment, 113, 345, 346
Demographic vigor, 50, 337
Denmark, 124, 125
Density-dependence, 202, 219, 222, 300, 328, 331, 333, 335
Desertification, 135, 248, 255
Disease, 225, 264, 266, 270, 285, 341
Dispersal, 124, 130, 140, 143, 146, 147, 243, 283, 286, 293, 294, 298, 299, 319, 322, 327, 337
Dolphin, 220
Domestic animals, 14, 49, 59, 91, 93, 96, 116, 186, 222, 225, 233, 234, 237, 239, 241, 242, 248, 259, 261, 263, 264, 265, 269, 274, 277, 278, 285, 317, 320, 322, 323, 324, 328, 345
Drought, 39, 100, 135, 146, 148, 224, 229, 265, 286, 333
Dugong, 185, 221
Duiker
 grey, 33
 red, 37, 44

E

East Africa, 69, 70, 183, 186, 187, 211, 212, 230
Eland, 32, 33, 35, 36, 37, 38, 39, 40, 44, 46, 73, 75
Elephant
 Addo population, 100, 223
 African, 1, 2, 14, 23, 31, 32, 36, 42, 43, 72, 73, 75, 76, 78, 83–86, 88, 91–118, 143, 146, 185, 210, 218, 219–224, 226, 227, 229, 230, 248, 253, 254, 255, 286, 308, 311, 319, 320, 322, 329, 330, 345–349, 350, 351
 Asiatic, 185
Elk, 8, 10, 251, 255, 343, 344, 349
Emigration, *see* Dispersal
Equilibria
 dynamic, 22, 308, 342
 K, 305, 306
 stable, 58, 62, 63, 64, 65, 66
 unstable, 60, 61, 62, 65, 66, 330
  see also Thresholds
Eragrostis plana, 63
Erosion, 16, 103, 135, 161–164, 226, 238, 240, 248, 263, 264, 319, 321, 322, 323, 324, 326, 328, 334, 336, 345, 346, 350, 351
Eruption/crash, 146, 243, 254, 264, 280, 303, 304, 308, 312, 323, 324, 325, 335, 338
Euclea divinorum, 148

F

Farallon Islands, 291
 southeast, 293, 295, 298, 299, 300
Farne Islands, 161–164, 320, 322
Fat reserves, 50, 266, 337
Feeding strategies, 217–232, 268, 269
Festuca, 261
Fiber, fiber production, 79, 80, 81
Fire, 22, 38, 42, 58, 59, 66, 75, 81, 82, 83, 87, 88, 96, 120, 130, 135, 148, 229, 238, 264, 329, 333, 339, 341, 343, 345, 346, 347, 348, 349
Fishing gear, 152, 153, 155, 156, 171, 201, 321, 322

G

Gemsbok, 33, 35, 36, 37, 38, 39, 40
Genetic variability, 16, 50, 272
Gerenuk, 220
Giraffe, 32, 36, 37, 38, 40, 42, 44, 221

Gonarezhou National Park, 94, 95, 112, 113, 114
Grazing, *see* Overgrazing, Undergrazing
Grewia, 111
Grey seals, *see* Seals
Gulls, 162
  herring, 161
  lesser black-backed, 161
Gynozis, 261

H

Hare, 33
Hartebeest, 73, 87
  red, 32, 35, 36, 37, 38, 39, 40, 44, 46
  Lichtenstein's, 75, 84
Herbage, dead, *see* Litter
Herring, 154
Hippopotamus, 2, 31, 32, 37, 43, 44, 46, 73, 113, 114, 115, 219, 221, 224, 225–229
Hluhluwe Game Reserve, 24, 129–150
Horses, feral, 221
Humpback whales, *see* Whales
Hyena, 8, 73, 132, 141, 181, 218
  spotted, 31
Hyparrhenia, 75

I

Ice dynamics, 305, 306, 311
Immigration, *see* Dispersal
Impala, 17, 31, 32, 33, 35, 36, 37, 38, 40, 44, 46, 49, 113, 114, 134, 148, 220, 319, 320
"Increaser" plant species, 237, 241
International Fund for Animal Welfare (IFAW), 1, 279, 285, 286
International Union for the Conservation of Nature and Natural Resources (IUCN), 1, 129, 182, 186, 278, 284, 286, 287
International Whaling Commission (IWC), 155, 211, 213
"Invader" plant species, 237, 254, 255
Isla de Guadalupe, 295, 298, 299
Island biogeography, 22, 119, 121, 122, 123, 124, 126, 211, 227, 228, 251, 319
Islas San Benitos, 169, 299
Isle Royale National Park, 181, 340–343, 349
Ivory, 91, 93, 94, 107, 146, 157, 278, 304, 322

J

Jackal, black-backed, 32
Julbernadia, 70, 72, 347

K

K, 210, 218–232, 273, 303, 304, 305, 306, 307, 308, 312
Kabalega Falls National Park, 26, 229, 230
Kaibab Plateau, 280
Kariba Lake, 103, 107, 347
Kasungu National Park, 16, 69–89, 329
Kelp, 172, 173, 174, 176, 177, 178, 218
  Macrocystic pyrifera, 172
  Nereocystic leutkeava, 172
Kenya, 286
Kidepo Park, 253
Kivu National Park, 227
Knysna Forest, 223
Kob, Uganda, 220
Krill, 154
Kruger National Park, 8, 17, 23, 24, 28, 29, 31, 32, 36, 39, 40, 42, 43, 46, 47, 65, 66, 114, 253, 256, 319, 324
Kudu, 32, 33, 35, 36, 37, 38, 39, 44, 46, 220
  lesser, 220
Kurile Islands, 168, 193
Kyle Game Park, 133

L

Lake Manyara National Park, 225
Leaf dynamics, 234–236
Lemming, 218
Leopard, 73, 181–191, 320, 321, 322, 331, 349
Lepidophyllum, 261, 264
Lichen, 218, 220, 227, 230, 239
Lion, 8, 31, 32, 43, 65, 66, 73, 132, 324
Litter, 238
Live capture, *see* Translocation
Los Coronados, 299
Lundi River, 112, 115

M

Mackerel, 154
Malawi, 16, 69–89, 247
Mana Pools National Park, 93, 95, 114
Manatee, 221
Marine Mammal Protection Act, 201, 202
Matusadona National Park, 95, 107, 109, 110, 111, 112, 114
Maximum sustainable yield (MSY), 10, 197, 213, 241, 261, 265, 270, 278, 284, 288, 309, 324, 329, 352
Meru National Park, 187, 331
Mexico, 291

Mkuzi Game Reserve, 24, 41
Models, modeling, 99, 100, 136, 142, 148, 207, 248–254, 255, 274, 278, 281–283, 304, 310, 312, 323, 324, 351, 353
Monkey, vervet, 32
Moose, 11, 13, 181, 218, 221, 340–343
Multiple stable states, 207, 254, 336, 338
Muskoxen, 119
Mweya peninsula, 226

N

Namibia, 45, 114
Natal, 319, 320, 330
Natal Parks, Game, and Fish Board, 24, 29, 31, 32, 34, 35, 37, 41, 43, 44, 46, 47, 49, 50, 142
New South Wales, 17
New Zealand, 13, 136
Newfoundland, 152–155
Nile River, 226
Nonintervention, 324, 339, 340, 342, 343, 349
North Pacific Fur Seal Commission, 197
Nuanetsi, 112, 113
Nursing, 295
Nutrition, 224, 225, 240, 268
Nyala, 2, 32, 37, 44, 46, 148, 220

O

Oil spill, 178, 179
Oman, 190
Orange Free State Division of Nature Conservation, 25, 29, 33, 35, 36, 37, 40, 41, 42, 43, 48
Oribi, 38
Otter, European, 185
Overbrowsing, 224, 341, 343, 344, *see also* Overgrazing, Undergrazing
Overgrazing, 65, 120, 123, 124, 125, 126, 134, 148, 223, 225, 234, 235, 236, 238, 240, 241, 242, 243, 248, 255, 264, 265, 319, 328, 343, 350, *see also* Overbrowsing, Undergrazing

P

Pacific walrus, *see* Walrus
Pampa Galeras National Reserve, 259–275, 277–289, 328
Panicum coloratum, 133
Parasitism, 341
Pastures, 4, 234
"Patch dynamics," 206, 208–212
Pericopsis, 75
Perturbations, man-made, 256, 309, 333, 352
Phytoplankton, 218
Plant resilience, 243, 244
Poa flabellata, 228
Poaching, 69, 72, 81, 84, 85, 86, 87, 107, 183, 186, 228, 229, 230, 248, 261, 266, 272, 278, 333
Polar bear, 151
Polilepis, 264
Predators, 252, 254, 266, 270, 280, 322, 333
Prey switching, 176, 177
Pribilof Islands, 193–204, 251, 280
Prince William Sound, 168
Protein, protein production, 78, 79, 80, 134
Puffins, 161, 162, 322
Puna, 261, 262, 263, 264, 273, 274, 286
Pycnophyllum, 264

Q

Queen Charlotte Islands, 169
Queen Elizabeth National Park, *see* Rwenzori National Park

R

r, 50, 132, 210, 218–232, 273, 307
Ranching, *see* Domestic animals
Reedbuck, 17, 73
  mountain, 33, 35, 36, 37, 38, 39, 40
  southern, 32, 37, 44, 46
Reindeer, 119, 124, 219, 221, 227–229, 230, 251, 280, 307, 312, 319, 333
Reintroduction, *see* Translocation
Rhinoceros, 73, 210
  black, 36, 37, 44, 141, 185, 221
  white, 37, 40, 41, 43, 44, 129–150, 221, 319, 320, 324, 329, 350
Rift Valley, 225
Rinderpest, 91, 96, 103, 207, 252, 322, 345
Roan antelope, 36, 38, 73, 75, 84, 87
Robben Island, 193
Rwenzori National Park, 134, 225, 226, 227

S

Sabi-Lundi, 112
Sable antelope, 33, 36, 38, 73, 75, 77, 84, 87, 348
St. George Island, *see* Pribilof Islands
St. Lawrence Island, 160

St. Matthew Island, 124, 227, 307, 312
St. Paul Island, *see* Pribilof Islands
San Miguel Island, 193, 299
San Nicolas Island, 299, 300
Savannas, 57–67, 148, 237, 323, 345
Scare device, 321
Scotland, 164
Sea otters, 151, 167–180, 221, 320, 322
Sea urchins, 168, 172, 173, 174, 176, 178
Seals
  Alaskan fur, *see* Seals, northern fur
  Antarctic fur, 220
  bearded, 157, 159
  crabeater, 220
  elephant, 220
    northern, 291–301, 327, 329
  grey, 2, 151, 152, 161, 162, 163, 220, 320, 322
  harbor, 202
  harp, 154
  leopard, 220
  northern fur, 193–204, 221, 352
  ringed, 220
  sea lion, 202
    Steller's, 221
  Weddell, 220
Sebungwe Region, 94, 95, 103, 104, 105, 107, 108, 109, 110, 112, 347–349
Seed dispersal, 239
Seed reserve, 336
Senecio, 261, 264
Sengwa Wildlife Research Area, 109, 110, 111
Serengeti National Park, 134, 252, 280
Shaba National Reserve, 189
Snow, 123, 342, 343
Soil erosion, *see* Erosion
Soil nutrients, 130, 134, 135, 148, 218, 219, 225, 264, 325, 326, 328, 329, 332, 333, 335, 336, 338, 350
South Africa, 17, 21–55, 211, 212, 223, 247, 248, 252, 319, 320, 352
South Africa National Parks Board, 24, 29
Southern Africa, 16, 18, 57–67, 70
South-West Africa, *see* Namibia
Spatial heterogeneity, 208–210, 330, 332
Species diversity, 240, 241, 272, 331
Sporobolus smutsii, 133
Springbok, 33, 35, 36, 37, 38, 39, 40, 44, 49
Squid, 154
Stable limit cycles, 9, 146, 219, 243, 264
Staple, *see* Farne Islands
Steenbok, 33
Stipa, 261

T

Tahr, Arabian, 190, 221
Tanzania, 252, 280
Terminalis, 96
Terminatia, 75
Themeda spp., 135
  triandra, 133, 134
Thresholds, 58, 60, 61, 65, 66, 147, 206, 235, 336, 338, *see also* Equilibria
Tiger, 185
Time lag, 206, 230, 327, 332, 335
Time/space considerations, 330, 332, 333, 334
Tourism, 72, 75, 88, 103, 201, 229, 261, 301
Trampling, 239
Translocation, 2, 3, 4, 10, 11, 23, 31, 35, 36, 37, 40, 41, 42, 43, 48, 50, 113, 114, 115, 130, 142, 168, 178, 179, 187, 188, 208, 219, 224, 261, 272, 273, 274, 284, 286, 287, 288, 319, 321, 330, 331, 350, 352
Transvaal Nature Conservation Division, 25, 28, 29, 31, 33, 35, 36, 38, 40, 42, 43, 48
Trophic levels, 136, 143, 181, 335
Tsavo National Park, 84, 229, 230, 286
Tsessebe, 36, 38, 115, 348

U

Uganda, 134, 225, 253
Umfolozi Game Reserve, 24, 41, 129–150, 319 324
Undergrazing, 234, 235, 238, 240, 241, 242, 243, 325, *see also* Overbrowsing, Overgrazing
Urochloa mossambicensis, 133
U.S.S.R., 15, 157

V

Vacuum zones, 86, 130, 143, 147, 212, 227, 229, 273, 330, 352
Vicuna, 2, 213, 221, 259, 275, 277–289, 319, 328
Vleis, 345, 346

W

Wales, 164
Walrus, 151, 152, 156–160, 221, 303–315, 320, 322
  nutrient cycling, 309

Wankie National Park, 91, 94, 95, 96, 97, 99, 100, 101, 102, 113, 114, 345, 346
Wapiti, 220
Warning devices, 156
Warthog, 32, 37, 44, 46, 73, 114, 226, 320
Waterbuck, 35, 37, 38, 44, 73, 75, 114, 226, 320
Whales, 218, 219
  baleen, 221
  blue, 312
  humpback, 151–156, 320, 321, 322
  killer, 221
  minke, 220
  sperm, 221
Wildebeest, 8, 31, 65, 66, 114, 135, 181, 220, 253, 256, 280, 319, 320, 324, 333, 350
  black, 32, 33, 35, 36, 37, 38, 39, 40, 42, 44, 46, 135
  blue, 31, 32, 33, 35, 36, 37, 38, 44, 46
Willow, 343, 344
Wolves, 121, 181, 185, 189, 218, 340–343
World Conservation Strategy, 48
World Wildlife Fund (WWF), 93, 122, 123, 126, 284, 285

Y

Yellowstone National Park, 8, 251, 255, 343, 344, 349

Z

Zambesi, River and Valley, 15, 93, 95, 103, 107, 347
Zebra, 8, 31, 36, 43, 73, 75, 84, 113, 114, 135, 148, 221, 320, 345, 346
  Burchell's, 32, 33, 36, 37, 38, 40, 44, 46
  mountain, 40
Zimbabwe, 45, 91–118, 248, 320, 322, 329, 345–346, 347–349
Zooplankton, 220
Zoos, 45, 130, 188, 208, 223, 247, 273, 331, 350

WORKSHOP PARTICIPANTS

* John R. Beddington
Daniel B. Botkin
Graeme Caughley
Stephen Cobb
David H. M. Cumming
S. K. Eltringham
James A. Estes
John Hanks
* John Harwood
* Sidney Holt
* Peter A. Jewell
Brian H. Walker
David R. Klein
David M. Lavigne
* Richard M. Laws
Burney J. Le Boeuf
Robert M. May
Imanuel Noy-Meir
K. C. Otte
Norman Owen-Smith
G. Carleton Ray
Peter Scott
A. R. E. Sinclair

ATTENDEES

Brian D. Davies
Donna Hart
W. J. Jordan
Ruth Walicki
Ian MacPhail
Daniel J. Morast
Gloria C. Toland

* Member, Steering Committee.